Botanical Gardens and Their Role in Plant Conservation

Approaching the contributions of a world-wide sector of scientific institutions to addressing the extinction crisis, *Botanical Gardens and Their Role in Plant Conservation* brings together a diversity of perspectives. There are more than 3,600 botanical gardens worldwide, where trees, shrubs, herbs, and other plants are studied and managed in collections. They are foremost among efforts to conserve the diversity of living plant species and ensure that crucial biodiversity is available for the future of humanity.

This book is a showcase for plant conservation, restoration, biodiversity, and related scientific and educational work of botanical gardens around the world, featuring both thematic overview chapters and numerous case studies that illustrate the critical role these institutions play in fighting extinction and ensuring plant diversity is available for sustainable use.

FEATURES

- A wide range of case studies derived from practical experience in a diversity of institutional, national, and biogeographical settings,
- Reviews of topics such as networking amongst institutions, the importance of global policy agreements such as the Convention on Biological Diversity and the Global Strategy for Plant Conservation,
- Profiles of botanical gardens contributions at the national level to conservation priorities,
- Real-world examples of programs in plant conservation for both critically endangered wild plant diversity and unique horticultural or cultural germplasm.

Botanical Gardens and Their Role in Plant Conservation includes contributions from institutions from Africa, Asia, Australia, Europe, and the Americas, and institutions of all sizes and histories, from long-established national gardens to new gardens offering their perspectives on developing their roles in this vital undertaking.

Botanical Gardens and Their Role in Plant Conservation

Asian Botanical Gardens, Volume 2

Edited by T. Pullaiah and David A. Galbraith

CRC Press is an imprint of the
Taylor & Francis Group, an **informa** business

First edition published 2024
by CRC Press
6000 Broken Sound Parkway NW, Suite 300, Boca Raton, FL 33487–2742

and by CRC Press
4 Park Square, Milton Park, Abingdon, Oxon, OX14 4RN

CRC Press is an imprint of Taylor & Francis Group, LLC

© 2024 selection and editorial matter, T. Pullaiah and David A. Galbraith; individual chapters, the contributors

Reasonable efforts have been made to publish reliable data and information, but the author and publisher cannot assume responsibility for the validity of all materials or the consequences of their use. The authors and publishers have attempted to trace the copyright holders of all material reproduced in this publication and apologize to copyright holders if permission to publish in this form has not been obtained. If any copyright material has not been acknowledged please write and let us know so we may rectify in any future reprint.

Except as permitted under U.S. Copyright Law, no part of this book may be reprinted, reproduced, transmitted, or utilized in any form by any electronic, mechanical, or other means, now known or hereafter invented, including photocopying, microfilming, and recording, or in any information storage or retrieval system, without written permission from the publishers.

For permission to photocopy or use material electronically from this work, access www.copyright.com or contact the Copyright Clearance Center, Inc. (CCC), 222 Rosewood Drive, Danvers, MA 01923, 978–750–8400. For works that are not available on CCC please contact mpkbookspermissions@tandf.co.uk

Trademark notice: Product or corporate names may be trademarks or registered trademarks and are used only for identification and explanation without intent to infringe.

ISBN: 978-1-032-25039-7 (hbk)
ISBN: 978-1-032-25040-3 (pbk)
ISBN: 978-1-003-28125-2 (ebk)

DOI: 10.1201/9781003281252

Typeset in Times
by Apex CoVantage, LLC

Contents

Preface vii
Editor Biographies ix
List of Contributors xi

Chapter 1 Botanical Gardens and Plant Conservation Initiatives in Nepal 1

Krishna Kumar Shrestha and Yadav Uprety

Chapter 2 Role of Indonesian Botanic Gardens in Plant Conservation 25

Joko Ridho Witono and Irawati

Chapter 3 Vietnam Botanic Gardens and Their Role in Plant Conservation 51

Hong Truong Luu, Huu Dang Tran, Ke Loc Phan, Van Sam Hoang, Quoc Binh Nguyen, The Cuong Nguyen, and Trong Nhan Pham

Chapter 4 Botanical Gardens in Malaysia and Their Role in Plant Conservation 65

L.G. Saw, A. Latiff, and J. Sang

Chapter 5 Ancillary Botanic Gardens: A Case Study of the American University of Beirut 81

Salma N. Talhouk, Ranim Abi Ali, Alan Forrest, and Yaser Abunnasr

Chapter 6 Conservation of Threatened Plant Species and Protected Areas in Korean Botanical Gardens and Arboreta 95

Yong-Shik Kim, Hyun-Tak Shin, and Sungwon Son

Chapter 7 Mongolian Botanical Gardens – Modern Plant Biodiversity Conservation Resources in Mongolia 103

Nanjidsuren Ochgerel, Luvsanbaldan Enkhtuyaa, and Victor Ya. Kuzevanov

Chapter 8 Jawaharlal Nehru Tropical Botanic Garden and Research Institute, a Treasure House of Tropical Plant Germplasm, Blends into the Western Ghats, the Biodiversity Hotspot in Indian Region 135

R. Prakashkumar and R. Raj Vikraman

Chapter 9 Botanical Gardens and Their Role in Education, Research, Conservation, and Bioprospecting of Plant Diversity: Lead Botanical Garden (LBG), Shivaji University, Kolhapur – a Case Study 149

M.M. Lekhak and S.R. Yadav

Chapter 10 Lead Botanical Garden of Yogi Vemana University, Kadapa, India, and Its Role in Plant Conservation 173

A. Madhusudhana Reddy and C. Nagendra

Chapter 11 Dhanikhari Experimental Garden-Cum-Arboretum 187

Lal Ji Singh, C.P. Vivek, Bishnu Charan Dey, C.S. Purohit, Gautam Anuj Ekka, Mudavath Chennakesavulu Naik, and Fouziya Saleem

Chapter 12 M. S. Swaminathan Botanical Garden – A Community Conservation Initiative in the Western Ghats of India 203

Nadesa Panicker Anil Kumar, V. Shakeela, Merlin Lopus, and Sarada Krishnan

Chapter 13 Calicut University Botanical Garden (CUBG) and Its Role in Plant Conservation 217

A.K. Pradeep and Santhosh Nampy

Chapter 14 Role of Botanical Garden in Conservation and Citizen Science – A Case Study from Mahatma Gandhi Botanical Garden, University of Agricultural Sciences, Bangalore 227

A.N. Sringeswara and Sahana Vishwanath

Chapter 15 The Role of the Kuzbass Botanical Garden in Solving Environmental Problems and in Plant Conservation *in situ* and *ex situ* 233

Andrey Kupriyanov

Chapter 16 Contribution of Botanic Garden to Plant Conservation: 233 Years of Conservation History and Actions of CSIR-NBRI Botanic Garden 247

J.S. Khuraijam, S.K. Tewari, and S.K. Barik

Index 261

Preface

The conservation and sustainable use of the world's diversity of plants is critical to the future of humanity. Of the approximately 400,000 species of plants, at least a quarter are of direct medicinal, economic, and food security importance, and many are at risk of extinction. Botanical gardens constitute a global network of research, educational, and conservation institutions that direct their efforts to plants. There are more than 3,000 botanical gardens worldwide, where trees, shrubs, herbs, and other plants are studied and managed in collections. They are foremost among the institutions around the world that are leading efforts to conserve the diversity of living plant species, document them, and ensure that this crucial biodiversity is available for the future of humanity.

Botanical gardens are a remarkably heterogenous group of institutions, with great ranges in size, programs, and capacities. Since the early 1980s, many have taken on the conservation of biological diversity as central to their missions. They are engaging in direct conservation programs focused on at-risk species (both *ex situ* and *in situ*), combating invasive species, supporting public education about conservation issues, and other actions directed at stemming the tide of extinction. More than 80,000 plant species are already protected to some degree in the collections of just 1,200 of these institutions.

As botanical gardens play a pivotal role in plant conservation, we have undertaken to bring examples of these conservation efforts together and showcase their scope, importance, and their own diversity. With this aim, we invited contributed chapters for this multivolume set of books from plant conservation practitioners working at botanical gardens worldwide. The goal of these edited volumes is to bring together experiences and case studies of the contributions of botanical gardens worldwide to the study, preservation, and sustainable use of plant diversity.

This three-volume set positions the work of botanical gardens in light of global needs and initiatives, highlighting the responses of individual institutions to their local circumstances, priorities, and objectives. Forty-five chapters contributed by 183 authors from 23 countries span the three volumes of this set. Nine introductory chapters present overviews of key issues by experts, including summaries of networking among botanical gardens, the role of botanical gardens in protecting food plant species, the state of development of seed gene banks, access and benefits-sharing regimes, plant propagation, and conservation strategies involving botanical gardens. Thirty-six chapters represent programs, collections, and contributions by individual botanical gardens on all continents except Antarctica. No single book or set of volumes could present all of the work of botanical gardens from all countries. It is our intention to that these chapters provide a wide range of examples of how botanical gardens are making contributions and also a survey of the kinds of issues botanical gardens involved in conservation face.

The conservation of plant diversity is along the most important environmental, social, and economic concerns of the twenty-first century. As the human population has exceeded eight billion and economic growth remains a high priority for many governments, the pressures that this large civilization places on remaining natural areas, the cradles and reservoirs of biological diversity, are ever increasing. Conversion of natural areas to cultivated, urbanized, or industrialized landscapes; fragmentation of remaining habitats; loss of ecological connectivity and ecological partners; and the global effects of climate change area all immediate and well-characterized threats. At the same time, humanity's ability to study, characterize, and conserve plant species diversity has never been greater. Globe-spanning databases and communications allow for information to be shared and accumulated, and analyzed across the planet, and molecular genetic techniques are expanding our ability to detect, read, and understand the DNA underlying that biological diversity like never before. The responses to threats to plant diversity are numerous too, including the full gamut of *in situ* and *ex situ* approaches to conservation.

Standing in the midst of all of this – of efforts to document diversity, protect it through a range of methods and actions, and critically interpret it to the public, are the world's 3,000+ botanical gardens. Having their origins nearly 500 years ago in the emergence of botany as a science during the Italian Renascence, botanical gardens have grown in their missions, capabilities, and collections and have diversified greatly from their start as small university teaching and research gardens to everything from community-led centers for well-being and beautification to major scientific institutions that have taken leading roles in organizing and implementing plant conservation at every level.

Just what botanical gardens can do to address plant conservation issues is under-appreciated. While at their hearts are the development of living plant collections, and thus *ex situ* conservation does Figure large in their programs and contributions, *ex situ* conservation (living plant collections and facilities like seed gene banks) is just one tool in an expanding tool-box. While no single institution might undertake all of these, the range of tools botanical gardens are deploying also include in-vitro conservation (preservation of viable tissue or specimens in artificial growth media), cryogenic preservation of germ-plasm (banking tissue in ultracold repositories), propagation, population supplementation, and reintroduction of target taxa into wild habitats, as well as the gathering and analysis of associated data relevant to conservation, plant taxonomy, and field studies of floristics. Increasingly, partnerships with local and Indigenous peoples to bring important new knowledge and approaches to conservation are being seen. These address historical issues of colonialism and recognize that plants are at the heart of human culture.

Above all, botanical gardens focus their attention, and the attention of their many visitors, stakeholders, and partners, on the world of whole-organism plant science. Reductionistic approaches to biology that employ DNA and other sequence data have made staggering advances in our scientific and applied knowledge about plants since the 1960s, but understanding and conserving plants, and being able to relate to them on a human level, remains firmly with the whole organism. Botanical gardens are uniquely equipped to bring that whole organism perspective to the awareness of the public and decision-makers.

These volumes were written during the SARS-COVID-2 pandemic that began in 2020. Following a call for any interested botanical gardens to contribute accounts of their roles in conservation, we found that not only were large institutions involved, but smaller botanical gardens in many countries were already undertaking a variety of plant conservation initiatives, some many years old, and some were just launching their own contributions.

In some ways, this project has been a response to the COVID epidemic, too. As authors were being identified and volume contents organized, what emerged was a symposium in print form, a chance for many voices to come together and be heard on this vital issue. The resulting three volume set provides both topical and geographic coverage. We hope that readers will take away a profound sense of hope and encouragement in the diversity and distribution of these efforts.

Plant conservation is not necessarily an "easy sell" to the public or to decision-makers. Although there are profound economic, social, and even spiritual reasons to conserve the diversity of plant species, other taxa often take precedent in action and conservation funding. A significant part of this is the perception of conservation value and the emotional responses to threats to charismatic megafauna, particularly threatened mammals and birds. These taxa are also of great importance, and conservation should not be viewed as a zero-sum game.

Faced with a multitude of urgent environmental and conservation matters ranging from local damage to climate change, the world's ability to take action has numerous constraints, ranging from the practical to the political. With the growing recognition that the threats to humanity posed by declines in biological diversity are as profound as the threat of climate change and its numerous consequences, however, not acting is unacceptable. It is our hope that this three-volume set will inspire botanical gardens to further contribute to the conservation of plant diversity, worldwide and close to home.

We would like to extend our profound thanks to all the authors who have contributed to these volumes.

T. Pullaiah,
Anantapur, Andhra Pradesh, India

David A. Galbraith,
Hamilton, Ontario, Canada

Editor Biographies

Prof. T. Pullaiah obtained his MSc and PhD degrees in Botany from Andhra University. He was a post-doctoral fellow at Moscow State University, Russia, during 1976–1978. He traveled widely in Europe and visited Universities and Botanic Gardens in about 17 countries. He joined Sri Krishnadevaraya University as Lecturer in 1979 and became a professor in 1993. He has published 120 books, 345 research papers, and 35 popular articles. His books have been published by reputed international publishers like Elsevier, Springer, CRC Press, Taylor & Francis, Apple Academic Press, Scientific Publishers, Astral International, CBS Publishers, etc. Under his guidance, 54 students obtained their PhD degrees and 34 students their M Phil degrees. He is the recipient of P. Maheshwari Gold Medal, Prof. P.C. Trivedi medal for Editorial excellence and Dr. G. Panigrahi Memorial Award of Indian Botanical Society, and Prof. Y. D. Tiagi Gold Medal of the Indian Association for Angiosperm Taxonomy. He was President of the Indian Association for Angiosperm Taxonomy (2013) and President of the Indian Botanical Society (2014). He was a member of the Species Survival Commission of International Union for Conservation of Nature and Natural Resources (IUCN).

David A. Galbraith completed his BSc and MSc at University of Guelph and PhD at Queen's University at Kingston, in Canada. His early research focused on evolutionary ecology of aquatic vertebrates. Following a post-doctoral fellowship in Canterbury, England, he served as executive director and curator of a small AZA-accredited center for endangered wildlife species conservation. In 1995, he joined Royal Botanical Gardens (Canada) to develop biodiversity projects among botanical gardens across Canada in response to the Convention on Biological Diversity. In 2006, he was appointed RBG's Head of Science, overseeing library, archives, and herbarium research and use of RBG resources by outside researchers. He has published many contributions on conservation policy, management, and history of botanical gardens. In 2002, Dr. Galbraith was honored by the American Public Gardens Association with their annual Professional Citation for his innovative work in public horticulture. He was named Hamilton Environmentalist of the Year in 2010 for his efforts to protect nature. Dr. Galbraith has always been passionately engaged in biology, history, cultural heritage, and the arts, and is fascinated by how all of these intersect within botanical gardens. He is an adjunct biology professor at McMaster University, a Fellow of the Royal Canadian Geographical Society, and a Fellow International of the Explorers Club.

Contributors

Ranim Abi Ali
AUBotanic – Botanic Garden of the American University of Beirut, Beirut, Lebanon

Yaser Abunnasr
Department of Landscape Design and Ecosystem Management, Faculty of Agricultural and Food Sciences, American University of Beirut, Beirut, Lebanon

S.K. Barik
CSIR-National Botanical Research Institute, Lucknow, India

Bishnu Charan Dey
Botanical Survey of India, Andaman & Nicobar Regional Centre, Port Blair, Andaman and Nicobar Islands, India

Gautam Anuj Ekka
Botanical Survey of India, Andaman & Nicobar Regional Centre, Port Blair, Andaman and Nicobar Islands, India

Luvsanbaldan Enkhtuyaa
Botanical Garden and Research Institute of the Mongolian Academy of Sciences, Ulaanbaatar, Mongolia

Alan Forrest
Centre for Middle Eastern Plants, Royal Botanic Garden Edinburgh, Scotland, UK

Van Sam Hoang
Vietnam National University of Forestry, Xuan Mai, Hanoi, Vietnam

Irawati
Research Center for Plant Conservation and Botanic Gardens, Indonesian Institute of Sciences (LIPI), Jalan Ir. H. Juanda 13, Bogor, Indonesia

J.S. Khuraijam
CSIR-National Botanical Research Institute, Lucknow, India

Yong-Shik Kim
Yeungnam University, Gyeongsan-si, Gyeongsangbuk-do, Korea

Sarada Krishnan
Global Crop Diversity Trust, Bonn, Germany

Nadesa Panicker Anil Kumar
M. S. Swaminathan Research Foundation, Community Agrobiodiversity Centre, Kalpetta, Wayanad, Kerala, India

Andrey Kupriyanov
Kuzbass Botanical Garden, The Federal Research Centre of Coal and Coal Chemistry of SB RAS, Kemerovo, the Russian Federation

Victor Ya. Kuzevanov
Baikal State University, Irkutsk, Russia

A. Latiff
Faculty of Science and Technology, Universiti Kebangsaan, Malaysia, Bangi, Selangor, Malaysia

M.M. Lekhak
Department of Botany, Shivaji University, Kolhapur, India

Merlin Lopus
M. S. Swaminathan Research Foundation, Community Agrobiodiversity Centre, Kalpetta, Wayanad, Kerala, India

Hong Truong Luu
Southern Institute of Ecology, Institute of Applied Materials Science & Graduate University of Science and Technology, Vietnam Academy of Science and Technology, Ho Chi Minh City, Vietnam

A. Madhusudana Reddy
Department of Botany, Yogi Vemana University, Kadapa, Andhra Pradesh, India

C. Nagendra
Department of Botany, Yogi Vemana University, Kadapa, Andhra Pradesh, India

Mudavath Chennakesavulu Naik
Botanical Survey of India, Andaman & Nicobar Regional Centre, Port Blair, Andaman and Nicobar Islands, India

Santhosh Nampy
Department of Botany, Calicut University, Kerala, India

Quoc Binh Nguyen
Vietnam National Museum of Nature, Vietnam Academy of Science and Technology, Hanoi, Vietnam

The Cuong Nguyen
Me Linh Station for Biodiversity, Institute of Ecology and Biological Resource, Vietnam Academy of Science and Technology, Hanoi, Vietnam

Nanjidsuren Ochgerel
Botanical Garden and Research Institute of the Mongolian Academy of Sciences, Ulaanbaatar, Mongolia

Ke Loc Phan
Hanoi University of Science, Vietnam National University Hanoi, Hanoi, Vietnam

Trong Nhan Pham
Forest Science Institute of Central Highlands and South of Central Vietnam (FSIH), Da Lat City, Lam Dong Province, Vietnam

A.K. Pradeep
Department of Botany, Calicut University, Kerala, India

R. Prakashkumar
KSCSTE-Jawaharlal Nehru Tropical Botanic Garden and Research Institute, Palode, Thiruvananthapuram, Kerala, India

C.S. Purohit
Botanical Survey of India, Andaman & Nicobar Regional Centre, Port Blair, Andaman and Nicobar Islands, India

Fouziya Saleem
Botanical Survey of India, Andaman & Nicobar Regional Centre, Port Blair, Andaman and Nicobar Islands, India

J. Sang
Forest Research Centre, Kuching, Sarawak, Malaysia

L.G. Saw
Penang Botanic Gardens, Jalan Kebun Bunga, Penang, Malaysia

V. Shakeela
M. S. Swaminathan Research Foundation, Community Agrobiodiversity Centre, Kalpetta, Wayanad, Kerala, India

Krishna Kumar Shrestha
Central Department of Botany, Tribhuvan University, Kirtipur, Kathmandu, Nepal

Hyun-Tak Shin
Korea National Arboretum, Pocheon-si, Gyeonggi-do, Korea

Lal Ji Singh
Botanical Survey of India, Andaman & Nicobar Regional Centre, Port Blair, Andaman and Nicobar Islands, India

Sungwon Son
Korea National Arboretum, Pocheon-si, Gyeonggi-do, Korea

A.N. Sringeswara
Mahatma Gandhi Botanical Garden, University of Agricultural Sciences, Bangalore, India

Salma N. Talhouk
Department of Landscape Design and Ecosystem Management, Faculty of Agricultural and Food Sciences, American University of Beirut, Beirut, Lebanon

S.K. Tewari
CSIR-National Botanical Research Institute, Lucknow, India

Huu Dang Tran
Becamex Institute of Research and Development, Becamex IDC Corp., Thu Dau Mot City, Binh Duong Province, Vietnam

Contributors

Yadav Uprety
Central Department of Botany, Tribhuvan University, Kirtipur, Kathmandu, Nepal

R. Raj Vikraman
KSCSTE-Jawaharlal Nehru Tropical Botanic Garden and Research Institute, Palode, Thiruvananthapuram, Kerala, India

Sahana Vishwanath
Mahatma Gandhi Botanical Garden, University of Agricultural Sciences, Bangalore, India

C.P. Vivek
Botanical Survey of India, Andaman & Nicobar Regional Centre, Port Blair, Andaman and Nicobar Islands, India

Joko Ridho Witono
Research Center for Plant Conservation and Botanic Gardens, Indonesian Institute of Sciences (LIPI), Bogor, Indonesia

S.R. Yadav
Department of Botany, Shivaji University, Kolhapur, India

1 Botanical Gardens and Plant Conservation Initiatives in Nepal

Krishna Kumar Shrestha and Yadav Uprety

CONTENTS

1.1 Introduction 1
1.2 Historical Background 2
1.3 Botanical Gardens in Nepal 3
1.4 National Policies for Botanical Gardens 3
1.5 The National Botanical Garden and Regional Botanical Gardens 10
1.6 The National Botanical Garden 13
    1.6.1 Salient Features of the National Botanical Garden 13
        1.6.1.1 Botanical Information and Exhibition Centre 13
        1.6.1.2 Biodiversity Education Garden 13
        1.6.1.3 Herbal Garden 14
        1.6.1.4 Fern Garden 14
        1.6.1.5 Lily Garden 14
        1.6.1.6 Rock Garden 14
        1.6.1.7 Systematic (or Taxonomy) Garden 14
    1.6.2 Research and Development Initiatives Associated with the National Botanical Garden 14
        1.6.2.1 National Herbarium and Plant Laboratories 14
        1.6.2.2 Conservation and Education Garden 15
1.7 ICIMOD Knowledge Park 15
1.8 Tribhuvan University Botanical Gardens 15
1.9 Public Gardens and Parks 15
1.10 Opportunities and Challenges 16
    1.10.1 *Ex Situ* Conservation of Rare, Threatened, and Endemic Species 20
    1.10.2 Developing an Online Portal of Botanical Gardens Database 21
1.11 Conclusion and Recommendations 21
Acknowledgments 22
References 22

1.1 INTRODUCTION

Plant biodiversity is highly important for securing different fundamental human needs. These valuable resources are, however, threatened due to various factors such as overexploitation, land use change, climate change, invasive species, and pollution (IPBES, 2019). The challenge posed by these anthropogenic factors calls for effective conservation approaches, thus there is a need to increase efforts to develop integrative approaches for plant biodiversity conservation (Chen and Sun, 2018). It is estimated that 391,000 species of vascular plants are found in the world (RBG Kew, 2016), and every year nearly 2,000 species of vascular plants are added as new to science (RBG Kew, 2016;

Smith, 2019). According to the International Union for the Conservation of Nature (IUCN) Red List Criteria, more than 20% of the global plant species are currently threatened with extinction (RBG Kew, 2016; Smith, 2019). The IUCN predicts extinction of more than one-third of plant species if the threatened species are not preserved (IUCN, 2010; Sharrock, 2011). This is where botanical gardens can contribute to plant biodiversity conservation.

Botanical gardens are playing an important role in conservation, preservation, and maintenance of plant biodiversity, specifically that of endemic, endangered, and rare plants (Volis, 2017). More than 105,000 species of different plant species, nearly 31% of all known plant species, representing 93% vascular plant families, are conserved and managed in the combined collections of the botanical gardens of the world (Sharrock, 2011; Mounce et al., 2017; Smith, 2019; Marquardt et al., 2020). The contribution of botanical gardens in conserving threatened species is praiseworthy as living laboratories, by conserving more than 41% of the known threatened species of the world in the form of living collections and seed banks (Mounce et al., 2017; www.bgci.org).

Thanks are due to the 60,000 plant scientists and horticulturists who are in the network of Botanic Gardens Conservation International (BGCI) for their contribution in maintaining and conserving biodiversity in botanical gardens (Mounce et al., 2017; Smith, 2019). More than 350 botanical gardens representing 74 countries have seedbanks, preserving about 57,000 taxa; among which, the Millennium Seed Bank, managed by the Royal Botanic Gardens Kew, is the largest and an outstanding example of *ex situ* conservation, comprising 13% of the world's wild species (Justice, 2015; Kovacs et al., 2021).

Most botanical gardens are centers of attraction for visitors, especially due to the large collections of varieties of exotic ornamental plants, enhancing threats for the introduction of alien invasive species in botanical gardens, parks, and home gardens. Less emphasis is given to the conservation and display of indigenous species, ignoring the value of respective country's flora. The Botanical Gardens Conservation Strategy (1989) has defined a botanical garden as the scientifically ordered and maintained collection of living plants, usually documented and labeled and open to the public for the purposes of recreation, education, and research. These gardens also play a central role in promoting awareness for meeting human needs and providing well-being, as well as research and conservation of plant species diversity (Pautasso and Parmentier, 2007; Chen and Sun, 2018). The Global Strategy for Plant Conservation (GSPC) has set an ambitious target to have 75% of the world's threatened plant species conserved *ex situ* by 2020 (CBD, 2012). The importance of botanical gardens is likely to increase as species are threatened with extinction because of climate change (Primack and Miller-Rushing, 2009).

1.2 HISTORICAL BACKGROUND

The Orto Botanico of Pisa (Pisa Botanical Garden) (est. 1543) and University of Padova Botanical Garden (est. 1545) in Italy are considered the oldest botanical gardens in the world. Since the sixteenth century, several botanical gardens were initiated throughout the world, such as Leiden Botanical Garden, Netherlands (1587); Leipzig Botanical Garden, Germany (1590); Oxford Botanical Garden, UK (1621); Paris Botanical Garden, France (1635); Royal Botanic Gardens Edinburgh, UK (1670); Chelsea Physic Garden, UK (1673); Tokyo Botanical Garden, Japan (1684); Petersburg Botanical Garden, Russia (1713); and many others (Bown, 1992; Xia et al., 2021). The Royal Botanic Garden Kew (UK), founded in 1759, is considered one of the largest and most popular botanical gardens in the world, comprising an area of 130 hectares, with ca. 20,000 taxa displayed in the garden, with 108,663 accession of living plants (www.kew.org). Today, there are 1,775 botanic gardens and arboreta in 148 countries, attracting an estimated 500 million visitors each year (www.bgci.org). These gardens comprise more than 80,000 taxa in cultivation and more than 6 million accession of living plants (Pautasso and Parmentier, 2007).

Botanical Gardens and Plant Conservation Initiatives in Nepal

The history of botanical gardens in Nepal dates back to the late nineteenth century, especially by the initiation of the Rana rulers, then Prime Ministers of the country. They had maintained small private botanical gardens in the palaces that were not open for the public. Such gardens include Singha Durbar, Keshar Mahal, Bag Durbar, Narayanhiti Palace, etc. After the commencement of democracy in Nepal during the early 1950s, a few public gardens in the form of parks were initiated in Kathmandu Valley, such as Bhugol Park, Ratna Park, Tribhuvan Park, etc. These gardens were, however, not aimed at preserving plant species as are botanical gardens; rather, they were more for aesthetic purposes.

The first naturalist to visit Nepal was Francis Buchanan-Hamilton, who visited Nepal during the early nineteenth century (1802–1803), en-route from the India–Nepal border (Raxaul to Kathmandu Valley). He collected more than 1,200 specimens and described 114 species of vascular plants from Narainhetty as new to science. The type locality of Narainhetty includes Narayan Hiti Royal Palace garden and nearby hills in the Kathmandu Valley (Devkota, 2022). Although the garden in the palace is still existing, the species described by Hamilton may not exist. The assessment of the status of vascular plants described by Hamilton from the Narainhetty (Narayanhiti) type locality should be initiated.

1.3 BOTANICAL GARDENS IN NEPAL

As a result of more than 200 years of botanical explorations, plant collections, and documentation, it is known that Nepal comprises nearly 7,000 species of vascular plants, ranking 35th in the world in plant diversity richness. Furthermore, in terms of floristic diversity, Nepal ranks tenth in Asia and second in South Asia (Shrestha and Bajracharya, 2019; Shrestha et al., 2022). The botanical gardens in Nepal are playing a significant role in the *ex situ* conservation of plant diversity, especially the rare, threatened, and high-value species growing from the lowland to the high hills of Nepal.

Nepal has undertaken plant conservation efforts through botanical gardens since the first botanical garden was established in the early 1960s. Since then, 12 botanical gardens are officially recognized and managed under the Ministry of Forests and Environment, Department of Plant Resources. In recent years, as the provincial and local governments function under the new federal structure of the country, many provincial and local governments have also made attempts to establish botanical gardens under their jurisdictions (Plates 1.1–1.6). There is scattered information at the moment on such botanical gardens, and they are in need of proper documentation. The motive behind these new formations is more for recreation rather than a scientifically ordered and maintained collection of living plants. There are also recognizable botanical gardens within university premises and old palaces where important collections are available.

In this chapter, we aim to provide a brief history of botanical garden establishment in Nepal, their distribution in the country, major plant collections, and management challenges. We also provide management recommendations and highlight the need for cooperation and collaboration among botanical gardens, at least at an Asian country level, for the development of botanical gardens in the region.

1.4 NATIONAL POLICIES FOR BOTANICAL GARDENS

In Nepal, plant biodiversity conservation initiatives have been started lately compared to faunal biodiversity. Both *in situ* and *ex situ* conservation priorities were set, but there was no specific program for botanical gardens. Nepal's Forestry Sector Strategy (2016–2025) has aimed to establish 20 botanical gardens all over Nepal, covering all physiographic regions, by 2025. Similarly, the Nepal National Biodiversity Strategy and Action Plan (2014–2020) has set action plans to strengthen conservation of threatened and rare plant species through the network of botanical

PLATE 1.1 National Botanical Garden, Godavari, Lalitpur.

Photo credits: AM Kayastha (Photos 1–6).

Botanical Gardens and Plant Conservation Initiatives in Nepal

PLATE 1.2 National Botanical Garden.

Photo credits: AM Kayastha (Photos 7–11).

12. Visitor looking the Way point in NBG

13. View of stream water fall in NBG

14. Japanese Style Garden, one of the thematic gardens of National Botanical Garden

15. ICIMOD Knowledge Park, Godavari

16. Herbal garden in ICIMOD Knowledge Park

PLATE 1.3 National Botanical Garden and ICIMOD Knowledge Park Godavari.

Photo credits: AM Kayastha (Photos 12–16).

Botanical Gardens and Plant Conservation Initiatives in Nepal

32. Southern view of Manjushree Park, Chovar

33. Central view of Manjushree Park

34. Panoramic view of Manjushree Park, Chovar, South of Kathmandu Valley

35. Plantation by VIP guests in Dewariya Botanical Garden, Dhangadi, Sudurpaschim Province

36. *Ex-situ* conservation of *Rauvolfia serpentina* at Dewariya Botanical Garden

PLATE 1.4 Manjushree Park, Chovar, and Dewariya Botanical Garden, Dhangadi.

Photo credits: AM Kayastha (Photos 32–34); Y. Uprety (Photos 35–36).

37. Entrance of Brindaban Botanical Garden

38. Brindaban botanical garden: Public charter

39. Brindaban Botanical Garden: Panoramic view of central part of the garden

40. Daman Botanical garden: winter view

41. Daman Botanical Garden: Nursery site

PLATE 1.5 Brindaban Botanical Garden, Hetauda and Mountain Botanical Garden, Daman, Makwanpur.

Photo credits: RR Parajuli (Photos: 37–41).

Botanical Gardens and Plant Conservation Initiatives in Nepal

PLATE 1.6 Botanical Gardens in Ilam, Salyan, Banke and Rupandehi districts.

Photo credits: KK Shrestha (Photos 42–43), Jeevan Pandey (Photos 44–45), Y. Uprety (46–47).

gardens and other means. Forest Policy (2015) has also focused on *in situ* and *ex situ* conservation of rare, endangered, and threatened plant species. These legal frameworks have provided legal backstopping for the development of botanical gardens, but the need for specific legal instruments for establishment and management of botanical gardens could be instrumental for better networking of botanical gardens in the country (Lamichhane and Rai, 2021). It is likely to be possible to meet the target of Forestry Sector Strategy as the provincial and local governments are also mandated to establish and manage the botanical gardens; nevertheless, technical knowhow for management and selection of the species to preserve could be the issues that need technical support and capacity enhancement.

1.5 THE NATIONAL BOTANICAL GARDEN AND REGIONAL BOTANICAL GARDENS

The history of establishment of botanical gardens in Nepal dates back to the 1960s when the late King Mahendra visited the Royal Botanic Garden Edinburgh (Lamichhane and Rai, 2021). Dr. Geoffrey Herklots, who was the landscape designer and orchid expert, was asked to provide technical support to Nepal by the British Queen for the establishment of a botanical garden after learning of the keen interest of King Mahendra. Then the Department of Medicinal Plants (renamed as Department of Plant Resources [DPR] in 1994) selected the present location of the National Botanical Garden (formerly established as the Royal Botanical Garden) with support from Dr. Herklots; it was inaugurated by King Mahendra on 28 October 1962 (Hughes and Lamichhane, 2017; Lamichhane and Rai, 2021). Since then, another 11 botanical gardens have been established and managed by the Department of Plant Resources and its Plant Research Centers across the country.

Altogether, 12 botanical gardens are spread over Nepal in different ecological regions, covering some 794.7 hectares of land, and they are managed under the Department of Plant Resources (DPR, 2017; Lamichhane, 2018; Lamichhane and Rai, 2021) (Figure 1.1 and Table 1.1). Of the total

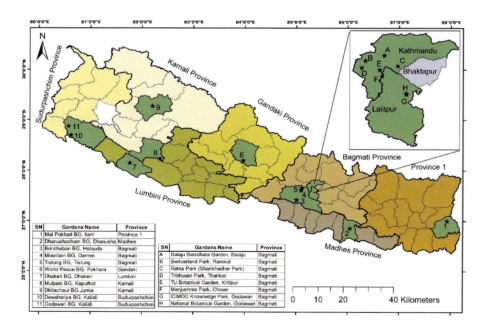

FIGURE 1.1 Distribution map of the botanical gardens and other parks and gardens in Nepal.

TABLE 1.1
Botanical Gardens of Nepal

Province District	Name of the Site	Year of Establishment	Area (Ha)	Physiographic Region	Elevation	Key Features
Koshi Province (Ilam)	Mai Pokhari Botanical Garden, Mai Pokhari	1992	9	Temperate	2,121m	• Habitat of endemic moss (*Sphagnum nepalense*) • 300 socio-economically valuable plant species conserved, including some medicinal plants • Included Mai Pokhari Ramsar site in the premises
Madhesh Province (Dhanusha)	Dhanushadham Botanical Garden, Dhanusha	1998	107	Tropical	107m	• Religious garden has been established • 103 socio-economically valuable plant species conserved
Bagmati Province (Lalitpur)	National Botanical Garden, Godawari	1962	82	Sub-tropical	1,515m	• The first national botanical garden and only member of BGCI from Nepal • *In situ* (456 species) and *ex situ* (580 species) conservation of socio-economically and ecologically valuable plant species • Rock garden, orchid house, cactus house, ethnobotanical garden, lily garden, taxonomic family garden, exhibition center, VVIP plantation area
(Makwanpur)	Brindavan Botanical Garden, Hetauda	1962	57	Tropical	405m	• Chure exhibition plots where species of Chure are conserved • *In situ* and *ex situ* conservation of 315 socio-economically valuable plant species
(Makwanpur)	Mountain Botanical Garden, Daman	1962	65	Temperate	2,320m	• *Rhododendron* species from different places are conserved • 220 socio-economically valuable plant species conserved
(Makwanpur)	Tistung Botanical Garden, Tistung	1962	45	Temperate	1,900m	• 212 socio-economically valuable plant species conserved
Gandaki Province (Kaski)	World Peace Biodiversity Garden, Pokhara	2010	164.76	Sub-tropical	775–1,078m	• Managed directly by the Department of Plant Resources • Situated on the southwest of famous Phewa lake; pristine forest near settlement • 160 socio-economically valuable plant species conserved
Lumbini Province (Banke)	Dhakeri Botanical Garden, Dhakeri	1998	5.29	Tropical	170m	• 400 socio-economically valuable plant species conserved
Karnali Province (Salyan)	Mulpani Botanical Garden, Kapurkot	1998	5.65	Sub-tropical	1,420m	• 375 socio-economically valuable plant species conserved
(Jumla)	Dhitachaur Botanical Garden, Jumla	1998	4.5	Temperate	2,500m	• High Himalayan species are conserved • 47 socio-economically valuable plant species conserved

(Continued)

TABLE 1.1 (Continued)

Province District	Name of the Site	Year of Establishment	Area (Ha)	Physiographic Region	Elevation	Key Features
Sudurpaschhim Province (Kailali)	Dewahariya Botanical Garden, Dhangadhi	1998	149.5	Tropical	170m	• Jokhar lake inside garden is the major attraction • 500 socio-economically valuable plant species conserved
(Kailali)	Godawari Botanical Garden, Godawari*	1998	100	Tropical	185m	• *In situ* conservation of rare, endangered, and native species in Chure region
	Total		**794.7**			

* Godawari Botanical Garden in Kailali does not have any physical infrastructure required to manage the garden. Its existence as a garden or as a Plant Conservation Area is also ambiguous. However, it has been listed as a botanical garden on the departmental website of DPR (see https://dpr.gov.np/).

Sources: Chaudhary et al. (2020), DPR (2021), Lamichhane and Rai (2021).

12 botanical gardens managed by the Department of Plant Resources (DPR) in seven provinces, one botanical garden each is located in Koshi Province, Madhesh Province, Gandaki Province, and Lumbini Province; two botanical gardens in Karnali Province and Sudurpachhim Province; and four botanical gardens in Bagmati Province. Similarly, five botanical gardens are located in the tropical region, four in the subtropical region, and three in the temperate region (Figure 1.1). The largest botanical garden of Nepal is the World Peace Biodiversity Garden (165 ha) situated in Pokhara, which is conserving about 1,000 species of plants, while the smallest one is Dhitachaur Botanical Garden (4.5 ha) situated in Jumla (DPR, 2021).

Thematic gardens within these botanical gardens are among the major features of the botanical gardens in Nepal. These thematic gardens were developed with some novel objectives and have contributed to enhancing the interest and ownership of the general public toward botanical gardens. For example, the Plant Research Center of Makwanpur has developed a part of the Brindaban Botanical Garden as a ceremonial garden, where the general public can choose and plant a species on the occasion of special days, such as birthdays, marriage anniversaries, and in memory of their beloved ones. They are requested to pay voluntarily some donation, which has generated some funds to be utilized for the management of the garden.

1.6 THE NATIONAL BOTANICAL GARDEN

The first national botanical garden and the only member of BGCI of Nepal, the National Botanical Garden, is located in the federal capital. With its enriched collection of *in situ* (plants that were already growing on the site when the garden was established are preserved) and *ex situ* plants, the garden is the attraction for thousands of people annually. It has recorded some 300,000–450,000 visitors per year. The garden includes a water fountain, orchid house, cactus house, rock garden, fern garden, physic garden, Japanese style garden, rare and endangered plants garden, lily garden, taxonomic family garden, and Rhododendron garden, among other attractions, for students and visitors (http://kath.gov.np/). The number of visitors could be increased in this garden if more information is disseminated to the public and the schools.

1.6.1 Salient Features of the National Botanical Garden

1.6.1.1 Botanical Information and Exhibition Centre

One of the major attractions of the National Botanical Garden is the Botanical Information and Exhibition Center, which is located close to the main entrance gate of the National Botanical Garden. The exhibition center displays pictorial and demonstration guides on the physiography, vegetation, plant diversity, and associated information related to the flora of Nepal.

1.6.1.2 Biodiversity Education Garden

The Biodiversity Education Garden is one of the major attractions, among others (see Table 1.1); it was established in 2016 to mark 200 years of the British–Nepal diplomatic relationship and reflects the collaboration between the Government of Nepal, the British Embassy, and the Royal Botanic Garden Edinburgh. The Biodiversity Education Garden hosts some 120 species of native plants from tropical, temperate, and alpine ecological regions of Nepal on 1.3 ha. Land (DPR, 2021). Each species is scientifically labeled with photographs and a QR code.

The Biodiversity Education Garden is enriched with a diverse group of plants, including *Abies spectabilis* (Fir), *Acer oblongum* (Maple), *Alnus nepalensis* (Alder), *Bombax ceiba* (Silk-cotton tree), *Choerospondias axillaris* (Nepalese hog-plum), *Pandanus nepalensis* (Screwpine), *Podocarpus neriifolius* (Mountain teak), *Rhododendron arboreum* (Rhododendron), *Syzygium jambos* (Rose apple), and *Taxus mairei* (Himalayan Yew), among others.

1.6.1.3 Herbal Garden

The Herbal Garden (*Physic Garden*), located near the entrance of the garden, comprises more than 100 species of popular medicinal and aromatic plants. Some of the common plants grown in this garden are *Acorus calamus* (Sweet flag), *Asparagus filicinus* (Wild asparagus), *Curcuma aromatica* (Yellow zedoary), *Ephedra gerardiana* (Ephedra), *Mahonia napaulensis* (Nepali Mahonia), *Rauvolfia serpentina* (Serpentine root), *Valeriana jatamansi* (Indian Valerian), and *Zanthoxylum armatum* (Prickly ash).

1.6.1.4 Fern Garden

The Fern Garden comprises a collection of native ferns and fern allies, ranging from the species of *Dryopteris*, *Polystichum*, and *Pteris* to *Cyathea spinulosa* (tree fern).

1.6.1.5 Lily Garden

The Lily Garden is home to indigenous and exotic species of Amaryllidaceae (*Agapanthus africanus*, *Narcissus tazetta*), Colchicaceae (*Gloriosa superba*), Iridaceae (*Iris pallida*, *Neomarckia gracilis*), Liliaceae (*Lilium oxypetalum*), and Melanthiaceae (*Paris polyphylla*).

1.6.1.6 Rock Garden

Located in the center of the left wing of the botanical garden, the attractive Rock Garden exhibits 28 species of diverse group of angiosperms, for example, *Agave cantula* (Century plant), *Cupressus torulosa* (Himalayan cypress), *Euphorbia royleana* (Royle's spurge), *Juniperus horizontalis* (Creeping juniper), *Opuntia ficus-indica* (Barbary fig), *Thysanolaena latifolia* (Nepalese broom grass), and *Yucca gloriosa* (Spanish dagger).

1.6.1.7 Systematic (or Taxonomy) Garden

The Systematic (or Taxonomy) Garden was initiated in 2016, and in it, the plants are arranged according to the order of families, based on Bentham and Hooker's natural system of classification. It is realized that one section of the systematic garden should be allocated for the arrangement of plant species according to the Angiosperm Phylogeny Group (APG) system of classification.

One of the important sites in the open spaces of the botanical Garden is the VVIP Plantation Area, which is allocated for the plantation of indigenous and threatened trees of Nepal by Heads of State during visits to Nepal. In addition, the attraction of the botanical garden is enhanced by the presence of the Tropical House, a typical glass house built in 1974 and the site for the preservation of lowland species. Similarly, other attractions are the Cactus House, Temperate Glass House, Rock Garden, Coronation Pond, and more.

1.6.2 RESEARCH AND DEVELOPMENT INITIATIVES ASSOCIATED WITH THE NATIONAL BOTANICAL GARDEN

1.6.2.1 National Herbarium and Plant Laboratories

Established in 1961 as the Botanical Survey and Herbarium, later named the National Herbarium and Plant Laboratories (KATH), this is one of the notable landmarks of the Department of Plant Resources. As a result of extensive expeditions throughout the country for the exploration of plant resources, the Herbarium houses 165,000 specimens, representing nearly 60% species of vascular plants reported from Nepal. The Herbarium is associated with the Economic Botany Museum, Phanerogams Section, Cryptogams Section (Algae, Fungi, Lichens, Bryophytes, and Pteridophytes), Digitation and Publicity Section, Ecology Section, Cytology Section, Xylarium (Wood Anatomy) Section, and Plant Protection Section.

1.6.2.2 Conservation and Education Garden

The Conservation and Education Garden (Arboretum) is located between the Botanical Garden and National Herbarium. The site is allotted with 90 species of trees, especially for education, research, and development purposes. Emphasis is given to the plantation and transformation of tree species representing several families from different ecological zones. Some noteworthy trees in the Conservation and Education Garden include *Aesculus indica* (Horse Chestnut), *Bauhinia variegata* (Orchid-tree), *Castanopsis tribuloides* (Chestnut), *Cedrus deodara* (Himalayan cedar), *Magnolia champaca* (Champak), *Quercus semecarpifolia* (Brown oak), *Sapindus mukorossii* (Soap nut), and *Senegalia catechu* (Cutch tree).

1.7 ICIMOD KNOWLEDGE PARK

The ICIMOD Knowledge Park (established in 1993 by the International Centre for Integrated Mountain Development) is located near the National Botanical Garden Godavari, in the foothills of the north-facing slope of Phulchoki Hill, the highest peak in Kathmandu valley (2,782 m). The park comprises mixed deciduous and evergreen broad-leaved forests at an elevation of 1,540–1,800 m, within an area of 30 hectares.

The objective of the knowledge park is the conservation and management of natural resources, such as vegetation management, water and soil management, and biodiversity management, using sustainable technologies. The park comprises nearly 700 species of indigenous, naturalized, and introduced flowering plants and maintains well-managed herbal gardens, with emphasis on the cultivation of indigenous and economically viable medicinal and aromatic plants by demonstrating low-cost agroforestry technologies (www.icimod.org/initiative/godavari-knowledge-park/).

1.8 TRIBHUVAN UNIVERSITY BOTANICAL GARDENS

Apart from the botanical gardens managed by the Department of Plant Resources, there are also gardens managed by the Tribhuvan University (TU) in Kirtipur.

The Botanical Garden of the Central Department of Botany was initiated in 1974. Emphasis was given to conserving indigenous species as well as exotic species of well-known plants for education and research purposes. This garden houses some unique collections of plant species, including *Butea buteiformis, Cassia fistula, Cedrus deodara, Choerospondias axillaris, Cinnamomum tamala, Dalbergia sissoo, Elaeocarpus sphaericus, Ginkgo biloba, Magnolia champaca,* and *Podocarpus neriifolius.* Besides native species, several species of useful plant resources, as well as introduced and exotic taxa, are also planted in the premises of the garden for educational and aesthetic purposes.

Tribhuvan University Coronation Garden (54 hectares) was established in 1975 to commemorate the coronation of the former King Birendra Bir Bikram Shah (Bajracharya et al., 1997). The university has also authorized the Central Department of Botany to develop the Golden Jubilee Garden (then called Coronation Garden) on the university premises. The Golden Jubilee Garden could serve as an important site for research and education.

Zakir Hussain Rose Garden, which is situated just in front of TU Central Library, was established during the state visit of Dr. Zakir Hussein (Late President of India) and comprises the richest collection of roses in the country (Bajracharya et al., 1997).

1.9 PUBLIC GARDENS AND PARKS

Some gardens and parks with the potential to conserve plant biodiversity are spread throughout the federal capital (Figure 1.1). There are such gardens also outside the capital in different parts of the country (Plates 7–9), but at present, they are not well managed, decorated only with ornamental plants that are mostly exotic; if proper guidance was provided for the management of these gardens and parks, they could serve as important conservation sites for many species. Nevertheless, the difference between botanical gardens and parks should be realized. As there is a growing trend to establish parks and gardens at the local level, they should be properly designed, and guidance should be provided for selection of the proper species.

A few of these parks are as follows (Source: https://trip101.com/article; https://english.online khabar.com/public-parks):

Balaju Water Garden, located in Balaju, the foothill of Nagarjun Hill (2,096 m), is a famous recreational site with beautiful flower gardens. It was built in the eighteenth century and is famous for its 22 stone water spouts, which are well-known as "Baees dhara Udhyan."

Bhugol Park (est. 1934), the first public park in Nepal, is located in the heart of the city, New Road, Kathmandu.

Ratna Park (est. 1964), recently renamed as Shankhadhar Park, is located between Rani Pokhari and Tundikhel.

Garden of Dreams, Keshar Mahal, Kathmandu (built in the 1920s), is located at the eastern edge of Thamel and covers 9.7 hectares.

United Nations Park (UN Park, est. 1997) is situated along the bank of Bagmati River, stretching from Shankhamul Ghat to Teku Dobhan in Lalitpur, covering 4.2 hectares. The park was established to commemorate the 50th anniversary of the United Nations, but construction is still undergoing.

In addition, several public parks and gardens are located in different corners of Kathmandu Valley, including the following:

Tribhuvan Park (est. 1972) is located in Thankot, the foothill of Chandragiri Hill, on the way to Prithvi Highway.

Manjushree Park is located near the Manjushree cave, Chovar, on the way to Dakhsinkali, south of Kathmandu.

Shankhamul Park is located in the Shankhamul Ghat area along Bagmati River, and is managed by the NRNA (Non-Resident Nepali Association).

Amideva Buddha Park, popularly known as Swayambhu Park, is located in the foothill of Swayambhu Nath temple, one of the popular World Heritage sites in the Kathmandu Valley.

One of the popular destinations in Kathmandu Valley for hiking is **Switzerland Park** (Indra Daha), located in Ramkot, Chandragiri, at an elevation of 1,830 m. The park is constructed around the cemetery of Kaji Kalu Pande, the Chief Commander of the Late King Prithvi Narayan Shah. Besides a historic pond, which is called "Indra daha," the park is also a popular pilgrim site destination due to the Hindu temples of Bindabasini and Mankamana (https://www.nepaladventureteam.com/indra-daha-switzerland-park-day-hiking).

1.10 OPPORTUNITIES AND CHALLENGES

Existing policies and current practices have contributed in many ways to the establishment of botanical gardens in the country. However, inadequate technical knowledge and funding for the extension and management of botanical gardens and the lack of trained staff, international exposure, and networking are some of the challenges to be addressed. Networking and collaboration with gardens in other countries would be useful to exchange ideas, and the exposure of people managing botanical gardens would help to bring new ideas for better management of the gardens. One of the opportunities is to invite corporations to support the management of the gardens where companies

Botanical Gardens and Plant Conservation Initiatives in Nepal

PLATE 7 Balaju Baeesdhara Park (Water Garden).

Photo credits: AM Kayastha (Photos 17–21).

22. Hindu temple inside Balaju Water Garden

23. 'Chautari', a rest place in the garden

24. Panoramic view of the central part of the Balaju Water Garden

25. Northern part of the water garden

26. Statue of Bhanu Bhakta, the renowned poet of Nepal, near the entrance of the garden

PLATE 8   Balaju Water Garden (cont.).

Photo credits: AM Kayastha (Photos 22–26).

Botanical Gardens and Plant Conservation Initiatives in Nepal

27. CDB TU Botanical Garden: Central view

28. Right wing of TU Botanical Garden

29. Panoramic view of Central Department of Botany, Tribhuvan University, Kirtipur

30. Dr. Zakir Hussain Rose Garden

31. Rose yard in Zakir Hussain Rose Garden

PLATE 9 TU Botanical garden and Zakir Hussain Rose Garden.

Photo credits: AM Kayastha (Photos 27–31).

such as commercial banks can easily support financial resources for management of the gardens. Inadequate human resources for maintaining the gardens is often discussed by managers. It is high time to call for better management of existing gardens with trained human resources and adequate financial resources while extending the networks of botanical gardens in the country to include major biodiversity hotspots. This is more relevant in the context of climate change and also in the present age of rapid loss of biodiversity.

As new botanical gardens are also being established in different parts of the country, it is important to know the objectives and role of botanical gardens. These new gardens are more parks for recreation, and the objectives of research and education are missing. Therefore, it is also important to develop guidelines for the establishment of such gardens. The Department of Plant Resources would have additional roles in this aspect, and studies are required to understand how representative are the present distribution of botanical gardens in the country and how many species are protected *in situ* and *ex situ* in these networks. As the Global Plant Council rightly noted, the importance of botanical gardens should not be underestimated (Global Plant Council, 2016) . The role of botanical gardens is not outdated – they are more relevant than ever – but these institutions should also be able to show their dynamic role by conducting cutting-edge research. Botanical gardens are not only visitor attractions but also scientific institutions (Smith, 2019).

1.10.1　*Ex Situ* Conservation of Rare, Threatened, and Endemic Species

The population of several species is decreasing in the natural habitat due to deforestation, overgrazing, slash and burn practices, expansion of agricultural lands, and habitat loss for infrastructure development. In this context, it is high time to emphasize the conservation of wild plants by *ex situ* conservation (off-site, or botanical gardens) to mitigate the risk of extinction of the species (Kovacs et al., 2021), and by *in situ* conservation (in natural habitat) through the active participation of institutions and communities.

Learning lessons from successful botanical gardens, the National Botanical Garden, as well as other botanical gardens and parks of Nepal, should focus on the maintenance of seedbanks by collecting seeds from the natural habitat of targeted species. Attempts should be made for the *ex situ* conservation of endemic species; plants belonging to the IUCN Red List Categories, such as endangered, threatened, and vulnerable species; and economically viable medicinal plants, representing species from lowland Tarai to the mid hills and high hills of all seven provinces of Nepal. In general, for effective *ex situ* conservation, propagules (viable seeds, seedlings, cuttings, *in-vitro* propagated plants, and planting materials) of the selective species should be collected from the natural habitat from the edaphic, geographical, and altitudinal range (Justice, 2015; Kovacs et al., 2021). It is suggested to conserve the following species of rare and threatened plants in the botanical gardens:

- Lowland species: *Butea monosperma* (Palans), *Cycas pectinata* (Kalbal), *Dalbergia latifolia* (Sati sal), *Gnetum montanum* (Bhote lahara), *Horsfieldia kingii* (Ban supari), *Pterocarpus marsupium* (Bijay sal), *Putranjiva roxburghii* (Pittamari), and *Senegalia catechu* (Khayar).
- Middle-altitude species: *Corylus ferox* (Lek katus), *Elaeocarpus angustifolius* (Rudrakshya), *Magnolia hodgsonii* (Bhalu kath), *Oroxylum indicum* (Tatelo), *Pandanus furcatus* (Bantari), *Podocarpus neriifolius* (Gunsi), *Taxus mairei* (Singi), and *Ulmus wallichiana* (Dhamina).
- Higher-altitude species: *Betula utilis* (Bhojpatra), *Dactylorhiza hatagirea* (Paanchaunle), *Ephedra gerardiana* (Somlata), *Podophyllum hexandrum* (Laghupatra), *Rheum nobile* (Kyanjo), *Rhododendron cowanianum*, and *Rhododendron lowndesii* (Barjhum mhendo)

1.10.2 Developing an Online Portal of Botanical Gardens Database

It is widely realized that emphasis should be given to the comprehensive documentation of all the species growing in the botanical gardens for user-friendly access to their collections for education, research, and conservation. Online databases are widely used to access the identification of living collections. The PlantSearch Database managed by BGCI comprises information about 640,000 taxa, contributed by nearly 1,200 institutions. It provides data on IUCN Red List threatened species, medicinal plants, wild crop relatives, and other plant species in the living collections of botanical gardens, facilitating education, research, and conservation efforts (Marquardt et al., 2020; https://tools.bgci.org/plant_search.php).

Identical to the online database of BGCI, the National Botanical Garden of Nepal may develop an online database of the living collections conserved in the botanical gardens managed by the Department of Plant Resources throughout the country; this could also be linked with the living collections managed by the public gardens and parks of the country. Collaboration is needed between the representatives of the gardens, taxonomists, GIS experts, and database experts to develop a user-friendly online database of botanical gardens of Nepal.

Networking of all the botanical gardens of Nepal and sharing their records of living collections should be initiated. For uniformity, a central database system could be designed, incorporating the following information per taxon: family name, accepted name with authority, most common synonyms, Nepali name (in Devanagari and English translation), common English name, phenology (flowering/fruiting), distribution range, elevation range, origin (native country), IUCN threat categories/CITES appendices, and use value. Furthermore, barcodes of each collection, GIS location, and photographs (habitat, inflorescence and closeup of flowers/fruit) should be incorporated. A distribution map of the taxa, as well as a location map of each collection, could be linked within the database. Besides this, additional information on horticultural practices for propagation and conservation measures should be appended.

1.11 CONCLUSION AND RECOMMENDATIONS

The initiation of *in situ* and *ex situ* conservation of plant species through botanical gardens is a comparatively recent practice in Nepal. Though there are only 12 such gardens officially recognized, other parks and gardens could also play important roles in plant conservation. The representativeness of these gardens to include various ecoregions should be further assessed. The National Botanical Garden of Nepal, with networking of other botanical gardens situated in various ecological regions of Nepal, managed under the umbrella of the Department of Plant Resources, is playing a key role in the display of native as well as introduced exotic species, attracting the general public and students. The mission of these gardens is to promote the *ex situ* conservation of native and threatened species, representing various ecological zones of Nepal. In general, other public botanical gardens and parks are not initiating display and management of native species. Therefore, it is recommended to form a broader network by inviting all the public gardens and parks to achieve the goal of conservation of plant species diversity.

The National Botanical Gardens should launch education and research initiatives by developing and/or strengthening various sections, such as a herbal garden section with a display of a wide variety of medicinal and aromatic plants; horticulture sections, focusing on the display of varieties of horticultural crops, with emphasis on fruit orchards; and floriculture and agronomy sections, among others. Besides these sections, emphasis should be given to upgrading the systematic garden, demonstrating evolution of classification systems such as the artificial or sexual system (eighteenth century), Carl Linnaeus's system; the natural system of classification (Bentham and Hooker); the phylogenetic system (Engler and Prantl's, and Takhtajan's systems); and the Angiosperm Phylogeny Group IV system. Obviously, such an upgraded National Botanical Garden will attract more visitors, especially school and college students, enhancing practical knowledge of plant biodiversity.

ACKNOWLEDGMENTS

The authors are grateful to the concerned authorities of the botanical gardens, public gardens, and parks for giving us permission to visit and photograph, especially to Dipak Lamichhane (Chief, Senior Garden Officer) and Jeevan Pandey (Garden Officer), National Botanical Garden, Godavari, Lalitpur, for their cooperation during the tour of the garden and for providing adequate information of the garden. Special thanks are also due to AM Kayastha (Ayush Man Kayastha), Smart Pix Studio (P.) Ltd., Bijeshwari, Kathmandu, for the photography of the gardens in the Kathmandu Valley; RR Parajuli (Raghu Ram Parajuli), Senior Scientific Officer (Chief, Plant Research Centre), Makwanpur; and Jeevan Pandey (former Chief, Mulpani Botanical Garden, Kapurkot, Salyan, and Dibas Shrestha (Central Department of Hydrology and Meteorology, Tribhuvan University, Kirtipur) for sharing the photographs and the map of Botanical gardens in Nepal, respectively.

REFERENCES

Bajracharya, D., K.K. Shrestha and R.P. Chaudhary. 1997. *Garden Flowers: An Illustrated Guide to Indoor and Outdoor Garden Plants in Nepal*. The King Mahendra Trust for Nature Conservation (KMTNC), Jawalakhel, Lalitpur.

Bown, D. 1992. *Four Gardens in One*. The Royal Botanic Garden Edinburgh, HMSO Publications, Edinburgh.

Convention on Biological Diversity (CBD). 2012. *Global Strategy for Plant Conservation: 2011–2020*. Botanic Gardens Conservation International, Richmond, UK.

Chaudhary, R.P., Y. Uprety, S. Devkota, S. Adhikari, S.K. Rai and S.P. Joshi. 2020. Plant biodiversity in Nepal: Status, conservation approaches, and legal instruments under new federal structure. In: M. Siwakoti et al. (eds.), *Plant Diversity in Nepal*. Botanical Society of Nepal, Kathmandu, pp. 167–206.

Chen, G. and W. Sun. 2018. The role of botanical gardens in scientific research, conservation, and citizen science. *Plant Diversity* 40(4): 181–188.

Devkota, S. 2022. History of endangered Nepalese plants in the Narayanhiti Museum (*Narayanhiti Sangrahalayama lop bhaeko Nepali Banaspatiko Itihas*, in Nepali). *Shilapatra Utkhanan Part 20*. https://shilapatra.com/detail/75049

DPR. 2017. *Annual Progress Report of FY 2073/074)*. Department of Plant Resources, Kathmandu, Nepal.

DPR. 2021. *Explore the Beauty within the Floral Diversity*. Department of Plant Resources, Kathmandu.

Hughes, K. and D. Lamichhane. 2017. Garden profile: The national botanical garden of Nepal. *Sibbaldia* (15): 9–30.

IPBES. 2019. Summary for policymakers of the global assessment report on biodiversity and ecosystem services of the Intergovernmental Science-Policy Platform on Biodiversity and Ecosystem Services.

IUCN. 2010. IUCN Red List of Threatened Species, Version 2010.4, IUCN.

Justice, D. 2015. The role of botanical gardens in plant conservation. Proceedings of the 2015 Annual Meeting of the International Plant Propagator's Society, pp. 127–129, Bellefonte, PA, USA.

Kovacs, Z., A.M. Csergo, P. Csontos and M. Hohn. 2021. Ex-situ conservation in botanical gardens: Challenges and scientific potential preserving plant biodiversity. *Notulae Botanicae Horti Agrobotanici Cluj-Napoca* 49(2): 12334. DOI: 10.15835/nbha49212334

Lamichhane, D. 2018. Ex-situ conservation of plants and botanical gardens in Nepal. In: Dhakal, M. et al. (eds.), *25 Years of Achievement on Biodiversity Conservation in Nepal*. Ministry of Forests and Environment, Kathmandu, pp. 61–65.

Lamichhane, D. and S.K. Rai. 2021. Role of botanical gardens of Nepal in biodiversity conservation, education and research. In: Siwakoti, M. et al. (eds.), *Integrating Biological Resources for Prosperity*. Botanical Society of Nepal, Nepal Biological Society and Department of Plant Resources, Kathmandu, Nepal.

Marquardt, J., N. Koster, G. Droege, J. Holetschek, A. Guntsch, F. Bratzel, G. Zizka, M.A. Koch and T. Borsch. 2020. What grows where? Towards an infrastructure to connect collections of Botanic gardens for research and Conservation. In: Espirito-Santo, M.D., A.L. Soares and M. Veloso (eds.), *Botanic Gardens, People and Plants for a Sustainable World*. IsaPress, Lisboa (Lisbon), p. ????.

Mounce, R., P. Smith and S. Brockington. 2017. Ex situ conservation of plant diversity in the world's botanic gardens. *Nature Plants* 3(10): 795–802. http://doi:10.1038/s41477-017-0019-3

Pautasso, M. and I. Parmentier. 2007. Are the living collections of the World's botanical gardens following species richness patterns observed in natural ecosystems? *Bot. Helv.* 117: 15–28. DOI: 10.1007/s00035-007-0786-y

Primack, R.B. and A.J. Miller-Rushing. 2009. The role of botanical gardens in climate change research. *New Phytol.* 182: 303–313.

RBG Kew. 2016. *The State of the World's Plants Report-2016*. Royal Botanic Gardens, Kew.

Sharrock, S.L. 2011. The biodiversity benefits of botanic gardens. *Trends Ecol. Evol.* 26: 433.

Shrestha, K.K., P. Bhandari and S. Bhattarai. 2022. *Plants of Nepal* (Gymnosperms and Angiosperms). Heritage Publishers & Distributors, Kathmandu, Nepal.

Shrestha, K.K. and S.B. Bajracharya. 2019. Biodiversity in Nepal. In: Pullaiah, T (ed.), *Global Biodiversity*, Vol. 1 (Asia). Apple Academic Press Inc., Burlington, Canada and Palm Bay, USA, pp. 427–472.

Smith, P. 2019. The challenge for botanic garden science. *Plants People Planet* 1: 38–43. DOI: 10.1002/ppp3.10

Volis, S. 2017. Conservation utility of botanic garden living collections: Setting a strategy and appropriate methodology. *Plant Diversity* 39(6): 365–372. https://doi.org/10.1016/j.pld.2017.11.006

Xia B., R. Chu, H. Yu, Z. Zhang and Y. Gu. 2021. The history of botanical gardens. In: *Phytohortology*. EDP Sciences, Les Ulis, pp. 10–22. https://doi.org/10.1051/978-27598-2531-8.c008

ONLINE RESOURCES

http://kath.gov.np/
https://dpr.gov.np/
https://globalplantcouncil.org/the-importance-of-botanic-gardens-in-the-21st-century/
https://tools.bgci.org/plant_search.php
The Switzerland Park. https://nepaladventureteam.com/indra-daha-switzerland-park-day-hiking
Ten Public Parks in Kathmandu. https://english.onlinekhabar.com/public-parks-in-kathmandu.html
Top 10 Gardens and Parks in Kathmandu. https://trip101.com/article/garden-of-dreams
www.bgci.org
www.icimod.org/initiative/godavari-knowledge-park/
www.kew.org

2 Role of Indonesian Botanic Gardens in Plant Conservation

Joko Ridho Witono and Irawati

CONTENTS

2.1 Introduction 25
2.2 Indonesian Botanic Gardens and Their Important Contributions throughout the Ages 26
    2.2.1 Bogor Botanic Gardens 26
        2.2.1.1 Cibodas Botanic Gardens 28
        2.2.1.2 Purwodadi Botanic Gardens 28
        2.2.1.3 Eka Karya Bali Botanic Gardens 28
        2.2.1.4 Cibinong Science Center and Botanic Gardens
              (Cibinong Botanic Gardens) 28
2.3 Role and Function of Indonesian Botanic Gardens 29
    2.3.1 Plant Conservation 29
    2.3.2 Research 30
    2.3.3 Education 31
    2.3.4 Tourism 31
    2.3.5 Environmental Services 31
    2.3.6 Plant Collection Enrichment in the Gardens 32
        2.3.6.1 Exploration 32
        2.3.6.2 Seed Exchange 34
        2.3.6.3 Seed Bank 34
        2.3.6.4 Plant Propagation 35
        2.3.6.5 Donation 35
2.4 Priority Setting for Plant Conservation in Indonesian Botanic Gardens 35
2.5 Developing New Botanic Gardens 38
2.6 Future Programs of Bogor Botanic Gardens 44
    2.6.1 Development and Supervision of Local Botanic Gardens 44
    2.6.2 National Focal Point GSPC 46
    2.6.3 Research Program 47
Acknowledgments 48
References 48

2.1 INTRODUCTION

For more than 185 years, Indonesia only had four botanic gardens: Bogor, Cibodas, Purwodadi, and Eka Karya Bali. At the same time, in the early twenty-first century, the USA, China, Australia, India, Russia, and the UK had more than 100 botanic gardens. As a tropical country with great biodiversity, Indonesia faces great pressure on its natural vegetation. Deforestation for development was unavoidable when industry developed, agriculture and estate crops extended, settlement for people increased, and exploitation of natural resources grew.

When people were unable to find some important plants in the wild and they found them in the botanic gardens, people realized that botanic gardens are not only collection gardens but also have a more important role as conservation gardens. Decreasing vegetation not only occurred in the forest but also in the cities during development and consequently reduced the quality of air in urban areas. People would need open green spaces with fresh air, therefore botanic gardens were also suitable for developing into better recreation areas, and at the same time, people also learn more about vegetation and environment at them. When plant conservation became an action priority, the existing botanic gardens were not enough to save Indonesian plants, and now botanic gardens have a strategic position as *ex situ* plant conservation areas in Indonesia. Indonesia has been blessed with the opportunity to develop more botanic gardens in the country since 1999.

In this chapter, contributions throughout the ages of Indonesian botanic gardens and their multiple tasks, plant collection enrichment, priority setting on plant conservation, development of new botanic gardens, and future programs of Indonesian Botanic Gardens, mainly Bogor Botanic Gardens, are presented.

2.2 INDONESIAN BOTANIC GARDENS AND THEIR IMPORTANT CONTRIBUTIONS THROUGHOUT THE AGES

The diversity of living organisms in Indonesia and the surrounding areas has been a great attraction to many Europeans wanting to know more. Therefore, in the fifteenth century, Europeans began traveling to this area to satisfy their curiosity. Irregularly, Swedish, French, and Spanish explorers visited East Hindia. The famous Georg Eberhard Rumphius visited and stayed in Ambon (1652–1702) and handed down his masterpiece *Herbarium Amboinense*. Alfred Russel Wallace's travels through the Malay Archipelago (1854–1913) is another example of a great naturalist that inspired the theory of natural selection of Charles Darwin. The diversity of Indonesian plants and the need for locations for the introduction of economic plants prompted the establishment of botanic gardens in Indonesia. The following are five botanic gardens managed by the Indonesian Institute of Sciences (LIPI) and their characteristic plant collections.

2.2.1 Bogor Botanic Gardens

King Willem I of the Netherlands appointed Caspar Georg Carl Reinwardt to discover the potential of his colony in the Dutch East Indies (now Indonesia) in 1815 (Goss, 2004). A garden was then established on 18 May 1817. Many interesting plants had been collected by C.G.C. Reinwardt, and the governor general allowed him to use a piece of land for planting his collection that initiated *s'Lands Plantentuin te Buitenzorg* (Bogor Botanic Gardens). This garden is 87 ha, beside the 28.4 ha Governor General Residence and now Bogor Presidential Palace. The garden and the palace are located at the center of Bogor city (Safarinanugraha et al., 2017).

Bogor Botanic Gardens celebrated its 200th anniversary in 2017. This garden has a long history. It was established and passed through different government authorities, managed by different citizenships from generation to generation, and continues to serve the country through plant conservation, science research, reference gardens, and a recreation area. It also provides a better environment for the surrounding areas. In the beginning, the garden was an important place to acclimatize economic plants from all over the world. More than 40 species of plants were introduced to Indonesia through Bogor Botanic Gardens, such as rubber (*Hevea brasiliensis*) from Brazil; oil palm (*Elaeis guineensis*) from Africa; quinine (*Cinchona calisaya*) from Peru; coffee (*Coffea canephora*) from Ethiopia; cacao (*Theobroma cacao*) from Brazil; vegetables, such as potato (*Solanum tuberosum*), eggplant (*Solanum melongena*), carrot (*Daucus carota*), cabbage (*Brassica oleracea*), chilli (*Capsicum* spp.), cassava (*Manihot utilissima*), and corn (*Zea mays*); and also fruits, such as strawberry (*Fragaria x ananassa*), apple (*Malus domestica*), pear (*Pyrus communis*), and so on (Kartawinata, 2010b).

At present, the plant collection of Bogor Botanic Gardens consists of 23,541 accessions and 4,273 cultivated taxa. Special collections include palms, orchids (more than 500 species), *Rafflesia*, dipterocarps, aroids, bamboo, ferns, pandanus, water plants, cacti and succulents, medicinal plants, climbers, shrubs, and an extensive collection of tropical woody species (Maryanto et al., 2013).

Herbarium collections, which have been a part of the garden since it was established in 1844, later became an independent new institution, Herbarium Bogoriense (BO). Since 2006, the garden collection herbaria (Hortus Botanicus Bogoriense) and voucher specimens of the garden's living collection were kept in a herbarium inside Bogor Botanic Gardens, consisting of about 50,000 specimens, including type specimens, dry herbaria, spirit, seed, and wood collections. A virtual visit to the herbarium is available through Hortus Botanicus Bogoriense Virtual Herbarium (HBBVH) (Jasa Ilmiah – KRB-LIPI, 2020).

In the history of Bogor Botanic Gardens, the garden officially has had several names. At the beginning, as a place to plant the collection, its first name was *s'Lands Plantentuin te Buitenzorg* (National Botanic Gardens, Bogor). Later, when it was beautifully designed and the collections were arranged, this big garden was famous as *Kebun Raya Bogor* or Bogor Botanic Gardens. Bogor Botanic Gardens is also known as Hortus Botanicus Bogoriensis and Jawatan Penyelidikan Alam, and it then became the Lembaga Pusat Penyelidikan Alam (LLPA), Kebun Raya Indonesia (Indonesia Botanic Garden), Shokubutsuen (Botanic Gardens), Unit Pelaksana Teknis Balai Pengembangan Kebun Raya (Botanic Garden Development Division), Pusat Konservasi Tumbuhan Kebun Raya Bogor (Center for Plant Conservation Bogor Botanic Gardens), and the latest, Pusat Penelitian Konservasi Tumbuhan dan Kebun Raya – LIPI (Research Center for Plant Conservation and Botanic Gardens, Indonesian Institute of Sciences).

Bogor Botanic Gardens is a forerunner of scientific institutions in Indonesia. Many institutions in Indonesia were initiated by Bogor Botanic Gardens, then later separated, developed, and reorganized into new institutions, such as Bibliotheca Bogoriensis (1842), Herbarium Bogoriense (1844), Cibodas Botanic Gardens (1852), Cikeumeuh Economic Gardens (1876), Photography and Illustration Institutions (1878), Museum Zoologicum Bogoriense (1894), Ocean Research Institution (1904), Sibolangit Hortus Botanicus (1914), Nature Preservation Institution (1937), Purwodadi Botanic Gardens (1941), Flora Malesiana Foundation (1950), Setia Mulia Nature Science Institution (1955), Biology Academy (1956), and Microbiological Institution (1956) (Sukarya and Witono, 2017). However, the old botanic garden has remained unchanged, even through the European and Japanese occupation of this country until Indonesia became an independent country. The garden has a strong foundation that easily implemented the development of government programs, and Bogor Botanic Gardens is the most important garden in Indonesia.

The herbarium collection of Bogor Botanic Gardens, with its important collections of C.L. Blume, J.K. Hasskarl, C.G.C. Reinwardt, J.J. Smith, and J.E. Teijsmann, in 1844 separated from the Botanic Garden and became an independent institution, namely Herbarium Bogoriense (BO). At present, BO is the National Herbarium of Indonesia with the biggest herbarium collections in Southeast Asia. BO hosts nearly 2 million herbarium specimens (more than 13,000 type specimens, 1.28 million dry specimens, 50,000 spirit collections, fungi, fossils, woods, etc.). The BO specimens represent not only all major islands in Indonesia but also the Malesia region. Historical collections from C.L. Blume, J.K. Hasskarl, J.E. Teysmann, S.H. Kooders, C.A. Baker, A.G.O. Penzig, K.B. Boedijn, A.J.G.H. Koostermans, J.J. Smith, and R.E. Holttum are also kept in BO. Herbarium collections at BO are constantly augmented through field expeditions by succeeding herbarium staff, gifts, and exchanges with other institutions in Indonesia and abroad. About 5,000 new specimens are added yearly to the BO collection. The care of the collections, or curation, is undertaken with great precision by the curators in the herbarium. All plant groups, flowering plants, gymnosperms, ferns and lycophyte, mosses, liverworts, and fungi are represented in the collection (Herbarium Bogoriense, 2020).

The BO building was formerly next to Bogor Botanic Gardens, but it has been moved to a new location in Cibinong with a new building supported by the World Bank through the Global Environmental Facility (GEF). Improvement of the management system, newly recruited staff, new facilities for improving the collection, and digitalization of the collection give better services to scientists visiting Herbarium Bogoriense, as well for loans and exchange activities (Irawati, 2003).

There are four branches of Bogor Botanic Gardens, as described in the following sections.

2.2.1.1 Cibodas Botanic Gardens

Cibodas Botanic Gardens was established in 1852. It is located in Cianjur, West Jawa, to conserve montane plants in wet climates in the western part of Indonesia. The garden covers 84.99 ha (210.0 acres), situated on the slopes of Mount Gede, at 1,300–1,425 m asl. (Sujarwo et al., 2019). This garden is famous as a place to introduce a high-value economic plant, quinine (*Cinchona calisaya*), and also for research on mycorrhiza that was initiated in the laboratory of Cibodas Botanic Gardens in 1890–1897 (Immamudin et al., 2006).

At present, Cibodas Botanic Gardens has 11,746 specimens (2,040 species) in the garden, including 701 specimens (122 species) of orchids, 1,004 specimens (95 species) of cacti, 1,287 specimens (68 species) of succulent plants, 180 specimens (101 species) of ferns, 222 specimens (46 species) of *Nepenthes*, and 67 specimens (18 species) of Gesneriads (Kebun Raya Cibodas, 2023a). The garden collections are mainly obtained from exploration, materials exchange with other botanic gardens, and donation. Additional collection between 2008–2018 added 473 specimens, of which 248 are from exploration, 217 from donation, 7 from propagation, and 1 from seed exchange (Hidayat et al., 2019). Important collections that attract a lot of people, such as *Amorphophallus titanum*, have been successfully propagated by seeds as well as *in vitro* culture. Generative propagation of the species results in interesting character variations. The garden also has important collections such as *Nepenthes*, moss, and tree fern species.

Cibodas Botanic Gardens Herbarium (Cianjur Herbarium Hortus Botanicus Tjibodasensis or CHTJ) was established in 1891 and now hosts more than 10,000 specimens. The main herbarium collections are dry and spirit collections and moss and seed collections (Kebun Raya Cibodas, 2023b).

2.2.1.2 Purwodadi Botanic Gardens

Purwodadi Botanic Gardens is located in Purwodadi, Pasuruan, East Java. The garden was established in early 1941 by Dr. Gerhard Lourens Baas Becking Marinus as a division of Bogor Botanic Gardens. It covers an area of 85 ha (210 acres) at an altitude of 300 m asl. Initially, the gardens were used for plantation crop research activities, then in 1954, the basics of a botanic garden began to be applied with the establishment of patches of plant collections. Most of the plant collections were reorganized according to the Engler classification system from 1980.

The garden is a major center of research on plant conservation for lowland plants in dry climates (Kebun Raya Purwodadi, 2023). At present, Purwodadi BG hosts more than 13,760 plant collections belonging to 179 families, 980 genera, and 2,207 species. Purwodadi BG herbarium preserves 3,531 herbarium specimens from 1,016 species, 669 genera, and 199 families in 2020 (Mudiana et al., 2020).

2.2.1.3 Eka Karya Bali Botanic Gardens

Eka Karya Bali Botanic Gardens was established in 1959 at an altitude of 1,300 m asl. Garden collections are conserving montane plants in wet climates in the eastern part of Indonesia. The 157.5 ha area of the garden is located at the central highland of Bali Island, and part of it is natural forest. Eka Karya Bali Botanic gardens has more than 21,000 living specimens from 2,400 species of trees, shrubs, orchids, ferns, begonias, cacti, bamboo, carnivorous plants, plants used for Balinese ceremony, and medicinal plants.

2.2.1.4 Cibinong Science Center and Botanic Gardens (Cibinong Botanic Gardens)

Cibinong Botanic Gardens covers 32 ha at an altitude of 165 m asl. And is located in the middle of a scientific institution Cibinong Science Center and next to Herbarium Bogoriense. It is not only greening and cooling the Science Center, but at the same time also became an ideal place for ecological studies. This garden is also known as Ecopark. Unlike most botanic gardens, Cibinong Botanic Gardens collections are arranged not in family groups but to represent seven Indonesian bioregions of lowland tropical rain forest, from Sumatra, Jawa and Bali, Kalimantan, Sulawesi, Mollucas, Lesser Sunda Islands, and Papua (Nafar and Gunawan, 2017). The garden has collected 1,147 specimens belonging to 86 families, 328 genera, and 733 species (Ariati et al., 2019; Hutabarat et al., 2020).

2.3 ROLE AND FUNCTION OF INDONESIAN BOTANIC GARDENS

The botanic garden plays a very important role in plant conservation, especially endemic and threatened plants in Indonesia. As a complement of *in situ* conservation managed by the Ministry of Environment and Forestry, some definitions of botanic garden are as follows:

> A botanical garden is a collection of growing plants, the primary purpose of which is the advancement and diffusion of botanical knowledge.
>
> *(Lawrence, 1969)*

> Botanic gardens are institutions holding documented collections of living plants for the purposes of scientific research, conservation, display, and education.
>
> *(Wyse Jackson and Sutherland, 2000)*

Based on these definitions, laws, and regulations, and experience in managing the Indonesian Botanical Gardens, the Bogor Botanic Gardens team formulated the definition of a botanic garden in Indonesia. The definition is very important as a legal basis for developing new botanic gardens in the future, which is later expected to become a national program involving various Ministries and Institutions in Indonesia. Regulation related to a botanic garden was established through Presidential Regulation Number 93 of 2011, which was stipulated on 27 December 2011. A botanic garden is defined as an *ex situ* plant conservation area that has a documented collection of plants and is organized based on taxonomic classification, bioregions, thematics, or a combination of these patterns for the purpose of conservation, research, education, tourism, and environmental services.

Based on the definition, botanic gardens in Indonesia have five main functions, namely plant conservation, research, education, tourism, and environmental services.

2.3.1 PLANT CONSERVATION

Conservation in the botanic gardens is carried out through preserving the diversity of plant species outside their habitat and conducting function studies for sustainable use. The wealth of plant diversity is the main research material and a source of propagation in the context of species reintroduction and ecosystem restoration. In this context, the meaning of conservation in a botanic garden is not only to save the existence of plants but also to explore the uses of plants and utilize them for human welfare. Currently, various research activities on the use of Indonesian plants are being carried out, such as *Amorphophallus* spp., fruit, and ornamental plants.

Botanic gardens in Indonesia strive to be the foremost botanical gardens in increasing the number of plant collections, especially endemic and threatened plants native to Indonesia. The increase in the number of collections is carried out through plant collection activities throughout Indonesia's tropical forests, material exchange, and donations from various research institutions and individuals who are concerned about the importance of plant conservation in Indonesia.

2.3.2 Research

Bogor Botanic Gardens is the first botanical and agricultural research institution in Indonesia. The research of Bogor Botanic Gardens is published in various world-leading journals. Also, in 1876, the Bogor Botanic Gardens published the scientific journal *Annales du Jardin Botanique de Buitenzorg*, which was managed by Dr. R.H.C.C. Scheffer and became an important reference for researchers in botany and agriculture. Its role as a research center was strengthened by the establishment of a laboratory in the garden that focused on tropical research in 1884 during the leadership era of Dr. Melchior Treub. Currently, the laboratory is still functioning as a center for experimental botany research and was given the name Treub Laboratory in appreciation of Dr. Melchior Treub (Sukarya and Witono, 2017).

Research in the Bogor Botanic Gardens that was conducted by foreign and Indonesian researchers has become a reference in botanical and agricultural research to date, such as the discovery of geotropism in orchids (Tischler, 1905), the hormone auxin by Frits W. Went in Treub Laboratory (Went, 1935), mycorrhiza in orchids found in the Bogor Botanical Gardens branch laboratory in Cibodas (Burgeff, 1959), and flower induction on *Dendrobium crumenatum* orchids (Rutgers and Went, 1916), which is now used commercially.

At present, all botanic gardens in Indonesia carry out and facilitate various research and development activities in the fields of botany, conservation, cultivation, and plant utilization. The wealth of plant collections is the main asset in carrying out research functions in botanic gardens. Research in Indonesian Botanic Gardens is directed at the fields of plant conservation, domestication, reintroduction of rare plants, and ecosystem restoration. Plant conservation research includes floristic studies, characterization, and plant systematics using molecular, chromosomal, and macromorphological characters.

Research on plant domestication is focused on food crops, such as *Amorphophallus* spp.; fruit crops, such as *Baccaurea* spp. and *Musa* spp.; and ornamental plants, particularly *Hoya* spp., *Aeschynanthus* spp., *Begonia* spp., and *Nepenthes* spp. Research on the reintroduction of rare plants has been carried out on 12 species (Table 2.1). The main obstacle faced in the rare plant

TABLE 2.1
The Rare Plant Reintroduction Program Carried Out by the Indonesian Botanic Gardens

No.	Species	Conservation Status	Total Seedlings	Location and Year of Planted
1.	*Pinanga javana* Blume	Endangered (WCMC, 1997); Endemic to Jawa	5,200	Gunung Halimun Salak NP, West Java (2005)
2.	*Calamus manan* Miq.	Vulnerable (WCMC, 1997)	670	Bukit Duabelas NP, Jambi (2006)
3.	*Alstonia scholaris* (L.) R.Br.	Low risk (IUCN, 2013)	1,000	Ujung Kulon NP, Banten (2007)
4.	*Parkia timoriana* (DC.) Merr.	Rare (Zuhud et al., 2003)	1,000	Meru Betiri NP, East Java (2007)
5.	*Intsia bijuga* (Colebr.) O. Kuntze	Vulnerable A1cd (WCMC, 1998)	500	Ujung Kulon NP, Banten (2009)
6.	*Diospyros macrophylla* Blume	Vulnerable A1c (Mogea et al., 2001)	100	Ujung Kulon NP, Banten (2009 and 2014)
7.	*Stelechocarpus burahol* (Blume) Hook.f. & Thomson	Low risk (Mogea et al., 2001)	400	Ujung Kulon NP, Banten (2009)
8.	*Vatica bantamensis* (Hassk.) Benth. & Hook.ex Miq.	Endangered A1c, D (Ashton, 1998); Endemic to Banten	100	Ujung Kulon NP, Banten (2014)
9.	*Heritiera percoriacea* Kosterm.	Endangered B1+2c (WCMC, 1998); Endemic to Banten	500	Ujung Kulon NP, Banten (2014)
10.	*Dacrycarpus imbricatus* (Blume) de Laubenf.	Least concern (Thomas, 2013)	2,000	Batukaru Protection Forest, Bali (2005–2012)
11.	*Elaeocarpus grandiflorus* J.E.Sm.	Near threatened (www.mybis.gov.my/, 2018)	15	Batukaru Protection Forest, Bali (2005–2012)
12.	*Michelia champaca* L.	Least concern (www.gbif.org/, 2021)	2,000	Batukaru Protection Forest, Bali (2005–2012)

reintroduction program is limited resources, especially funds, because reintroduction requires a very high budget. Collaboration with *in situ* management authorities and other research institutions is crucial to the success of this program in the future.

2.3.3 EDUCATION

Bogor Botanic Gardens and all Indonesian botanic gardens provide information to increase knowledge in the fields of botany, conservation, environment, and plant utilization and to stimulate the growth and development of awareness, concern, responsibility, and commitment of the wider community toward plant conservation and environmental sustainability. The botanic garden is a library and plant laboratory that has very complete value and information compared to other conservation and education places. Plant collections and data collections are important sources of information for students, teachers, researchers, and the wider community to know more about the plant world.

In general, the location of botanic gardens in Indonesia is closer to community activities compared to *in situ* conservation sites, such as national parks, wildlife reserves, and natural reserves. Various species and uses of plant collections are ideal educational facilities to increase people's appreciation of the role of plants in human life. Plants have two main functions for human beings, namely, economic and ecological functions. In the economic function, plants are used as food, fruits, vegetables, building materials, fiber materials, and so on. In terms of ecology, plants are producers of oxygen for other living things, shade, erosion control, microclimate improving, and so on.

2.3.4 TOURISM

A botanic garden is a comfortable and healthy tourist destination and also has educational value for the community. Thematic and aesthetic parks and lawns in the botanic garden areas are public spaces where people can gather and socialize while enjoying the beauty of plant collections and well-maintained areas. The development of a botanic garden should adopt the social and cultural aspects of the local community. Parks and infrastructure built in the botanic garden area can reflect the diversity of arts and culture of the local community. The botanic garden is also equipped with supporting infrastructure, such as information centers, ethnobotany museums, libraries, and herbarium buildings that function to provide plantation information and display the arts and culture of the local community.

Each botanic garden in Indonesia has a unique concept in its development, both related to the theme of collections and supporting infrastructure. Determination of the theme of the plant collection is based on the environmental and socio-cultural conditions of the community. For example, the Eka Karya Bali Botanic Gardens has a collection theme of montane plants in wet climates in the eastern part of Indonesia and gives a touch of Balinese ethnicity to the entire building, such as the gate, guest house, meeting hall, and so on.

2.3.5 ENVIRONMENTAL SERVICES

A botanic garden has an ecological impact on improving the quality of the environment in and around the area, which includes aspects of water management, biodiversity, carbon sequestration, and landscape. The following are some examples of botanic gardens having an important contribution to environmental services:

1. Cibinong Science Center and Botanic Gardens (CSC-BG) is the only botanic garden in Indonesia whose plant collections are arranged based on the concept of Indonesian bioregion. This botanic garden provides a green open space area and is the main oxygen producer for the environment at the Indonesian Institute of Sciences (LIPI) research center in Cibinong and Bogor Regency. The lake built within the area conserves 22 springs. These

water sources have very important functions for maintaining plant collections and microclimate control, and they have become an attraction for visitors.
2. The construction of the reservoir in the Balikpapan Botanic Gardens plays an important role in providing water for plant collections during the dry season, increasing tourist attraction, and providing water needs of the community around the area.
3. Bogor Botanic Gardens is one of the largest green open spaces in the city of Bogor, and it is able to significantly improve the microclimate since it has the highest number and density of trees compared to the surrounding locations. According to Asiani (2007), the temperature inside the Bogor Botanic Gardens area is lower and the humidity is higher than the surrounding area.

2.3.6 PLANT COLLECTION ENRICHMENT IN THE GARDENS

Despite its limitations, *ex situ* conservation has a major role to play as part of an integrated approach in many conservation procedures, such as species recovery, reintroduction, and ecological restoration (Heywood, 2017a, 2017b). Continued loss of Indonesian forests raises a question about how to save plants from extinction. Cribb (1994) suggested that there are many techniques necessary to protect, propagate, and even reintroduce plants, and botanic gardens have a major role to play to solve this problem through education as well as providing facilities.

Botanic gardens collections as *ex situ* conservation are backups to *in situ* conservation and serve as good representative support, and therefore, they need to be as complete as possible. In general, plant collection enrichment in Indonesian botanic gardens was conducted by plant exploration, seed exchange, seed bank, propagation, and donation.

2.3.6.1 Exploration

The tropical forests of Indonesia attracted the founder of the Bogor Botanic Gardens, Dr. Reinwardt, to collect plants during his expeditions. The plants collected became a repository of living plants and inspiration for the birth of a botanical institution. Later, many new plants were described and other observations were supported with scientific equipment in Java. It is amazing to read the records of early plant explorers satisfying their curiosity in a *Paradise Lost* of unspoiled nature. Plant exploration followed by scientific research activities to find out the potential of the collected plant resources became an attraction for scientists around the world to visit and work at Treub laboratory in this garden. For more than one century, Bogor Botanic Gardens has been a world-famous tropical biological research center (Goss, 2004).

Plant exploration made by many different nationalities was recorded, not only for Indonesian plants but also the Botanical region Malesia, including Malay peninsula, Singapore, Sarawak, Sabah, Brunei Darussalam, Philippines, East Timor, and New Guinea. The technique of plant collecting and preservation in the humid tropical region was written up by van Steenis and Kruseman (1950) in detail, including planning, equipment, processing of herbarium specimens and how to collect living plants, labeling, recording, and documentation. Duplicates of herbarium specimens were also sent to other herbaria, such as Leiden, Kew, Brisbane, Manila, Singapore, British Museum, Berlin-Dahlem, and Harvard, and additional ones were sent to world herbaria at Geneva, Paris, New York, Washington, Michigan, and Berkeley (van Steenis, 1950).

Records of collected specimens from different parts of Indonesia in periods 1817–1950 were a strong base to arrange further exploration. Photographic documentation was also taken during that time on glass plates and black and white films; they are valuable documents that are still admired. The number of collections planted in Bogor Botanic Gardens as well as the herbaria specimens have changed from time to time. In 1823, the first garden catalog was published; it contained 912 species of plants. At that time, *Litchi chinensis* was introduced from China to the garden. In 1844, the second catalog recorded 150 species of fern, 25 species of gymnosperm, 510 species of monocots, and 2,200 species of dicots. An important collection is *Elaeis guineensis* from Africa as this

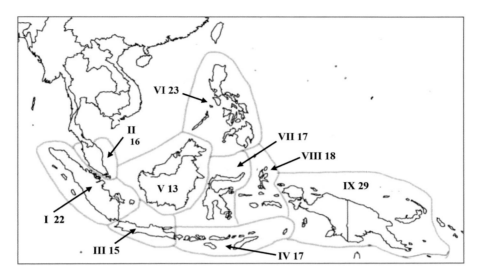

FIGURE 2.1 Plant exploration in Malesia region before 1840. Arabic numerals represent the number of expeditions in each subunit, and Roman numbers are artificial units for division.

Source: van Steenis-and Kruseman (1950).

plant is the parent of palm oil in Indonesia and Malaysia. Quinine (*Cinchona calisaya*) is also one of the most important medicinal plants; it was planted in 1852 at Cibodas Botanic Gardens, and the planting date of quinine has been selected as the time of establishment of Cibodas Botanic Gardens (Sukarya and Witono, 2017).

Plant explorations have been conducted by Indonesian botanic gardens mainly to enrich their collections, for horticultural research, or for other botanical studies. Exploration is one of the most important activities in Bogor Botanic Gardens and Herbarium Bogoriense. The number of specimens collected is shown in Table 2.2. For about 27 years, Herbarium Bogoriense was part of Bogor

TABLE 2.2
Number of Specimens Collected from 1817 to 1950 and from 1950 to 2008

		1817–1950		1950–2008	
Island	Area (km²)	Number Collected	Average Number Collected per 100 km²	Herbarium Specimens Collected	Living Specimens Collected
1. Sumatra	479.513	87,000	18	26,966	3,576
2. Jawa	132.474	247,522	187	7,455	2,351
3. Sunda Islands	98.625	24,545	25	4,363	3,638
4. Borneo	739.175	91,550	12	(Kal.*) 28,820	(Kal*.) 2,739
5. Sulawesi	82.870	32,350	18	15,420	1,834
6. Moluccas	63.575	27,525	43	22,216	1,173
7. Papua New Guinea	2,980.155	196,755	3.6	(Papua**) 2,150	(Papua**) 946
Total (estimation)	5,099.226	989,492	19	254,782	16,257

* Kal.: Indonesian Borneo
** Papua: Indonesian New Guinea

Source: Kartawinata (2010a).

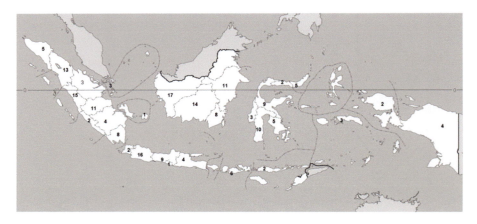

FIGURE 2.2 Plant exploration in periods 1992–2016 (numbers represent number of expeditions per province).

Botanic Gardens, therefore, these two institutions have much information to share and many exploration activities involving both institutions. Living plants are recorded at the registration section and planted in the botanic gardens. The herbaria information is noted in the herbarium sheets and spirit collections of the herbarium. Indonesian botanic gardens managed by LIPI have been actively collecting plants from all parts of Indonesia since 1992. A total of 200 expeditions had been conducted by 2016 (Hidayat et al., 2017) (Figure 2.2).

2.3.6.2 Seed Exchange

The seeds for exchange are from the garden collection. Living plant collections alone do not guarantee their sustainability, therefore, the collection of seeds, *in vitro* culture as well as seedlings, are other activities conducted at botanic gardens. To improve the collection, Bogor Botanic Gardens actively exchanges plants with other botanic gardens around the world. Every three years a list of plants for exchange is printed in "Index Seminum" and distributed to other botanic gardens. Seed exchange between botanic gardens began in the sixteenth century, and sharing collections is an important part of conservation activity. Havinga et al. (2016) proposed an online platform for seed exchange between botanic gardens to support the maintenance and development of living plant collections in botanic gardens. The botanic garden community needs to establish an online platform for the exchange of seeds to replace the centuries-old habit of circulating seed lists.

2.3.6.3 Seed Bank

The effort to conserve plants is also conducted by botanic gardens in Indonesia through seed conservation. In November 2016, Indonesian Botanic Gardens integrated with the Millennium Seed Bank Partnership (MSBP), the largest international program on *ex situ* conservation focused on wild species. It is coordinated by Millennium Seed Bank (MSB) – Royal Botanic Gardens, Kew, UK. Seed banks in the botanic gardens have three main functions for complementary, supplementary, and active collections. A complementary collection is a seed collection that duplicates a plant collection in the botanic gardens. A supplementary seed collection is an additional plant species that has not been collected and grown in the garden. An active collection means that the seed collection is distributed and used for research, reintroduction, and seed exchange based on orders, needs, and requests.

After two years, the MSBP project has successfully conserved 1,355 accessions of seeds from the garden collections and explorations; these are stored in the seed banks of four botanic gardens managed by LIPI. These results contribute significantly toward Target 8 of the Global Strategy for

Plant Conservation (Latifah et al., 2018). The MSBP project has joined Kew's Global Tree Seed Bank Project and is funded by the Garfield Weston Foundation (Hardwick et al., 2017).

2.3.6.4 Plant Propagation

Propagation of garden collections has been conducted since the gardens were established. Important collections, and especially rare species, were propagated by seeds as backup collections and are available to the public. Orchids, Nepenthes, ferns, and aroids are propagated in the nursery as well as through *in vitro* culture in the laboratory. Research in conservation by synthetic seeds was conducted for orchids. Propagation of several plants for reintroduction is conducted in four botanic gardens.

2.3.6.5 Donation

Some botanic gardens collections were received as donations, such as *Nepenthes*, cherry blossom (*Prunus* spp.) at Cibodas Botanic Gardens, orchids, *Begonia*, and other collections in Bogor, Cibodas, Purwodadi, and Eka Karya Bali Botanic Gardens.

2.4 PRIORITY SETTING FOR PLANT CONSERVATION IN INDONESIAN BOTANIC GARDENS

Plant conservation is the most important task for Bogor Botanic Gardens and other botanic gardens in Indonesia. This activity began more than 200 years ago and will continue in the future. Plant collections are the most valuable for botanic gardens. Limited resources, such as manpower, facilities, and funds for conserving Indonesian plants, require that the most is made of it through setting priorities or species to be conserved. Possingham et al. (2002) suggested additional criteria in the selection of biological conservation in creating tools for allocation of resources, including socio-economic cost and benefit as considerations to be taken. Bogor Botanic Gardens undertook to set the priorities of 100 species for conservation in 2009 (Risna et al., 2010).

Threatened plant species are those species classified by the IUCN Red List as critically endangered, endangered, and vulnerable categories. The IUCN Red List for threatened species evaluated 50,369 plant species, among them 674 endemic species of Indonesia, and the conservation status of 683 species needs to be updated (IUCN, 2020). Threatened species criteria are also often considered as a logical reason for the plant to be conserved. Processes that threaten the extinction of a species are varied, therefore, a red list of threatened plant species alone is not enough to set the priority species for conservation. Although not all threatened species can be saved from extinction, efforts should be made at least to minimize the loss of species. According to Risna et al. (2010), methods of setting priorities are as follows:

1. Determine the group of plants (plant families) to be assessed, and select the local species. Bogor Botanic Gardens selected four families, Arecaceae, Cyatheaceae, Nepenthaceae, and Orchidaceae, consisting of 191 species.
2. Determine the criteria in the priority setting process, focusing on taxonomic and geographic distinctiveness, population status, threats, vulnerability, propagation potential, and use value of the target species. This method is the modified Molloy and Davis (1992) system. The criteria used for assessment in the form of a score are taxonomic uniqueness, geographical distribution, population size, population density, the largest population size, the condition of the population, population decline, legal protection in the natural habitat, estimation of the threatened status and habitat loss, effect of predators or exploitation on the survival of the species, competition between species, other pressures on the survival/existence of the species, habitat and/or nutrient specificity, reproductive specialization, propagation potential, and uses (Risna et al., 2010).
3. Evaluation of scores based on scientific, objective, and independent judgments. Expert judgments would increase the accuracy and reliability of the estimated values.

TABLE 2.3
Total Number of Priority Species for Conservation in Indonesia

Family	Category A	B	C
Arecaceae	14	19	18
Cyatheaceae	8	25	1
Nepenthaceae	34	15	2
Orchidaceae	44	0	0
Total	100	59	21

The scoring number ranged from 1 to 5, then the total number of individual scoring (genuine score) and compromised scoring values were recorded. The three categories for the resulting scores were as follows: A – total score of more than 50, for species with a priority for immediate conservation action; B – total score between 42–50, for species with lower priority for conservation; and C – total score less than 42, for species with less priority for conservation action. When individual scores were highly different, then compromised judgment was conducted. Among 191 Indonesian plant species selected for evaluation, 180 species fall into the three categories, and the remaining 11 species were excluded (Table 2.3). Species listed in category A are presented in Table 2.4.

TABLE 2.4
The List of Plants That Have High Priority for Conservation (Category A)

Priority Species for Conservation

No.	Arecaceae	Cyatheaceae	Nepenthaceae	Orchidaceae
1.	*Arenga distincta* Mogea	*Cyathea magnifolia* v.A.v.R.	*Nepenthes adnata* Tamin & M.Hotta ex J. Schlaue	*Ascocentrum aureum* J.J. Sm.
2.	*Arenga hastata* (Becc.) Whitmore	*Cyathea modesta* (Bak.) Copel.	*Nepenthes aristolochioides* Jebb & Cheek	*Arachnis hookeriana* (Rchb.f.) Rchb.f.
3.	*Arenga longipes* Mogea	*Cyathea pallidipaleata* Holttum	*Nepenthes bicalcarata* Hook.f.	*Bulbophyllum phalaenopsis* J.J. Sm.
4.	*Arenga talamauense* Mogea	*Cyathea punctulata* v.A.v.R.	*Nepenthes bongso* Korth.	*Cymbidium hartinahianum* J.B. Comber & Nasution
5.	*Calamus manan* Miq.	*Cyathea setifera* Holttum	*Nepenthes campanulata* Sh. Kurata	*Dendrobium ayubii* J.B. Comber & J.J. Wood
6.	*Ceratolobus glaucescens* Blume	*Cyathea strigosa* Christ.	*Nepenthes clipeata* Danser	*Dendrobium capra* J.J. Sm.
7.	*Ceratolobus pseudoconcolor* J. Dransf.	*Cyathea teysmannii* Copel.	*Nepenthes densiflora* Danser	*Dendrobium devosianum* J.J. Sm.
8.	*Daemonorops acomptostachys* Becc.	*Cyathea tripinnatifida* Roxb.	*Nepenthes dubia* Danser	*Dendrobium jacobsonii* J.J. Sm.
9.	*Hydriastele flabellata* (Becc.) W.J. Baker & Loo		*Nepenthes ephippiata* Danser	*Dendrobium laxiflorum* J.J. Sm.

TABLE 2.4
(Continued)

Priority Species for Conservation

No.	Arecaceae	Cyatheaceae	Nepenthaceae	Orchidaceae
10.	*Iguanura leucocarpa* Blume		*Nepenthes eustachya* Miq.	*Dendrobium militare* P.J. Cribb
11.	*Johannesteijsmannii altifrons* (Rchb.f. & Zoll.) H.E. Moore		*Nepenthes eimae* Sh. Kurata	*Dendrobium nindii* W. Hill
12.	*Licuala pumila* Blume		*Nepenthes fusca* Danser	*Dendrobium pseudoconanthum* J.J. Sm.
13.	*Pinanga javana* Blume		*Nepenthes hamata* J.R. Turnbull & A.T. Middleton	*Dendrobiium taurulinum* J.J. Sm.
14.	*Sommieria leucophylla* Becc.		*Nepenthes inermis* Danser	*Dendrobium tobaense* J.J. Wood & J.B. Comber
15.			*Nepenthes insignis* Danser	*Paphiopedilum gigantifolium* Braem, M.L. Baker & C.O. Baker
16.			*Nepenthes klossii* Ridley	*Paphiopedilum glaucophyllum* J.J. Sm.
17.			*Nepenthes lavicola* Wistuba & Rischer	*Paphiopedilum kolopakingii* Fowlie
18.			*Nepenthes mapuluensis* J.H. Adam & Wilcock	*Paphiopedilum mastersianum* (Rchb.f.) Stein
19.			*Nepenthes mikei* B.Salmon & Maulder	*Paphiopedilum moquettianum* (J.J.Sm.) Fowlie
20.			*Nepenthes mollis* Danser.	*Paphiopedilum niveum* (Rchb.f.) Stein
21.			*Nepenthes ovata* J.Nerz & A.Wistuba	*Paphiopedilum primulinum* M.W. Wood & P. Taylor
22.			*Nepenthes paniculata* Danser	*Paphiopedilum sangii* Braem
23.			*Nepenthes papuana* Danser	*Paphiopedilum schoseri* Braem & H. Mohr
24.			*Nepenthes pilosa* Danser	*Paphiopedilum supardii* Braem & Loeb
25.			*Nepenthes rhombicaulis* Sh. Kurata	*Paphiopedilum victoria-mariae* (Sander ex. Masters) Rolfe
26.			*Nepenthes singalana* Becc	*Paphiopedilum victoria-regina* (Sander) M.W. Wood
27.			*Nepenthes spathulata* Danser	*Paphiopedilum violascens* Schltr.
28.			*Nepenthes spectabillis* Danser	*Papilionanthe tricuspidata* (J.J.Sm.) Gray
29.			*Nepenthes stenophylla* Mast.	*Paraphalaenopsis denevei* (j.J.Sm.) A.D. Hawkes
30.			*Nepenthes sumatrana* (Miq.) Beck.	*Paraphalaenopsis labukensis* Shim, A.L. Lamb & C.L. Chan
31.			*Nepenthes talangensis* J.Nerz & A.Wistuba	*Paraphalaenopsis laycockii* (M.R. Hend.) A.D. Hawkes

(Continued)

TABLE 2.4
(Continued)

Priority Species for Conservation

No.	Arecaceae	Cyatheaceae	Nepenthaceae	Orchidaceae
32.			*Nepenthes tenuis* J.Nerz & A.Wistuba	*Paraphalaenopsis serpentilingua* (J.J.Sm.) A.D. Hawkes
33.			*Nepenthes treubiana* Warb.	*Phalaenopsis celebensis* Sweet
34.			*Nepenthes veitchii* Hook.f.	*Phalaenopsis floresensis* Fowlie
35.				*Phalaenopsis gigantea* J.J. Sm.
36.				*Phalaenopsis inscriptiosinensis* Fowlie
37.				*Phalaenopsis javanica* J.J. Sm.
38.				*Phalaenopsis modesta* J.J.Sm.
39.				*Phalaenopsis tetraspis* Rchb.f.
40.				*Phalaenopsis venosa* Shim & Fowlie
41.				*Phalaenopsis viridis* J.J. Sm.
42.				*Vanda devoogtii* J.J. Sm.
43.				*Vanda jennae* O'Brien & Vermeulen
44.				*Vanda sumatrana* Schltr.

Note: The list of the species is arranged alphabetically, not according to the highest to lowest scores.

Conservation activities of the botanic gardens are focused on these priority species. Not all of them are available in the botanic gardens, therefore, the prioritized species are realistic guidance in planning conservation activities. These selected species are from different islands in Indonesia, and exploration activities are focused on these priority species (Kusuma et al., 2011).

2.5 DEVELOPING NEW BOTANIC GARDENS

The development of botanic gardens in various regions in Indonesia is one of the outstanding programs of the Indonesian Institute of Sciences (LIPI). This program was initiated in 1999 based on the consideration that the four botanic gardens managed by LIPI, namely Bogor Botanic Garden, Cibodas Botanic Garden, Purwodadi Botanic Garden, and Eka Karya Bali Botanic Garden, are estimated to only be able to conserve 30%–40% of Indonesian plants. This is due to limited resources owned by LIPI, such as land area, human resource management, and budget. The development of botanic gardens in Indonesia is based on the ecoregion concept. According to Wikramanayake et al. (2001), Indonesia consists of 33 terrestrial ecoregions. Furthermore, the data were elaborated with RBI digital maps (2008) through application of the GeoProcessing Wizard ArcView GIS 3.2 and administrative area divisions so that Indonesia's ecoregions totaled 47 ecoregions (Figure 2.3) (Witono et al., 2012). This analysis was adopted to address that the number of botanic gardens developed is a minimum of 47 botanic gardens, representing the number of ecoregions in Indonesia.

The new botanic garden development program gained momentum after the President's speech at the commemoration of the National Technology Awakening Day on 11 August 2004 in Serpong, which emphasized the importance of developing botanic gardens in each province in Indonesia. Furthermore, it was followed up with the Minister of Research and Technology instruction to all

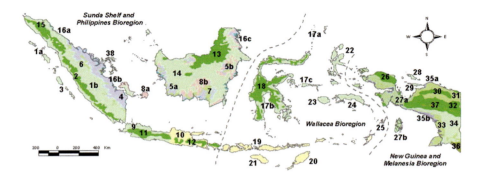

FIGURE 2.3 Map of Indonesian ecoregions.

(1a) Nias Islands lowland forest; (1b) Sumatran lowland forests; (2) Sumatran montane rainforests; (3) Mentawai Islands lowland forest; (4) Sumatran peat-swamp forests; (5a) Western Kalimantan peat-swamp forests; (5b) Eastern Kalimantan peat-swamp forests; (6) Sumatran freshwater swamp forests; (7) Southern Kalimantan freshwater swamp forests; (8a) Bangka Belitung heath forests; (8b) Kalimantan heath forests; (9) Western Java lowland rainforests; (10) Eastern Java-Bali lowland rainforests; (11) Western Java montane rainforests; (12) Eastern Java-Bali montane rainforests; (13) Kalimantan montane rainforests; (14) Kalimantan lowland rainforests; (15) Sumatran tropical pine forests; (16a) Northern Sumatran mangrove forests; (16b) Southern Sumatran mangrove forests; (16c) Eastern Kalimantan mangrove forests; (17a) Sangihe-Talaud Islands lowland rainforests; (17b) Sulawesi lowland rainforests; (17c) Banggai-Sula Islands lowland rainforests; (18) Sulawesi montane rainforests; (19) Lesser Sunda Islands deciduous forests; (20) Timor and Wetar deciduous forests; (21) Sumba deciduous forests; (22) Halmahera rainforests; (23) Buru rainforests; (24) Seram rainforests; (25) Banda Sea Islands moist deciduous forests; (26) Vogelkop montane forests; (27a) Vogelkop lowland forests; (27b) Aru lowland forests; (28) Biak-Numfoor rainforests; (29) Yapen rainforests; (30) Northern Papua montane rainforests; (31) Northern Papua lowland rainforests; (32) Central range montane rainforests; (33) Southern Papua freshwater swamp forests; (34) Southern Papua lowland forests; (35a) Northern Papua mangroves; (35b) Southern Papua mangroves; (36) Papuan savanna; (37) Central Papua subalpine savanna; (38) Riau islands rainforests.

Source: Witono et al. (2012).

Governors in Indonesia No. 77/M/VIII/2004 that each province seeks the development of at least one botanic garden and coordinates with LIPI for its implementation. The enthusiasm of the Regional Government in the development of the botanic gardens reached the highest level when 23 regional heads (governor and mayor) signed the Bedugul Declaration in 2009. This Bedugul Declaration contained a joint commitment of 23 Regional Governments, LIPI, and the Ministry of Public Works (the name of the Ministry of Public Works and Public Housing at the time) to accelerate the development of local botanic gardens. As the basis for the national policy for the development of local botanic gardens, Presidential Instruction No. 3 of 2009 was issued regarding the infrastructure development of the Presidential Palace, Botanic Gardens, and Certain Cultural Heritage, strengthening the importance of developing botanic gardens in Indonesia. The Presidential Instruction provides a role for technical ministries to be actively involved in the development of botanic gardens. Furthermore, with the development of the local botanic gardens, there must be a stronger national policy basis.

Presidential Regulation Number 93 of 2011 concerning Botanic Gardens was issued for supporting the program to develop new botanic gardens in Indonesia. This Presidential Regulation is the legal basis for LIPI and the Ministry of Public Works and Public Housing (PUPR) for developing local botanic gardens in accordance with their respective competencies. The development of the local botanical gardens has been established as a national program from the Government of

Note: ○ Botanic gardens managed by Indonesian Institute of Sciences (LIPI)
       ○ Botanic gardens managed by local governments
       ● Botanic gardens managed by universities

FIGURE 2.4 Location and distribution of botanic gardens in Indonesia by December 2020 (the number refers to Table 2.1).

Republic Indonesia after being included in the Medium Term Development Plan (RPJMN) 2010–2015, which is regulated in Presidential Regulation Number 5 of 2010. In the document, it is stated that there is a target of two botanic gardens to be launched and two botanic gardens to be initiated every year. Furthermore, Presidential Regulation Number 2 of 2015 concerning the RPJMN 2015–2019, the development of local botanic gardens, is included in the national priority for the science and technology sector with the same target (Presidential Regulation Number 2 of 2015).

By the end of 2020, there were 45 botanic gardens in Indonesia consisting of 5 botanic gardens managed by LIPI, 37 botanic gardens managed by local governments, and 3 botanic gardens managed by universities (Figure 2.4; Table 2.5). This number has covered 17 types of ecoregions so continued development of several new botanic gardens is necessary to be able to conserve the entire richness of plant species in Indonesia.

TABLE 2.5

Location, Area, and Theme of the Collection of Botanic Gardens in Indonesia as of December 2020

No.	Botanic Gardens (BG)/ Managed by	Location	Cover Area (ha)	Plant Theme/Ecoregion
1.	Bogor BG Indonesian Institute of Sciences	Bogor, West Java Province	87	Lowland plants in wet climates/ Western Java lowland rainforests
2.	Cibodas BG Indonesian Institute of Sciences	Cianjur, West Java Province	85	Montane plants in wet climates on western part of Indonesia/Western Java montane rainforests
3.	Purwodadi BG Indonesian Institute of Sciences	Pasuruan, East Java Province	85	Lowland plants in dry climates/ Eastern Java-Bali lowland rainforests
4.	Eka Karya Bali BG Indonesian Institute of Sciences	Tabanan, Bali Province	157.5	Montane plants in wet climates on eastern part of Indonesia/Eastern Java–Bali montane rainforests
5.	Cibinong Science Center and BG Indonesian Institute of Sciences	Bogor, West Java Province	189	Indonesian plants based on ecoregion/ Western Java lowland rainforests

TABLE 2.5
(Continued)

No.	Botanic Gardens (BG)/ Managed by	Location	Cover Area (ha)	Plant Theme/Ecoregion
6.	Balangan BG Balangan Regency Government	Balangan, South Kalimantan Province	7.6	Local plants of Balangan and lowland plants of Kalimantan/Kalimantan lowland rainforests
7.	Balikpapan BG Balikpapan Municipal Government	Balikpapan, East Kalimantan Province	309	Wood plants of Indonesia/Kalimantan heath forests
8.	Balingkang Bali BG Bangli Regency Government	Bangli, Bali Province	17.6	Hindu ceremonial plants in the mountains of Bali/Eastern Java–Bali montane rainforests
9.	Banua BG South Kalimantan Province Government	Banjarbaru, South Kalimantan Province	100	Medicinal plants of Kalimantan/Kalimantan lowland rainforests
10.	Batam BG Batam Municipal Government	Batam, Kepulauan Riau Province	86	Small island plants of Indonesia/Riau islands rainforests
11.	Baturraden BG Central Java Province Government	Baturraden, West Java Province	142	Montane plants of Java/Western Java montane rainforests
12.	Belitung Timur BG East Belitung Regency Government	Manggar, Bangka Belitung Province	96	Heath forest plants of Bangka Belitung/Bangka Belitung heath forests
13.	Bukit Sari Jambi BG Jambi Province Government	Batanghari and Tebo, Jambi Province	425	Lowland plants of Sumatra/Sumatran lowland forests
14.	Danau Lait BG West Kalimantan Province Government	Sanggau, West Kalimantan Province	328	Equatorial plants of Indonesia/Kalimantan lowland rainforests
15.	Gianyar BG Gianyar Regency Government	Gianyar, Bali Province	9.7	Native plants of Gianyar, medicinal and ceremonial plants of Bali/Eastern Java–Bali montane rainforests
16.	Gunung Tidar BG Magelang Regency Government	Magelang, Central Java Province	72	Ornamental plants in the lowland rain forest of Java/Eastern Java–Bali lowland rainforests
17.	Indrokilo BG Boyolali Regency Government	Boyolali, Central Java Province	8	Eastern Java Lowland Forest Plants of Eastern Java/Eastern Java–Bali lowland rainforests
18.	ITERA BG Sumatra Institute of Technology	Jati Agung, Lampung Province	75.5	Lowland plants of Sumatra/Sumatran lowland forests
19.	Jagatnatha BG Jembrana Regency Government	Negara, Bali Province	5.8	Medicinal and traditional ceremony plants of Bali/Eastern Java–Bali lowland rainforests
20.	Jompie BG Parepare Municipal Government	Parepare, South Sulawesi Province	13.5	Wallacea coastal plant/Sulawesi lowland rainforests
21.	Katingan BG Katingan Regency Government	Kasongan, Central Kalimantan Province	127	Fruit plants of Indonesia/Kalimantan heath forests

(Continued)

TABLE 2.5
(Continued)

No.	Botanic Gardens (BG)/ Managed by	Location	Cover Area (ha)	Plant Theme/Ecoregion
22.	Kolaka BG Kolaka Regency Government	Kolaka, Southeast Sulawesi Province	60	Mekongga forest plants/Sulawesi lowland rainforests
23.	Kendari BG Kendari Municipal Government	Kendari, Southeast Sulawesi Province	113	Ultra basic plants of Indonesia/ Sulawesi lowland rainforests
24.	Kuningan BG Kuningan Regency Government	Kuningan, West Java Province	172	Rocky ground plants and mount Ciremai plants/Western Java montane rainforests
25.	Liwa BG West Lampung Regency Government	Liwa, Lampung Province	116	Ornamental plants of Indonesia/ Sumatran montane rainforests
26.	Lemor BG East Lombok Regency Government	Suela, West Nusa Tenggara Province	130	Sunda Islands plant/Lesser Sunda Islands deciduous forests
27.	Mangrove Surabaya BG Surabaya Municipal Government	Surabaya, East Java Province	309	Mangrove plants of Indonesia/Eastern Java–Bali lowland rainforests
28.	Massenrempulu BG Enrekang Regency Government	Enrekang, South Sulawesi Province	300	Wallacean plants/Sulawesi lowland rainforests
29.	Megawati Soekarnoputri BG Southeast Minahasa Regency Government	Ratahan, North Sulawesi Province	221	Lowland wallacean plants/Sulawesi lowland rainforests
30.	Minahasa BG Minahasa Regency Government	Tondano, North Sulawesi Province	186	Montane Wallacean plants/Sulawesi montane rainforests
31.	Pelalawan BG Minahasa Regency Government	Pangkalan Kerinci, Riau Province	100	Peat swamp forest plants of Sumatra/ Sumatran peat-swamp forests
32.	Pucak BG Maros Regency Government	Maros, South Sulawesi	120	Economical plants of Indonesia/ Sulawesi lowland rainforests
33.	Rimbe Mambang BG Bangka Regency Government	Sungailiat, Bangka Belitung Province	55.7	Native plants of Bangka and lowland plants of Sumatra/Sumatran lowland forests
34.	Sambas BG Bangka Regency Government	Sambas, West Kalimantan Province	300	Riparian plants of Kalimantan/ Kalimantan lowland rainforests
35.	Samosir BG Samosir Regency Government	Simanindo, North Sumatra Province	100	Montane plants of North Sumatra/ Sumatran tropical pine forests
36.	Sampit BG East Kotawaringin Regency Government	Sampit, Central Kalimantan Province	501	Heath forest plants of Kalimantan/ Kalimantan heath forests
37.	Sigi BG Sigi Regency Government	Sigi, Southeast Sulawesi Province	78.8	Monsoon forest plants of Sulawesi/ Sulawesi lowland rainforests

TABLE 2.5
(Continued)

No.	Botanic Gardens (BG)/ Managed by	Location	Cover Area (ha)	Plant Theme/Ecoregion
38.	Sipirok BG South Tapanuli Regency Government	Sipirok, North Sumatra	80	Transitional plants of lowland forest and mountain forest plants of Sumatra/Sumatran lowland forests
39.	Solok BG Solok Regency Government	Solok, West Sumatra Province	112.6	Spice plants of Indonesia/Sumatran montane rainforests
40.	Sriwijaya BG South Sumatra Province Government	Indralaya, South Sumatra	100	Medicinal and wet plants of Sumatra/Sumatran peat-swamp forests
41.	Tanjungpuri BG Tabalong Regency Government	Tabalong, South Kalimantan Province	50	Lowland plants of Kalimantan/Kalimantan lowland rainforests
42.	UHO BG Halu Oleo University	Kendari, Southeast Sulawesi Province	22.9	Endemic plants of Sulawesi/Sulawesi lowland rainforests
43.	UPR BG Palangkaraya University	Palangkaraya, Central Kalimantan Province	363	Peat swamp forest plants of Kalimantan/Western Kalimantan peat-swamp forests
44.	Wamena BG Wamena Regency Government	Wamena, Papua Province	160	Montane plants of Central Papua/Central range montane rainforests
45.	Wolobobo BG Ngada Regency Government	Bajawa, East Nusa Tenggara Province	91.8	Monsoon forest and montane plants of Lesser Sunda Islands/Lesser Sunda Islands deciduous forests

According to Presidential Regulation Number 93 of 2011, the development of local botanic gardens is carried out through three stages: planning, development, and management. A botanic garden is categorized as being in the management stage if it has been launched to the public. The botanic gardens should have the following:

1. Legal land status
2. Permanent management agency
3. Five functions of botanic gardens (i.e., conservation, research, education, tourism, and environmental services)
4. Adequate infrastructure in the welcome, management, and collection zones.

(Decree of Chairman of LIPI No. 2/F/2015)

The accumulation of local botanic gardens launched in the 2013–2020 period amounted to 14 botanic gardens (Figures 2.5 and 2.6). In 2020, no local botanic gardens launched due to the COVID-19 pandemic, and the Government of Republic Indonesia has refocused the budget. In the following years, the target is for at least two local botanic gardens to be launched per year.

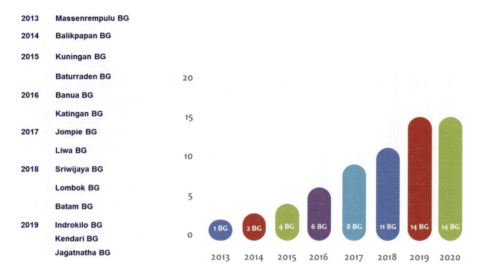

FIGURE 2.5 Launching of local botanic gardens (2013–2020).

2.6 FUTURE PROGRAMS OF BOGOR BOTANIC GARDENS

In general, three important programs will be developed by the Research Center for Plant Conservation and Botanic Gardens LIPI in the coming period.

2.6.1 Development and Supervision of Local Botanic Gardens

The number of new botanic gardens initiated has increased significantly from year to year. However, 45 botanic gardens in Indonesia (including botanic gardens managed by LIPI) only cover 17 of 47 Indonesian ecoregions, so acceleration in achieving the number of ecoregions is crucial. There are at least three strategies that can be done to achieve these goals: (1) prioritizing the location of new botanic gardens needs to be carried out in the future, especially for new ecoregions in the eastern part of Indonesia, such as Nusa Tenggara, Maluku, and Papua; (2) botanic gardens managed by local government or universities, except botanic gardens managed by LIPI, should be designed to conserve at least two types of ecoregions; and (3) initiation of the development of new botanic gardens managed by LIPI in new types of ecoregions.

Based on Presidential Regulation Number 18 of 2020 concerning the RPJMN 2020–2024, the Research Center for Plant Conservation and Botanic Gardens LIPI was given a mandate by the Government of Republic Indonesia to support the local botanic gardens in launching two botanic gardens every year. Thus, it is targeted that there will be an additional ten botanic gardens by 2024 that will be launched and managed by both local governments and universities. To achieve this target, the local botanic gardens development team carried out various strategies as follows: (1) facilitate infrastructure development support to the Ministry of Public Works and Housing; (2) facilitate policy support for ninistries/agencies related to the organization of land management institutions and legality; (3) develop human resources for local botanic gardens staff, both in terms of management and technical aspects; (4) support the enrichment of plant collections through plant exploration and donation activities; (5) support the preparation and development of thematic gardens and aesthetic gardens; and (6) support the improvement of the quality of the area and plant collections.

The Bogor Botanic Gardens development team (2020) has compiled a roadmap for the development of local botanic gardens for the period 2020–2024 that contains plans for infrastructure development and local botanic gardens that will be launched during that period. According to Purnomo et al. (2020), the botanic gardens that will be launched in the 2020–2024 period are as follows: in 2020: CSC-BG, ITERA BG, and Sipirok BG; 2021: Gianyar BG and Balangan BG;

Role of Indonesian Botanic Gardens in Plant Conservation

FIGURE 2.6 Fourteen local botanic gardens have been launched from 2013–2020.

(a) Massenrempulu BG, (b) Balikpapan BG, (c) Kuningan BG, (d) Baturraden BG, (e) Banua BG, (f) Katingan BG, (g) Jompie BG, (h) Liwa BG, (i) Sriwijaya BG, (j) Lombok BG, (k) Batam BG, (l) Indrokilo BG, (m) Kendari BG, (n) Jagatnatha BG.

Source: Data presented in Table 2.5.

2022: Pucak BG and Samosir BG; year 2023: Sambas BG and UHO BG; 2024: Sampit BG and Megawati Soekarnoputri BG. Since Indonesia and also almost all countries had pandemic COVID-19, BG launching planned were delayed. In 2021, only Sipirok BG has launched, and ITERA BG and CSC-BG have launched in 2022.

2.6.2 National Focal Point GSPC

The Convention on Biodiversity (CBD) is an international convention initiated by UNEP for the conservation of global biodiversity that is binding on countries that ratify the convention. Indonesia has ratified CBD through Law Number 5 of 1994 and so is obliged to contribute to the achievement of the implementation of CBD and all its derivatives. One of the implementations of the CBD is the establishment of the Global Strategy for Plant Conservation (GSPC), which consists of 16 targets (CBD, 2002) that were later updated with targets for 2011–2020.

As the national focal point of the GSPC, the Research Center for Plant Conservation and Botanic Gardens of LIPI is obliged to coordinate and compile a national report related to the implementation of the 16 GSPC targets. The targets adopted at the global level provide guidance for setting national plant conservation targets (Sharrock, 2012). There are several targets closely related to the role and functions of botanic gardens:

Target 1: An online flora of all known plants
Target 2: An assessment of the conservation status of all known plant species, as far as possible, to guide conservation action
Target 8: At least 75% of threatened plant species in *ex situ* collections, preferably in the country of origin, and at least 20% available for recovery and restoration programs
Target 11: No species of wild flora endangered by international trade
Target 14: The importance of plant diversity and the need for its conservation incorporated into communication, education, and public awareness programs
Target 15: The number of trained people working with appropriate facilities sufficient according to national needs, to achieve the targets of this strategy
Target 16: Institutions, networks, and partnerships for plant conservation established or strengthened at national, regional, and international levels to achieve the targets of this strategy

In this chapter, two targets that are directly related to the main role of Bogor Botanic Gardens on plant conservation in Indonesia, namely target 2 and target 8, will be discussed. Target 2 emphasizes the importance of assessment of the conservation status of plant species native to Indonesia. In this regard, the Bogor Botanic Gardens has carried out several activities: (1) conducted an assessment of priority species for conservation action using the point-scoring method and expert panels on species from four families, namely, Arecaceae, Cyatheaceae, Nepenthaceae, and Orchidaceae (Risna et al., 2010); (2) developed a strategy and conservation action plan (SRAK) with the Ministry of Environment and Forestry for Indonesian flora mascots, namely *Rafflesia* and *Amorphophallus* (Susmianto et al., 2015a, 2015b); and (3) conducted a conservation assessment of Indonesian tree species toward the global tree assessment in collaboration with Botanic Gardens Conservation International in 2020–2021. In this activity, the results of the conservation status assessment were carried out on 1,043 species belonging to 69 families that were submitted to the IUCN Red List. The preparation of assessments of priority species for conservation in botanic gardens and assessment of the conservation status of plants in Indonesia is mostly carried out using desk studies using herbarium data, publications, and field experience of the assessor team. The development of assessments in tropical countries, including Indonesia, is indeed slower than in subtropical and temperate countries because most natural forests are relatively difficult to access (Pitman and Jorgensen, 2002).

Target 8 emphasizes aspects of plant recovery and restoration programs. According to the IUCN Red List (2009), 386 Indonesian plants are included in the critically endangered (CR), endangered (EN), and vulnerable (VU) categories. Indonesia's number of threatened plant species ranks fourth with Brazil, after Ecuador (1,835 species), Malaysia (685 species), and China (446 species). Four botanic gardens managed by LIPI have been able to conserve 83 threatened species (21.5% of the total number of Indonesian plants that are threatened) (Purnomo et al., 2010). The addition of the number of botanic gardens in Indonesia to 25 in 2013 was able to significantly increase Indonesia's achievements in the conservation of threatened species. In 2012, the number of Indonesian threatened species amounted to 393 species (IUCN Red List, 2013). The number of species that were successfully collected in the Indonesian botanic gardens reached 97 species (24.01%), or 104 species (25.74%) if the species in the nursery were included (Purnomo et al., 2015). In 2018, with 37 Indonesian botanic gardens, they have collected 122 Indonesian threatened species or 28.5% of all threatened species in Indonesia (437 species). If the nursery collections in the botanic gardens are included, then 29.2% of Indonesia's threatened plants have been successfully conserved *ex situ* (Widyatmoko et al., 2018). This data shows a significant increase in the number of threatened species being conserved as a result of the increase in the number of botanical gardens. Efforts to enrich plant collections in botanic gardens are more focused on threatened species starting in 2021.

Based on target 8, the achievement of rare plant species that have been reintroduced by the Indonesian botanic gardens in natural habitats has not been significant. Up to 2020, as many as 11 plant species have been reintroduced in their natural habitats. A continuous monitoring program and improvement of reintroduction methods are urgently needed so that the survival of reintroduced seedlings can be further improved (Ahmad et al., 2013).

2.6.3 RESEARCH PROGRAM

Research has been focused on protecting and utilizing sustainably the value and potential of the Indonesian native plant diversity. In Bogor Botanic Gardens, there were three research groups in the last two decades focused on the following:

1. Plant Conservation
   Indonesia has long been known for its high diversity of tropical plant life, both at the species level and in terms of intra-species genetic diversity. So we have to highlight conservation action as the highest priority. First of all, we need to know how many species of plants we have, how many species are in danger, and what are the relevant biological features of endangered species, including seed biology, genetics, and so on. In general, this group conducted studies in various research areas, such as species diversity of specific habitats, conservation genetics of endangered plants, medicinal and food-source plant conservation, seed conservation/seed banking, pollen conservation/pollen banking, DNA barcoding of Indonesian Botanic Garden collections, and selected flagship species: *Amorphophallus titanum* and *Rafflesia* spp.
2. Plant Domestication
   Plant domestication is a long-term selection endeavor. Domestication is the process of hereditary reorganization of wild plants into domestic and cultivated forms according to the interests of people. The fundamental distinction of domesticated plants from their wild ancestors is that they are created by human labor to meet specific requirements or whims and are adapted to the conditions of continuous care and solicitude people maintain for them. In general, this group conducted studies in various research areas, such as early stages in plant domestication, crop wild relatives, wild ancestors and new varieties, underutilized species, plant breeding, and germplasm in gene banks providing research materials for understanding domestication and plant breeding.
3. Plant Reintroduction and Habitat Restoration
   Reintroduction of threatened species into their natural habitats has become an essential program for the conservation and management of species in natural, semi-natural, and wild populations, through procedures involving repeatable, scientifically validated, practical

methods. Successful implementation of reintroduction programs leads to the recovery of natural populations of endangered species, beginning with effective planting out of available population followed by a stable regeneration process (Center for Plant Conservation Botanic Gardens of Indonesia, 2015).

Those research groups in the Indonesian Botanic Garden were extended into 2020. Then, based on the Decree of the Deputy of Life Sciences LIPI No. B-137/II/HK.01.03/1/2021, research groups were composed of 13 groups: Bioconservation of Indonesian Orchid Mycorrhizae, Conservation Biology of Indonesian Mountain Flora, Conservation Biology of Indonesian Ferns, Biosystematics and Bioadaptation of Indonesian Orchids, Domestication of Indonesian Native Plants, Restoration Ecology for Biodiversity Recovery and Ecosystem Services, Phytotechnology and Environmental Management, Conservation of Indonesian Endangered Plants, Species Management Invasive Plants, Genetics of Plant Conservation, Utilization of Local Plants for Restoration of Watershed Areas, Reintroduction and Spatial Ecology, and Conservation of Seeds of Rare and Potentially Indonesian Plants.

ACKNOWLEDGMENTS

We thank the Research Center for Plant Conservation and Botanic Gardens, the Indonesian Institute of Sciences, for providing data and documentation related to the development of new botanic gardens.

REFERENCES

Ahmad, T.L.S., D. Setiadi and D. Widyatmoko. 2013. Selecting plant species for forest restoration based on their photosynthetic parameters. *Jurnal Biologi Indonesia* 9(2): 233–243. [Indonesian]

Ariati, S.R., R.S. Astuti, I. Supriyatna, A.Y. Yuswandi, A. Setiawan, D. Saftaningsih and D.O. Pribadi. 2019. *An Alphabetical List of Plant Species Cultivated in the Bogor Botanic Gardens*. Research Center for Plant Conservation and Botanic Gardens, Indonesian Institute of Sciences, Bogor.

Ashton, P. 1998. *Vatica bantamensis*. The IUCN Red List of Threatened Species 1998: e.T31319A9625224. DOI: 10.2305/IUCN.UK.1998.RLTS.T31319A9625224.en (Accessed 12 November 2014).

Asiani, Y. 2007. *The Effect of Green Open Space (RTH) Conditions on Microclimate in the City of Bogor*. Thesis, Indonesia University, Depok. http://lib.ui.ac.id/file?file=pdf/abstrak-109946.pdf. [Indonesian]

Burgeff, H. 1959. Mycorrhiza of orchids. In: Withner, C.L. (ed.), *The Orchids: A Scientific Survey*. Ronald Press, New York, pp. 361–395.

CBD. 2002. Global Strategy for Plant Conservation. The Secretariat of the Convention on Biological Diversity, Montreal.

Center for Plant Conservation Botanic Gardens of Indonesia. 2015. *Research Program 2015–2019*. Center for Plant Conservation Botanic Gardens of Indonesia, Indonesian Institute of Sciences, Bogor.

Cribb, P. 1994. New initiatives in the conservation of orchids. In: Suhirman, G., B. Fuaddini, J. Pfeiffer, M. Richardson and Suhendar (eds.), *Strategies for Flora Conservation in Asia*. The Kebun Raya Bogor Conference Proceedings, Bogor, pp. 85–90.

Goss, A.M. 2004. *The Floracrats: Civil Science, Bureaucracy, and Institutional Authority in the Netherlands East Indies and Indonesia, 1840–1970*. PhD Dissertation in the University of Michigan, Ann Arbor. UMI Number: 3138159.

Hardwick, K., D. Latifah, A.R. Gumilang and M. Zuhri. 2017. The biodiverse island nation of Indonesia joins the MSBP. *Samara* 32: 1–2.

Havinga, R., A. Kool, F. Achille, J. Bavcon, C. Berg, C. Bonomi, M. Burkart, D.D. Meyere, J. 't Hart, M. Havström, P. Keßler, B. Knickmann, N. Köster, R. Martinez, H. Ostgaard, B. Ravnjak, A.-C. Scheen, P. Smith, P. Smith, S.A. Socher and V. Vange. 2016. The index seminum: Seeds of change for seed exchange. *Taxon* 65(2): 333–336.

Herbarium Bogoriense. Research Center for Biology, Indonesian Institute of Sciences (LIPI), Bogor. www.biologi.lipi.go.id/botani/index.php/about-bo (Accessed 22 December 2020).

Heywood, V.H. 2017a. The future of plant conservation and the role of botanic gardens. *Plant Divers.* 39: 309–313.
Heywood, V.H. 2017b. Plant conservation in the anthropocene: Challenges and future prospects. *Plant Divers.* 39: 314–330.
Hidayat, I.W., N.I. Kurnita and D. Ardiyanto. 2019. The contribution of Cibodas Botanic Garden as an ex-situ conservation site for tropical mountainous plants: The last decade. *Jurnal Biologi Tropis* 19(2): 161–171.
Hidayat, S., D.M. Puspitaningtyas, S. Hartini, E. Munawaroh, I.P. Astuti and H. Wawangningrum. 2017. *The 25 Years of Flora Exploration: Exploring the Indonesian Jungle*. LIPI Press, Jakarta. [Indonesian]
Hutabarat, P.W.K., R.N. Zulkarnaen and M. Mulyani. 2020. Diversity of mistletoes in Ecopark, Cibinong Science Center-Botanic Gardens. *Al-Kauniyah* 13(2): 263–277. [Indonesian]
Immamudin, H., D.A. Nurdin, Solehuddin and T. Utomo (eds.). 2006. *History of Cibodas Botanic Gardens*. Cibodas Botanic Gardens, Indonesian Institute of Sciences, Cianjur. [Indonesian]
Irawati. 2003. Herbarium Bogoriense: Present and future activities. *Telopea* 10(1): 29–32.
IUCN. 2009. IUCN Red List of Threatened Species. www.redlist.org/ (Accessed 9 March 2010).
IUCN. 2013. IUCN Red List of Threatened Species. www.redlist.org (Accessed 9 November 2013).
IUCN. 2020. Rules of Procedure for IUCN Red List Assessments 2017–2020. Version 3.0 Approved by the IUCN SCC Steering Committee in September 2016. (Accessed 22 December 2020).
Jasa Ilmiah – KRB-LIPI. 2020. PubInfo. www.pubinfo.id/informasi-2668-jasa-ilmiah-krblipi-%C2%BB-herbarium.html (Accessed 22 December 2020).
Kartawinata, K. 2010a. Two centuries reveal diversity of Indonesian flora and ecosystem. Paper presented at Sarwono Prawirohardjo Memorial Lecture X on August, Indonesian Institute of Sciences, Jakarta. [Indonesian]
Kartawinata, K. 2010b. History of the Bogor Botanical Gardens. Paper presented at Botanical Garden International Workshop 2010, Bogor, Indonesia (unpublished).
Kebun Raya Cibodas. 2023a. https://krcibodas.brin.go.id/statistik-jumlah-koleksi-tanaman/. (Accessed 19 April 2023). [Indonesian]
Kebun Raya Cibodas. 2023b. https://sindata.krcibodas.brin.go.id/ (Accessed 19 April 2023). [Indonesian]
Kebun Raya Purwodadi. 2023. https://krpurwodadi.brin.go.id/koleksi/ (Accessed 19 April 2023). [Indonesian]
Kusuma, Y.W.C., Dodo and R. Hendrian. 2011. Propagation and transplanting of manau rattan *Calamus manan* in Bukit Duabelas National Park, Sumatra, Indonesia. *Conserv. Evid.* 8: 19–25.
Latifah, D., D. Widyatmoko, S.U. Rakhmawati, M. Zuhri, K. Hardwick, A.S. Darmayanti and P.K. Wardhani. 2018. The role of seed banking technology in the management of biodiversity in Indonesia. *IOP Conf. Series: Earth and Environmental Science* 298: 012006.
Lawrence, G.H.M. 1969. *Historical Roles of the Botanic Gardens*. The Longwood Seminars. University of Delaware, Newark.
Maryanto, I.J., S. Rahajoe, S.S. Munawar, W. Dwiyanto, D. Asikin, S.R. Ariati, Y. Sunarya and D. Susilaningsih. 2013. *Bioresources for Green Economy Development*. LIPI Press, Jakarta. [Indonesian]
Mogea, J.P., G. Djunaedi, H. Wiriadinata, R.E. Nautian and Irawati. 2001. *Indonesian Rare Plants*. Research and Development Center for Biology, Indonesian Institute of Sciences (LIPI), Bogor. [Indonesian]
Molloy, J. and A. Davis. 1992. *Setting Priorities for the Conservation of New Zealand's Threatened Plants and Animals*. Department of Conservation, Wellington.
Mudiana, D., E. Renjana, E.R. Firdiana, L.W. Ningrum, M.H. Angio and R. Irawanto. 2020. Plants collection enrichment of Purwodadi Botanic Garden through exploration in Alas Purwo National Park. *Jurnal Penelitian Kehutanan Wallacea* 9(2): 83–92. [Indonesian]
Nafar, S. and A. Gunawan. 2017. Ecological design of fernery based on bioregion classification system in Ecopark Cibinong Science Center Botanic Gardens, Indonesia: The 2nd International Symposium Sustaninable Landscape Development. *IOP Conf. Ser. Earth Environ. Sci.* 91: 012032.
Pitman, N.C.A. and P.M. Jorgensen. 2002. Estimating the size of the world's threatened flora. *Science* 298: 989.
Possingham, H.P., S.J. Andelman, M.A. Burgmam, R.A. Medellin, L.L. Master and D.A. Keith. 2002. Limits to the use of threatened species list. *Trends Ecol. Evol.* 17(11): 503–507.
Presidential Regulation Number 2 of 2015 concerning National Mid-Term Development Plan in periods 2015–2019. [Indonesian]
Presidential Regulation Number 18 of 2020 concerning National Mid-Term Development Plan in periods 2020–2024. [Indonesian]
Presidential Regulation Number 93 of 2011 concerning Botanic Gardens. [Indonesian]

Purnomo, D.W., R. Hendrian, J.R. Witono, Y.W.C. Kusuma, R.A. Risna and M. Siregar. 2010. Indonesian Botanic Gardens' achievement on target 8 of the global strategy for plant conservation (GSPC). *Buletin Kebun Raya* 13(2): 40–50. [Indonesian]

Purnomo, D.W., M. Magandhi, F. Kuswantoro, R.A. Risna and J.R. Witono. 2015. Developing plant collections on the regional botanic gardens in framework of Plant Conservation Strategy in Indonesia. *Buletin Kebun Raya* 18(2): 111–124 [Indonesian]

Purnomo, D.W., S. Wahyuni, D. Safarinanugraha, R.N. Zulkarnaen, D.M. Puspitaningtyas and J.R. Witono. 2020. Review of 10 years of botanical gardens development in Indonesia. *Warta Kebun Raya Special Edition* 18(1): 1–15. [Indonesian]

Risna, A.R., Y.W.C. Kusuma, D. Widyatmoko, R. Hendrian and D.O. Pribadi. 2010. *Priority Species for Indonesian Plant Conservation Seri 1: Arecaceae, Cyatheaceae, Nepenthaceae, and Orchidaceae*. Center for Plant Conservation Bogor Botanic Gardens, Indonesian Institute of Sciences, Bogor. [Indonesian]

Rutgers, A.A.L. and F.A.F.C. Went. 1916. Pertodische Erscheinungen bei den Bluten des *Dendrobium Crumenatum* Lindl. *Annales du Jardin botanique de Buitenzorg* 29: 129–160.

Safarinanugraha, D., A.A. Gunawan and W.Q. Mugnisjah. 2017. The development of Bogor Botanic Garden design from 1817 to 2017 base on spatial and functional. *IOP Conf. Ser. Earth Environ. Sci.* 179: 012026. DOI: 10.1088/1755-1315/179/1/012026

Sharrock, S. 2012. *Global Strategy for Plant Conservation a Guide to the GSPC All the Targets, Objectives and Facts*. Botanic Gardens Conservation International, Richmond.

Sujarwo, W., A.R. Gumilang and I.W. Hidayat. 2019. *List of Living Plants Collections Cultivated in Cibodas Botanic Gardens*. Cibodas Botanic Gardens, Indonesian Institute of Sciences, Cianjur.

Sukarya, D.G. and J.R. Witono. 2017. *Two Centuries of Sowing the Earth's Plant Diversity in Indonesia*. PT. Sukarya and Sukarya Pandetama, Jakarta.

Susmianto, A., D. Widyatmoko, B.D. Adji and P. Utama (eds.). 2015a. *Conservation Strategy and Action Plan of Rafflesiaceae 2015–2025*. Ministry of Environment and Forestry, Jakarta. [Indonesian]

Susmianto, A., D. Widyatmoko, B. Dahono and P. Utama (eds.), 2015b. *Conservation Strategy and Action Plan of Corpse Flower (Amorphophallus Titanum (Becc.) Becc. ex Arcang) 2015–2025*. Ministry of Environment and Forestry, Jakarta. [Indonesian]

Thomas, P. 2013. *Dacrycarpus imbricatus*. The IUCN Red List of Threatened Species 2013: e.T42445A2980614. https://dx.doi.org/10.2305/IUCN.UK.2013-1.RLTS.T42445A2980614.en (Accessed 5 August 2021).

Tischler, G. 1905. Uber das Workommen von Statolithen bei wenig oder garnicht geotropischen Wurzelen. *Flora* 94: 1–69.

van Steenis, C.G.G.J. 1950. *Flora Malesiana*, Series 1, Vol. 1. Noordhoff-Kolff N.V., Djakarta.

van Steenis, C.G.G.J., and Kruseman, M.J. 1950. Malaysian plant collectors and collections being a Cyclopaedia of botanical exploration in Malaysia and a guide to the concerned literature up to the year 1950. *Flora Malesiana – Series 1, Spermatophyta* 1(1): 2–639.

WCMC. 1997. Globally and Nationally Threatened Taxa of Indonesia (247 Records). World Conservation Monitoring Centre Plants Programme, Kew.

WCMC. 1998. Intsia bijuga. The IUCN Red List of Threatened Species 1998: e.T32310A9694485. DOI: 10.2305/IUCN.UK.1998.RLTS.T32310A9694485.en (Accessed 23 February 2016).

Went, F.W. 1935. Auxin, the plant growth-hormone. *Bot. Rev.* 1: 162–182.

Widyatmoko, D., R.A. Risna, D.W. Purnomo, D.O. Pribadi and S.R. Ariati. 2018. Implementation of the global strategy for plant conservation in Indonesia: The global partnership for plant conservation meeting. Conducted by Botanic Gardens Conservation International, Convention on Biological Diversity, and South African National Biodiversity Institute, Cape Town, South Africa, 28–30 August.

Wikramanayake, E.D., E. Dinerstein, C. Loucks, D. Olson, J. Morrison, J. Lamoreux, M. McKnight and P. Hedao. 2001. *Terrestrial Ecoregions of the Indo-Pacific: A Conservation Assessment*. Island Press, Washington, DC.

Witono, J.R., D.W. Purnomo, D. Usmadi, D.O. Pribadi, D. Asikin, M. Magandhi and S. Yuzammi. 2012. *Development Plan of Botanic Gardens in Indonesia*. Center for Plant Conservation Bogor Botanic Gardens. Indonesian Instutite of Sciences, Bogor. [Indonesian]

Wyse Jackson, P.S. and L.A. Sutherland. 2000. *International Agenda for Botanic Gardens in Conservation*. Botanic Gardens Conservation International, Kew, London.

Zuhud, E.A.M., L.B. Prasetyo, H. Dewi and H. Sumantri. 2003. *Study of Vegetation and Distribution Patterns of Medicinal Plants in Meru Betiri National Park, East Java*. Laboratory of Plants Conservation, IPB University, Bogor. [Indonesian]

3 Vietnam Botanic Gardens and Their Role in Plant Conservation

Hong Truong Luu, Huu Dang Tran, Ke Loc Phan, Van Sam Hoang, Quoc Binh Nguyen, The Cuong Nguyen, and Trong Nhan Pham

CONTENTS

3.1 Introduction 51
    3.1.1 Vietnam and Its Natural Conditions 51
    3.1.2 The Flora of Vietnam 52
    3.1.3 Threats to Vietnamese Plant Resources and Conservation Needs 53
3.2 Botanic Gardens in Vietnam: A Brief Introduction 54
    3.2.1 Saigon Zoo and Botanical Garden 54
    3.2.2 Hanoi Botanic Garden 56
    3.2.3 Trang Bom Arboretum 56
    3.2.4 Lang Hanh Arboretum 56
    3.2.5 Cu Chi Botanic Garden 57
    3.2.6 Cuc Phuong Botanic Garden 57
    3.2.7 Bu Gia Map Botanic Garden 57
    3.2.8 Bidoup – Nui Ba Botanic Garden 57
    3.2.9 Botanic Garden, Vietnam National University of Forestry 58
    3.2.10 Botanical Collections, Me Linh Station for Biodiversity 58
    3.2.11 Phu An Bamboo Village 58
3.3 Role of Botanic Gardens in Vietnamese Plant Conservation and Related Policy 60
3.4 Conclusions and Recommendations 62
Acknowledgments 62
References 62

3.1 INTRODUCTION

3.1.1 Vietnam and Its Natural Conditions

Vietnam is located at the eastern edge of the Indochinese Peninsula, between 8° 30' and 23° 30' N latitudes and 102° 10' and 109° 27' E longitudes (Figure 3.1). The member of ASEAN is bordered by China to the north, by Laos and Cambodia to the west, and by the Pacific Ocean to the south and east. Its land area of about 330,000 km² supports a population of more than 98.5 million inhabitants in 2021 (General Statistics Office of Vietnam, 2021).

Three-fourths of the country area are mountainous and hilly, with the largest mountain systems in Indochina. Most of the mountains are formed with granite, gneiss, shale, schist, and sandstone, but a considerable mountainous area (i.e., 60,000 km²) is composed of highly eroded karst (Do, 2010). The highest peak Fan Si Pan (3,143 m asl) and many others lie over 3,000 m asl

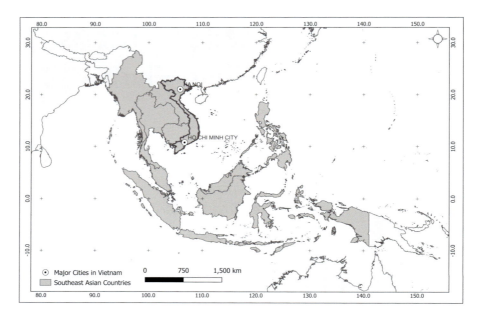

FIGURE 3.1 Location of Vietnam in Southeast Asia.

in the northwest. There are two main large deltas that provide the key source of rice for Southeast Asia: Red River Delta (15,000 km²) and Mekong Delta (40,000 km²). Between the mountainous regions and the lowlands are many dissected plateaus (500–1,500 m asl). The long shape (with 3,440 km-long coastline) and highly changing relief of its mainland make most of the country undergo tropical climate with high temperature and humidity all year around and subtropical in the high-elevated parts. The situation is more complicated by the regional monsoon regime that causes different seasons in other parts of the country (Averyanov et al., 2003). Besides, a number of islands are also included in its territory, including the UNESCO Biosphere Reserve Ha Long Bay. A river system of 41,000 km length in total flows in the country, including the two longest rivers in Southeast Asia: the Red River in the north and the Mekong, which makes a huge wetland complex in the south.

According to MARD (2021), Vietnam has 10.279.185 ha of natural forests and 4.398.030 ha of plantations covering 42% of the country's total area. The country has a total of 11,847,975 ha of wetlands, accounting for 37% of the country's total area (MONRE, 2019).

3.1.2 The Flora of Vietnam

Vietnam's special topography and climate harbors a high level of biodiversity with diverse vegetation types, from rainforests and dry forests to mangroves, which are home to an unusually rich array of plants. About 20,000 terrestrial and water plants were reported for Vietnam (UNDP, 2022). The vascular flora of Vietnam was estimated by Pham-Hoang (1991) to have about 12,000 species, although his work *Cây cỏ Việt Nam – An illustrated Flora of Vietnam*, which is Vietnam's first national flora, listed some 10,500 species. About ten years later, the number became 11,662 species in his updated version, including about 10,200 native ones (Pham-Hoang, 1999–2000). *Checklist of Plant Species of Vietnam*, in three volumes (CRES, 2001; Nguyen et al., 2003; Nguyen et al., 2005), enumerated 11,083 vascular plant species.

More and more botanical surveys are still in need to ascertain this number given the fact that more than 50 new (and therefore endemic mostly) species or new country records have been published from Vietnam every year of the last ten plus years (Middleton et al., 2014; Luong et al., 2015; Luu et al., 2014, 2016, 2017; Do et al., 2018; Tran et al., 2018; Hsu et al., 2020; Lin et al., 2021; Luu et al., 2022). A recent analysis shows that the actual total number of vascular plant species in Vietnam should well exceed 12,000 (Middleton et al., 2019). This is composed of a high rate of endemism, which is estimated to be around 30% countrywide and perhaps reaching 50% in northern Vietnam (Regalado Jr et al., 2005).

Nevertheless, the rich flora has been internationally recognized. According to IUCN, Vietnam has five centers of plant diversity (Davis and Heywood, 1995). The UNEP World Conservation Monitoring Centre (United Kingdom) considers the Hoang Lien Son Range, the Truong Son Range (Annamites), and the Tay Nguyen Plateau (Central Highlands) as the centers of plant diversity in Vietnam (Regalado Jr et al., 2005)

Together with those of birds and mammals, the rich biodiversity and high levels of regional and local endemism in plants makes Vietnam ranked among the world's 25th in species diversity per unit area (Groombridge and Jenkins, 2002). The country is considered as part of one of the world's 25 biodiversity hotspots (i.e., the Indo-Burma biodiversity hotspot) (Myers et al., 2000), with seven priority ecoregions, namely Northern Indochina Subtropical Moist Forests, Southeast China-Hainan Moist Forests, the Annamite Range Moist Forests, Cardamom Mountains Moist Forests, Indochina Dry Forests, Mekong River, and Xi Jiang Rivers and Streams (Olson & Dinerstein, 2002). Obviously, the flora is an important part of Vietnam's 11 biosphere reserves currently recognized by UNESCO (UNESCO, 2022).

3.1.3 THREATS TO VIETNAMESE PLANT RESOURCES AND CONSERVATION NEEDS

The quality of Vietnam's natural forests has degraded with poor and extremely poor forests (10–100 m^3 of timber per ha) accounting for 40.27%. The rich forest area (over 200 m^3 of timber per ha) account for only 8.71% of the country's total natural forest area (MARD, 2016). Threats to Vietnam's biodiversity, which is stored mainly in natural forests, include habitat fragmentation and above all the degradation of forest vegetation due to logging, agricultural conversion, deforestation of natural forests, and wildfires. Massive over-utilization, such as overgrazing, hunting, or collecting rare medicinal plants and timbers, has a major impact on the decline of biodiversity as well (MacKinnon, 1997; Rambaldi et al., 2001; Ngo et al., 2020). According to a recent report by WWF (Thuaire et al., 2021), the loss of natural habitats between 2000 and 2018 is caused by two main economic sectors: agriculture and forestry. For plants, biological resource uses, agriculture and aquaculture, and residential and commercial development are the three most significant threats.

Consequently, the flora becomes more threatened, and conservation is in need. The current IUCN Red List catalogued 227 globally threatened plants (including 69 Vulnerable, 94 Endangered and 64 Critically Endangered species). Most of them (at least 61%) have encountered decrease in population size and geographic range (Thuaire et al., 2021). Climate change and severe weather may cause negative impacts to plants, especially those narrowly endemic, but those are not often reported for individual plant species due to lack of research.

Vietnam's most recent *Red Data Book* (MOST and VAST, 2007) listed 428 threatened vascular plants, a 20% increase compared to its first version (MOSTE, 1996) with 356 threatened plants. The current situation is not known, but a higher figure would be foreseen. Therefore, efforts of conservation is increasingly urgently needed.

Conservation of biodiversity is generally done *in situ* and *ex situ*. Most threatened plants in Vietnam are presently conserved *in situ* in protected areas (such as nature reserves, national parks, or landscape conservation areas), which are expected to play an increasingly important part in the conservation and protection of the country's biodiversity. However, the efficiency of the protected areas is questioned as the magnitude of forested habitat loss in these areas between 2000 and 2018 was found even higher than that nationally (Thuaire et al., 2021). Many plants become rare and face risk of extinction in the field despite the fact they have not been assessed; these should be added to the global and national red lists. In this context, *ex situ* practices should be employed and promoted as significant measures, especially for those species that are narrowly endemic and/or critically endangered. In this regard, botanic gardens hold an important role.

3.2 BOTANIC GARDENS IN VIETNAM: A BRIEF INTRODUCTION

The first botanic garden in Vietnam was established in 1864 in Sai Gon (Ho Chi Minh City now) on a 12-ha land (extended to be 25 ha in 1865) along the Thi Nghe River (Chevalier, 1945); it is currently known as Saigon Zoo and Botanical Garden. The garden was intended specifically for plant and agricultural research and development in Indochina, although a first nursery was built to provide trees for the city's streets and parks (Chevalier, 1919, 1945). A short time later, Hanoi Botanic Garden was established in 1890 on a 33-ha land in the middle of Hanoi with collections of various plants from all over the world. During the French colonial time, two other large plant collections were made in Trang Bom and Lang Hanh, southern Vietnam.

Many botanic gardens have been or are planned to be established in the last decades, but a synthetic assessment has not been made (Hoang, 2020). Most protected areas, including national parks and nature reserves, have plans to develop botanic gardens for research, conservation, and ecotourism purposes, but only a few have been established. Besides, several have been seen in research institutions and universities for research and education. In general, most of them are at an early stage of development and will take time to be completed. Meanwhile, private gardens exist elsewhere, but no related information is available.

Vietnam's national strategy of biodiversity until 2020 with a vision for 2030 identified seven botanic gardens of a total 480 ha established in the country with fewer than 300 plant species each (MONRE, 2013). However, no list of botanic gardens was provided.

A search on the database of BGCI lists ten botanic gardens in Vietnam, namely Lam Vien Forestry Experimental Station, Forest Science Institute of Central Highlands, Hanoi Botanic Garden, Saigon Zoo and Botanical Garden, Bidoup–Nui Ba Botanic Garden, Hanoi University of Pharmacy, Vietnam National University of Forestry, Cuc Phuong Botanic Garden, Phuoc Binh National Park, and Tam Dao Botanic Garden (BGCI, 2022a). The last six are defined as national tropical botanic gardens. Of these, only Bidoup–Nui Ba Botanic Garden is presently a known member of BGCI.

The following is a brief introduction to 11 better known and/or typical botanic gardens in the country (eight of them are illustrated in Figure 3.2), including four established in the French colonial time, one established recently by a city's government, three within protected areas, one by a university, one by a research institution, and one by private scientists. Although we are trying to provide as much information as possible, this introduction simply reflects what we had access to during our preparation of this review rather than all botanic gardens of the country.

3.2.1 SAIGON ZOO AND BOTANICAL GARDEN

Currently, Saigon Zoo and Botanical Garden operates as a company owned 100% by the People's Committee of Ho Chi Minh City. An unpublished checklist of plants made by its own staff records a total of 308 plant species, including 7 gymnosperms and 301 angiosperms. This is a decrease compared with a report by Ly (1968), who enumerated 449 vascular plants, including 22 ornamental and 60 medicinal species. Prominent species are *Dipterocarpus alatus*, *Hopea odorata*, and *Khaya senegalensis*, which were planted from the day of its establishment. These plants are commonly used in streetscaping in Ho Chi Minh City nowadays. However, remarkable threatened species are *Adansonia grandidieri*, *Chukrasia tabularis*, *Dalbergia tonkinensis*, *Diospyros mollis*, *D. mun*, *Guaiacum officinale*, *Intsia bijuga*, and *Sophora tonkinensis*.

Vietnam Botanic Gardens and Their Role in Plant Conservation

FIGURE 3.2 Selected botanic gardens in Vietnam.

(A) Saigon Zoo and Botanical Garden, (B) Hanoi Botanic Garden, (C) Trang Bom Arboretum, (D) Lang Hanh Arboretum, (E) Cu Chi Botanic Garden, (F) Cuc Phuong Botanic Garden (G) Bu Gia Map Botanic Garden, (H) Botanical Collections, Me Linh Station for Biodiversity.

Sources: Photos by Thanh Truc Luu (A), Quoc Binh Nguyen (B), Hong Truong Luu (C, E), Trong Nhan Pham (D), Manh Cuong Nguyen (F), Huu Thang Khuong (G), and The Cuong Nguyen (H).

Owning a very good facility system, the garden defines its mission to develop propagation/breeding, conservation, and research on animals and plants and environmental education (Saigon Zoo and Botanical Garden, 2022). However, it is more well known for recreational and environmental education programs than for plant research and conservation. Its most recent activity in plant collection is establishing a garden of medicinal plants, which is aimed for educational and research purposes. Botanically, the garden is visited mostly by students of botany and pharmacology. In fact, botanical research appears not to be its management board's primary intention in the moment, although a herbarium of about 400 specimens of some 100 plant species is available for the public.

3.2.2 Hanoi Botanic Garden

Hanoi Botanic Garden was established in 1890 for plant experiments. Cultivated plants were those of use, such as timber, agricultural crop, ornament, cattle grass, etc., and most of them were introduced from abroad. They are still growing well. The original area for the garden was 33 ha, but it is 10 ha presently with 78 tree species grown (Hanoi Green Trees Park Limited Company, 1994).

After 1954, the botanic garden was renovated and added more mammals and birds, making it an important leisure place for people. After 1975, all animals were moved to Thu Le Zoo, and Hanoi Botanic Garden became its own function, i.e., hosting plants only.

The botanic garden is under the management of the Peoples' Committee of Hanoi Capital. Although it is not large and almost no research activities are run, more than 100 invaluable tree species find it as a safe area to continue their lives. The living collection includes precious timber trees, such as *Afzelia xylocarpa*, *Dalbergia tonkinensis*, *Parashorea stellata*, *Pterocarpus macrocarpus*, *Terminalia chebula*, etc. Other globally and nationally red-listed species being conserved are *Cleidiocarpon cavaleriei*, *Dipterocarpus retusus*, and *Millingtonia hortensis*. Some collaboration has been known for plant exchange and enrichment of its collections.

3.2.3 Trang Bom Arboretum

The Trang Bom Arboretum is currently located on 5 ha in the southeastern Dong Nai Province, where 189 tree species are grown. According to Ly (1969), it was originally established by the Service des Eaux et Forêts de l'Indochine in 1905 with 70 native tree species of 34 families growing on 316 ha. The arboretum was then under management of different agencies during the French colonial time, Vietnam War, and afterwards.

At present, it is managed by the Eastern South Vietnam Forest Scientific and Production Centre under the Forest Science Institute of South Vietnam. Its defined functions include collection and preservation of native and exotic plants, including rare and threatened species, providing an experiment environment for research institutions and universities and research of plant growth. One of its strengths is tree propagation and tree growth monitoring, especially in reforestation programs. Among threatened species being conserved are those of precious timber, such as *Anisoptera costata*, *Cephalotaxus hainanensis*, *Dalbergia oliveri*, *Keteleeria davidiana*, *Parashorea chinensis*, and *Sindora tonkinensis*.

3.2.4 Lang Hanh Arboretum

As part of the 106-ha Lang Hanh Forest Experiment Station, the Lang Hanh Arboretum is located in Lam Dong Province of the Central Highlands. It was established by the Institut des Recherches Agronomiques de l'Indochine and is currently managed by the Forest Science Institute of Central Highlands and South of Central Vietnam based in Da Lat City, which used to be a member of BGCI in 2021 and 2022.

Based on living collections and natural forest trees of 170 species, the arboretum functions mainly to produce seedlings for replantation besides conserving threatened plants. Like Trang

Bom Arboretum, it has extensive experience in propagating trees, especially threatened species, for reintroduction to the wild and reforestation. Very old collections are conserved for globally and nationally red-listed species, such as *Keteleeria evelyniana*, *Khaya senegalensis*, *Lithocarpus amygdalifolius*, *Pterocarpus indicus*, *Quercus setulosa*, and *Swietenia macrophylla*.

3.2.5 CU CHI BOTANIC GARDEN

This botanic garden was established by the People's Committee of Ho Chi Minh City in 2003 based on an existing protection forest named Ben Dinh. Its collections of plants started in 2009 with focus on rare, endemic, economically important, and typical species from all over Vietnam and especially from Southeast Vietnam.

Until December 2019, its area of 40 ha harbored collections of more than 500 tree species (with nearly 10,000 individuals), 52 orchid species, and dozens of others (lianas and succulents). They are cultivated in different thematic zones, such as rare plants, endemics, dipterocarps, palms, bamboos, cycads, evolution, orchids, etc. These include many globally and nationally threatened plants, such as *Afzelia xylocarpa*, *Dalbergia oliveri*, *Hopea cordata*, *Shorea falcata*, etc.

3.2.6 CUC PHUONG BOTANIC GARDEN

This 167-ha botanic garden was established by the Ministry of Forestry in 1988 and could be the first botanical collections within a protected area in Vietnam. Its living collections started in 1985 and presently include 811 plant species, among which 210 trees are native to the park, 85 trees from other parts of Vietnam, 5 from abroad, 25 aroids native to the park, 20 fruity plants, 15 bamboos, 17 cycads, 15 palms, 296 medicinal plants, and 140 orchids (Duwe et al., 2022). The garden is targeted to preserve rare plants and employs breeding experiments for scientific, learning, and tourism purposes. Currently, it holds the most global and national plant species among the 11 botanic gardens discussed in this chapter. Some examples are *Camellia cucphuongensis*, *Amorphophallus interruptus*, *Carya sinensis*, *Castanopsis lecomtei*, *Euphorbia prostrata*, *Gynostemma pentaphyllum*, *Paphiopedilum concolor*, etc.

The garden is part of Cuc Phuong National Park – Vietnam's first national park, which was established in 1962 and is about 120 km southeast of Hanoi. This well-known protected area is one of the world's five centers of plant diversity in Vietnam (Davis and Heywood, 1995).

3.2.7 BU GIA MAP BOTANIC GARDEN

This garden was established within Bu Gia Map National Park of Binh Phuoc Province in 2004. It is based on a 21.5-ha natural forest after human impacts, which serves as a convenient habitat to conserve rare and threatened plants collected from the park. Besides, the living collections are used for scientific research, environmental education, and ecotourism. Basic infrastructure has been invested thanks to the provincial People's Committee.

Naturally growing trees include 117 species of 42 families, among which, 6 are identified as threatened in the current IUCN Red List and 4 in the *Vietnam Red Data Book*. In the moment, the park is cooperating with the Southern Institute of Ecology to conserve its locally endemic camellia (*Camellia bugiamapensis*). Many other threatened species are being conserved, such as *Aquilaria crassna*, *Dalbergia cochinchinensis*, *Dipterocarpus intricatus*, *Pterocarpus macrocarpus*, *Shorea roxburghii*, etc.

3.2.8 BIDOUP–NUI BA BOTANIC GARDEN

This very new botanic garden is part of Bidoup–Nui Ba National Park, which is located in the Lang Biang Plateau – one of the world's centers of plant diversity (Davis and Heywood, 1995). Unlike most other protected areas where botanic gardens are built mainly by adding trails and simple instruction materials to a chosen area of natural forest without elaborated landscaping design and extensive living plant collections, Bidoup–Nui Ba National Park has an ambition to establish an internationally recognized botanic garden. Therefore, from the beginning, its management board has requested the Royal Botanic Gardens & Domain Trust (Australia) for help in designing. Its general planned shape reflects the structure of the cone of *Pinus krempfii* – a special two-flat leaved pine endemic to the southern Truong Son (Annamites). It is aimed to be a place to appreciate the beauty, uniqueness, and usefulness of Vietnamese plants in an accessible beautiful landscape, close to visitors and scientific facilities, and providing opportunities to further explore the natural forests.

At present, the provincial People's Committee and several national and international partners (e.g., Southern Institute of Ecology, Dr. Cecilia Koo Botanic Conservation Center from Taiwan, and WWF-Vietnam) are working to support this botanic garden, which covers an area of 2.2 ha of pure *Pinus kesiya* forest, which is planned for different specific plant collections and landscapes. Basic roads and facilities have been constructed, and some living collections are being developed. Besides the already-cultivated 100+ fern and orchid species native to the Lang Biang Plateau (Figure 3.3), a new collecting project is just starting with a target of 120 Vietnam-endemic and globally and nationally threatened species by the end of 2023. Some are already collected, such as camellias, slipper orchids, *Magnolia bidoupensis*, *Polyspora huongiana*, etc.

3.2.9 BOTANIC GARDEN, VIETNAM NATIONAL UNIVERSITY OF FORESTRY

This botanic garden was originally established at Vietnam National University of Forestry in 1984. It is located in Xuan Mai town, Ha Noi capital with an area of 100 ha and more than 400 native plants growing naturally or introduced, including 27 species endemic or listed in the current IUCN Red List and *Vietnam Red Data Book*. Examples of threatened threes are *Castanopsis cerebrina*, *Cinnamomum balansae*, *Lithocarpus elegans*, and *Quercus platycalyx*.

It is considered to be a living museum for training, scientific research, and exchange. It is being planned to upgrade to be a national botanic garden.

3.2.10 BOTANICAL COLLECTIONS, ME LINH STATION FOR BIODIVERSITY

Me Linh Station for Biodiversity (belonging to Institute of Ecology and Biological Resources) was established in 1999 by the National Center for Science and Technology (presently Vietnam Academy of Science and Technology) aiming to develop living collections of plants and animals to support biodiversity conservation and scientific research. In this regard, it appears to follow the original idea of the Saigon Zoo and Botanical Garden. Besides 1,220 plants growing naturally in a 170.3-ha area, Me Linh Station for Biodiversity has developed living collections of 46 medicinal plant, 90 timber, and more than 100 orchid species, including threatened species such as *Paphiopedilum malipoense*. More living collections are to be made.

3.2.11 PHU AN BAMBOO VILLAGE

This is one of the typical examples of increasing specialized gardens that focus on collections of specific plant groups, such as medicinal plants and bamboos. The Phu An Bamboo Village was established by a group of Vietnamese and French scientists in 1999 in Ben Cat District of the southeastern province of Binh Duong, which is about 35 km north of Ho Chi Minh City. As Vietnam's first bamboo living museum, it became more known after winning the UNDP Equator Prize in 2010. Its aim is to preserve bamboo diversity and promote public awareness in Vietnam and neighboring countries, with a strong vision for future development as the largest bamboo arboretum in Asia. More than 350 bamboo living collections have been grown on about 10 ha; however, its checklist of species is not available.

FIGURE 3.3 Collections of orchids and ferns at Bidoup–Nui Ba Botanic Garden. Front of the photo: *Paphiopedilum appletonianum*.

Source: Photo by Van Son Le.

The conservation activities aim to contribute to knowledge expansion on natural resources, to determine the adaptation conditions of bamboos in Phu An as well as proper techniques of propagation, and to study soil fertility conservation techniques and potential improvements in agricultural production. Local villagers are provided with bamboo cuttings from the garden to cultivate for handicraft production.

3.3 ROLE OF BOTANIC GARDENS IN VIETNAMESE PLANT CONSERVATION AND RELATED POLICY

Botanic gardens are obviously an effective measure of *ex situ* conservation of Vietnamese plant diversity, which is threatened due to various factors. According to BGCI, botanic gardens are institutions holding documented collections of living plants for the purpose of scientific research, conservation, display, and education. They should have a greater emphasis on conserving rare and threatened plants, compliance with international policies, and sustainability and ethical initiatives (BGCI, 2022b). In general, all existing and planned botanic gardens in Vietnam, including specialized ones (e.g., medicinal plant gardens and others), are defined to address these functions. In addition, ecotourism is also one of their key targets and embedded in their operations. However, the actual operation of the existing gardens indicates that they have emphasized either education and entertainment purposes or seedling production as these provide important capital for their operation besides the subsidiary from the State. Good examples are botanic gardens in the cities of Hanoi and Ho Chi Minh and the arboretums as mentioned in this chapter. This is also the case for all botanic gardens built or planned to be built in the protected areas, although research and conservation are indicated as their original key targets. In fact, those in protected areas are normally poorly designed and invested, and consequently, income from ecotourism appears to be minor.

Nevertheless, existing botanic gardens hold a considerable number of plants. Based on their unpublished checklists, we could only provide preliminary information on how many threatened plants are conserved in selected Vietnamese botanic gardens (Table 3.1) as there is no updating synthetic study yet. In total, there are 212 threatened species (including 113 globally and 131 nationally) currently being conserved in these botanic gardens, among which Cuc Phuong and Cu Chi Botanic Gardens hold the largest collections (82 and 80 species, respectively). Some conservation activities and threatened species being conserved are illustrated in Figure 3.4.

TABLE 3.1
Number of Threatened Plants Grown in Selected Botanic Gardens

Botanic Garden	Total	IUCN Red List (2022) CR	EN	VU	Vietnam Red Data Book (2007) CR	EN	VU
Saigon Zoo and Botanical Garden	30	2	10	12	0	10	5
Hanoi Botanic Garden	14	0	4	6	0	2	4
Trang Bom Arboretum	35	1	10	12	0	12	7
Lang Hanh Arboretum	14	0	5	5	1	3	3
Cu Chi Botanic Garden	80	7	22	26	2	23	23
Cuc Phuong Botanic Garden	82	5	11	15	1	15	43
Bu Gia Map Botanic Garden	32	2	11	10	1	9	8
Bidoup–Nui Ba Botanic Garden (on-going collection)	37	4	5	7	1	12	12
Botanic Garden, Vietnam National University of Forestry	27	1	11	6	0	11	8
Botanic Collections, Me Linh Station for Biodiversity	25	4	4	6	2	10	18
Total	**212**	**16**	**45**	**52**	**4**	**49**	**78**

Vietnam Botanic Gardens and Their Role in Plant Conservation

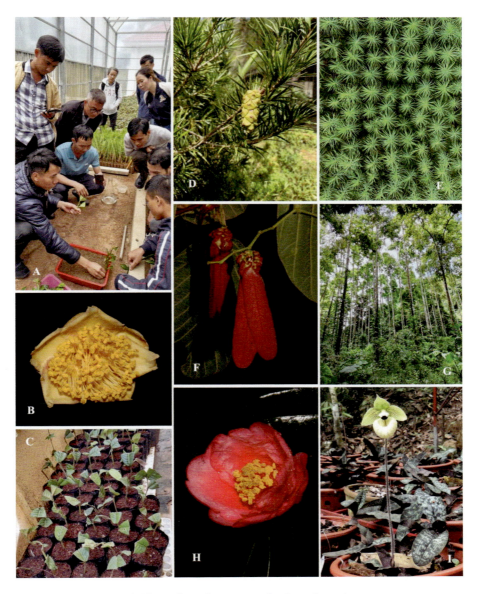

FIGURE 3.4 Conservation activities and species conserved at botanic gardens.

(A–C) Training on propagation of *Camellia dalatensis* funded by BGCI at FSIH (Lang Hanh Arboretum); (D–E) *Pinus krempfii* conserved at Bidoup–Nui Ba Botanic Garden; (F) *Dipterocarpus intricatus* (Cu Chi Botanic Garden); (G) *Hopea chinensis* (Cuc Phuong Botanic Garden); (H) *Polyspora huongiana* (Bidoup–Nui Ba Botanic Garden); (I) *Paphioedilum malipoense* (Me Linh Station for Biodiversity).

Source: Photos by Hong Truong Luu (A, C, F, H), Tran Quoc Trung Nguyen (B), Quang Cuong Truong (D, E), Manh Cuong Nguyen (G) and The Cuong Nguyen (I).

In spite of the role of botanic gardens, they are not mentioned in key national laws related to biodiversity, such as the Law on Biodiversity (2008) and the Law on Environment Protection (2020). The Law on Forestry (2017) is the first that uses the phrase "national botanic garden"; although this concept is not defined, the law regulates plant collection and research activities in such entities that are currently not known to exist. However, both Vietnam's national strategy of biodiversity until 2020 with a vision for 2030 (MONRE, 2013) and the updated version (Decision No. 149/QĐ-TTg by the Prime Minister dated 28 January 2022) (Prime Minister, 2022) appreciate the role of and encourage the development of botanic gardens for *ex situ* conservation of plants. Decision No. 45/QĐ-TTg by the Prime Minister dated 5 January 2014 (Prime Minister, 2014) plans to have 4 botanic gardens by 2020 and 11 by 2030 in Vietnam, which are typical for and distributed across the country's ecological regions. Besides, five and six medicinal plant gardens are planned to be developed by 2020 and 2030, respectively. Nevertheless, no detailed plan for botanic garden development has been known for Vietnam.

3.4 CONCLUSIONS AND RECOMMENDATIONS

Botanic gardens have been developed for nearly 150 years in Vietnam, and their role in plant conservation is known, with more or less ten existing and many more planned to be developed across the country. Their functions as defined by BGCI are integrated in their establishment and operation plans, but not all are well performed. Generally, environment education and entertainment rather than conservation of plants are emphasized in their current activities, although seedling production for replantation is seen more pronounced in arboretums run by research institutions.

The development of botanic gardens is encouraged by several national policies on biodiversity, but it needs a strategic national plan with clear targets, detailed actions, and consensus instruction for national and local policies and actions. To be feasible and effective, this should be based on a national assessment of their present status and actual related demands for development and operation of botanic gardens with careful consideration of both national and local plant conservation strategies and needs. Categories of botanic gardens should also be developed. Besides financial consideration, which is crucial for developing countries like Vietnam, personnel resources must be prepared and invested according to actual development demands. National resources are available, but they are presently scattered. More exposure to successful international experience appears to be critical for all stakeholders, especially policy makers and responsible governmental authorities.

ACKNOWLEDGMENTS

We thank Thanh Truc Luu, Manh Cuong Nguyen (Cuc Phuong National Park), Huu Thang Khuong (Bu Gia Map National Park), Thanh Truong Hoang (FSIH), Tran Quoc Trung Nguyen (SIE), Quang Cuong Truong, and Van Son Le (Bidoup–Nui Ba National Park) for allowing their photos to be used in the illustrations. We are grateful to Mr. Van Bang Tran, Mr. Thanh Luc Nguyen, Mr. Hieu Cuong Nguyen (SIE), and Prof. Dr. Ke Long Phan (VNMN) for their help in map and data preparation. The work is supported by the project "Field research and conservation of Camellias in Southern Vietnam" supported by Botanic Gardens Conservation International to SIE and Bidoup–Nui Ba National Park.

REFERENCES

Averyanov, L., K.L. Phan, T.H. Nguyen and D. Harder. 2003. Phytogeographic review of Vietnam and adjacent areas of Eastern Indochina. *Komarovia* 3: 1–83.

BGCI. 2022a. Advance Garden Search. https://tools.bgci.org/garden_advanced_search.php?action=Find&mode=&ftrCountry=VN&ftrInstitutionType=Botanic+Garden&ftrKeyword=&x=78&y=5#results

BGCI. 2022b. Botanic Gardens and Plant Conservation. www.bgci.org/about/botanic-gardens-and-plant-conservation
Chevalier, A. 1919. *Catalogue des plantes du Jardin botanique de Saigon*. Imprimerie nouvelle A. Portrail, Saigon.
Chevalier, A. 1945. Les améliorations scientifiques et techniques réalisées par la France en Indochine. *Journal d'agriculture traditionnelle et de botanique appliquée* 25(277): 133–162.
CRES. 2001. *Checklist of Plant Species of Vietnam*, Vol. 1. Agriculture Publishing House, Hanoi.
Davis, S.D. and V. Heywood. 1995. *Centres of Plant Diversity: A Guide and Strategy for Their Conservation*, Vol. 2. Asia, Australasia and the Pacific. IUCN Publications Unit, Cambridge, U.K.
Do, T. 2010. Characteristics of karst ecosystems of Vietnam and their vulnerability to human impact. *Acta Geologica Sinica-English Edition* 75(3): 325–329.
Do, V.T., T.T.H. Vu, H.T. Luu and T.T. Nguyen. 2018. *Aristolochia nuichuaensis* (subg. Siphisia, Aristolochiaceae), a new species, an updated key and a checklist to the species of *Siphisia* in Vietnam. *Annales Botanici Fennici* 56(1–3): 107–113.
Duwe, V.K., L. Van Vu, T. von Rintelen, E. von Raab-Straube, S. Schmidt, S. Van Nguyen, T.D. Vu, T. Van Do, T.H. Luu and V.B. Truong. 2022. Contributions to the biodiversity of Vietnam: Results of VIETBIO inventory work and field training in Cuc Phuong National Park. *Biodiversity Data Journal* 10: e77025.
General Statistics Office of Vietnam. 2021. Infographic Population, Labour and Employment in 2021. www.gso.gov.vn/en/data-and-statistics/2022/01/infographic-population-labour-and-employment-in-2021/
Groombridge, B. and M.D. Jenkins. 2002. *World Atlas of Biodiversity: Earth's Living Resources in the 21st Century*. University of California Press, Berkeley, USA.
Hanoi Green Trees Park Limited Company. 1994. *Feasibility Study for Hanoi Botanical Garden*. Unpublished Technical Report, Hanoi.
Hoang, V.S. 2020. Nghiên cứu đề xuất qui định và hướng dẫn quản lý vườn thực vật quốc gia [A study to recommend regulations and guidelines for management of national botanic gardens]. Technical report. GIZ, Hanoi.
Hsu, T.C., C.W. Chen, H.C. Hung, N.K.T. Tram, Q.C. Truong, H.T. Luu and C.W. Li. 2020. New and noteworthy orchids (Orchidaceae) discovered in Langbiang Plateau, Southern Vietnam 1. *Taiwania* 65(2): 237–248.
Lin, C.W., T.C. Hsu, H.T. Luu, I.L.P.T. Nguyen, T.Y.A. Yang and C.W. Li. 2021. Revision of *Begonia* (Begoniaceae) in Bidoup-Nui Ba National Park, Southern Vietnam, including two new species. *Phytotaxa* 496(1): 77–89.
Luong, V.D., H.T. Luu, T.Q.T. Nguyen and Q.D. Nguyen. 2015. *Camellia sonthaiensis* (Theaceae), a new species from Vietnam. *Annales Botanici Fennici* 52: 289–295.
Luu, H.T., H.C. Nguyen, T.Q.T. Nguyen and Q.B. Nguyen. 2022. *Arisaema vietnamense* (section Nepenthoidea, Araceae): A new species from Vietnam. *Academia Journal of Biology* 44(1): 1–9.
Luu, H.T., T.Q.T. Nguyen and Q.D. Nguyen. 2016. *Camellia luteopallida* (Theaceae), a new species from Vietnam. *Annales Botanici Fennici* 53(1–2): 135–138.
Luu H.T., T.T. Nguyen, G. Tran G., Q.D. Đinh, N.L. Vu, L.X.B. Nguyen, T.T.D. Ngo & T.T. Nguyen, 2014. *Thismia okhaensis* (Thismiaceae) – a new fairy lantern from Vietnam. Phytotaxa 164 (3): 161–200.
Luu, H.T., H.T. Van, T.T.D. Ngo, L.P. Nguyen and H.P. Le. 2017. *Typhonium thatsonense* (Araceae), a new species from Vietnam. *Novon* 25(4): 438–441.
Ly, V.H. 1968. Guide Botanique de la Ville de Saigon. *Bulletin de la Société des Etudes Indochinoises de Saigon* 18(4): 311–351.
Ly, V.H. 1969. *Natural Plants and Artificial Forests at Trang Bom Experiment Forest*. Institute of Agricultural Research, Saigon.
MacKinnon, J. 1997. *Protected Areas Systems Review of the Indo-Malayan Realm*. The Asian Bureau of Conservation Ltd. World Conservation Monitoring Centre, Cambridge.
MARD. 2016. Decision No. 4290/QD-BNNTCLN dated 21 October 2016 by the Ministry of Agriculture and Rural Development approving the results of forest mventory and inventory in 25 provinces in 2014–2015 under the "National Forest Inventory Period 2013–2016".
MARD. 2021. Decision No. 1558/QĐ-BNN-TCLN on announcing the current forest status nationwide in 2020.
Middleton, D., K. Armstrong, Y. Baba, H. Balslev, K. Chayamarit, R. Chung, B. Conn, E. Fernando, K. Fujikawa and R. Kiew. 2019. Progress on Southeast Asia's Flora projects. *Gardens' Bulletin Singapore* 71(2): 267–319.

Middleton, D., H. Atkins, L.H. Truong, K. Nishii and M. Moeller. 2014. *Billolivia*, a new genus of Gesneriaceae from Vietnam with five new species. *Phytotaxa* 161(4): 241–269.

MONRE. 2013. *Vietnam's National Strategy of Biodiversity Until 2020 with a Vision for 2030*. Ministry of Natural Resources and Environment, Hanoi.

MONRE. 2019. *The Sixth National Report of the United Nations Convention on Biological Diversity*. Ministry of Natural Resources and Environment, Hanoi.

MOST and VAST. 2007. *Vietnam Red Data Book. Part II. Plants*. Publishing House for Science and Technology, Hanoi.

MOSTE. 1996. *Vietnam Red Data Book. Volume 2 (Plants)*. Science and Technology Publishing House, Hanoi.

Myers, N., R.A. Mittermeier, C.G. Mittermeier, G.A. Da Fonseca and J. Kent. 2000. Biodiversity hotspots for conservation priorities. *Nature* 403(6772): 853–858.

Ngo, D.T., A.V. Le, H.T. Le, S.M. Stas, T.C. Le, H.D. Tran, T. Pham, T.T. Le, B.D. Spracklen and C. Langan. 2020. The potential for REDD+ to reduce forest degradation in Vietnam. *Environmental Research Letters* 15(7): 074025.

Nguyen, T.B., P.A. Tran, T.B. Tran, K.B. Le, Q.B. Nguyen, V.D. Nguyen, T.D. Nguyen, V.H. Vu, D.H. Duong, C.K. Tran, K.K. Nguyen, K.L. Tran, D.L. Tran, T.N. Nguyen, N.N. Tran, X.P. Vu, H.Q. Bui and L.V. Averyanov. 2005. *Checklist of Plant Species of Vietnam*, Vol. 3. Agriculture Publishing House, Hanoi.

Nguyen, T.B., T.P.A. Tran, K.B. Le, Q.B. Nguyen, T.D. Ha, V.D. Nguyen, D.D. Tran, K.D. Nguyen, T.D. Nguyen, H.H. Nguyen, T.H. Nguyen, V.H. Vu, D.H. Duong, C.K. Tran, D.K. Nguyen, K.K. Nguyen, K.L. Tran, K.L. Phan, D.L. Tran, N.N. Tran, X.P. Vu, M.T. Ha, N.T. Nguyen, T.X. Do, N.N. Arnautov, L.V. Averyanov, A.L. Budantsev, V.I. Dorofeev, M. Mikhailova, V.P. Serov and N.T. Skvortsova. 2003. *Checklist of Plant Species of Vietnam*, Vol. 2. Agriculture Publishing House, Hanoi.

Olson D.M. and E. Dinerstein. 2002. The Global 200: Priority ecoregions for global conservation. *Annals of the Missouri Botanical Garden* 89(2): 199–224.

Pham-Hoang, H. 1991. *Cây cỏ Việt Nam [An illustrated flora of Vietnam]*, Vol. 1. Mekong Printing, Montreal, Canada.

Pham-Hoang, H. 1999–2000. *Cây cỏ Việt Nam [An Illustrated Flora of Vietnam]*, Vol. 1–3. Youth Publishing House, Ho Chi Minh.

Prime Minister. 2014. Decision No. 45/QĐ-TTg on approving the master plan on nation-wide biodiversity conservation by 2020, with a vision to 2030.

Prime Minister. 2022. Decision No. 149/QĐ-TTg on approving the national strategy for biodiversity until 2030 and vision to 2050.

Rambaldi, G., S. Bugna and M. Geiger. 2001. Review of the protected area system of Vietnam. *Asean Biodiversity* 1(4): 43–51.

Regalado Jr, J.C., N.T. Hiep, P.K. Loc, L. Averyanov and D.K. Harder. 2005. New insights into the diversity of the Flora of Vietnam. *Biologiske Skrifter* 55: 189–197.

Saigon Zoo and Botanical Garden. 2022. https://saigonzoo.net/about-us/

Thuaire, B., A. Y, V.A. Hoang, K.Q. Le, H.T. Luu, T.C. Nguyen and T.T. Nguyen. 2021. *Assessing the Biodiversity in Viet Nam: Analysis of the Impacts from the Economic Sectors*. WWF-Viet Nam, Hanoi.

Tran, H.D., H.T. Luu, N.T. Tran, T.T. Nguyen, Q.B. Nguyen and J. Leong-Škorničková. 2018. Three new *Newmania* species (Zingiberaceae-Zingibereae) from central Vietnam. *Phytotaxa* 367(2): 145–157.

UNDP. 2022. Viet Nam's First National Ecosystem Assessment Warns of Steady Biodiversity Loss Upending the Economy. www.undp.org/nairobi-gc-red/news/viet-nam%E2%80%99s-first-national-ecosystem-assessment-warns-steady-biodiversity-loss-upending-economy

UNESCO. 2022. Biosphere Reserves in Asia and the Pacific. https://en.unesco.org/biosphere/aspac/

4 Botanical Gardens in Malaysia and Their Role in Plant Conservation

L.G. Saw, A. Latiff, and J. Sang

CONTENTS

4.1 Introduction 65
4.2 Spice Cultivation in Penang and the First Botanic Garden 66
4.3 Arboretum at the Forest Research Institute 67
4.4 Sabah 69
4.5 Sarawak 69
4.6 Botanical Gardens in Malaysia 70
    4.6.1 Johor Botanic Gardens 70
    4.6.2 Royal Palace Botanical Garden 71
    4.6.3 Perak Botanical Gardens 71
    4.6.4 Rimba Ilmu Botanic Gardens, University of Malaya 71
    4.6.5 Fernarium, Universiti Kebangsaan Malaysia 71
    4.6.6 Perdana Botanical Gardens, Kuala Lumpur 71
    4.6.7 Putrajaya Botanic Garden 72
    4.6.8 National Botanic Gardens, Shah Alam 72
    4.6.9 Melaka Botanic Garden 72
    4.6.10 Penang Botanic Gardens (Plates 4.2, 4.3) 72
    4.6.11 Suriana Botanic Conservation Gardens 72
    4.6.12 Rainforest Discovery Centre (Plate 4.4) 72
    4.6.13 Mount Kinabalu Botanical Garden 73
    4.6.14 Sabah Agriculture Park (Plate 4.5) 73
    4.6.15 Botanical Research Centre (BRC) 73
    4.6.16 Sarawak Botanical Garden 76
    4.6.17 Sarawak Biodiversity Centre 76
4.7 Role of Botanical Gardens in Malaysia 76
4.8 Plant Conservation in Malaysia 77
4.9 Malaysian Botanical Gardens Network (MYBGNet) 79
4.10 Conclusion 79
Acknowledgments 79
References 79

4.1 INTRODUCTION

Malaysia has among the richest flora in the world due to being in the humid tropic, and it has been recognized as 1 of the 17 biodiverse countries. Its species diversity is estimated to be about 15,000 species of vascular plants (Saw and Chung, 2015). The study and documentation of the country's flora have always been centered around its botanical institutions, such as research institutes, especially Forest Research Institute Malaysia, government departments such as Sabah Forestry

Department and Sarawak Forestry Department, and universities. Historically, the arrival of the British colonists in the nineteenth century started the interest in the flora of this region. However, both the Portuguese (1511–1641) and the Dutch (1641–1824) did not contribute much to the exploitation of the Malay Peninsular natural resources, including establishment of botanical gardens.

The settlement of Penang by the British East India company in 1786 fulfilled an immediate requirement for a convenient port along the trade route to China (Langdon, 2015). British interest in the sixteenth century in the Far East was spurred by trade of plant products from China, Malesian archipelago, and the Far East. Tea, nutmeg, mace, and clove were immensely valuable in Europe then, and the control of the trade of tea and spices strongly influenced the colonization and exploration of the Far East by the Europeans. The Portuguese, followed by the Dutch and British, made significant inroads toward the colonization of Southeast Asia and the Far East. The Dutch and British established the first botanical gardens in the region: The Dutch East India Co. founded Bogor Botanical Garden in 1817; the British founded the Singapore Botanic Gardens in 1859 and Penang Botanic Gardens in 1884.

When the early botanical gardens were first established, they played the role of the introduction and trial of new crop plants with the objective of broadening the agricultural exploitation and industries. They introduced and brought into cultivation many of these crops in the colonies they occupied. Through the East India Company, Royal Botanic Gardens, Kew coordinated the establishment of the many botanical gardens in the British colonies, and botanists stationed in these gardens took on the responsibility of introducing new crops. They also documented and studied the local flora, either for ethnobotanical purposes or to learn about the flora in general.

Subsequently, agricultural departments were established; they took over the role of crop development. Agricultural departments were able to widen the experiments of growing the crops, breeding, propagating, and improving them, and then manufacturing and analyzing the products. Botanical gardens became more inclined toward horticulture, the studying and documenting of native species, landscaping, and growing of plants for recreation and public enjoyment. Botanic gardens are often limited in space and could not do the kind of studies or crop trials that require much larger spaces. A botanical garden has many functions and roles, such as recreational, educational, scientific research, and conservation, and their roles evolve with changing needs and demands (Anon, 1980).

After the 1980s, many botanic gardens all over the world have changed in their functions and roles depending on the scope of conservation issues and problems (Bramwell et al., 1987). In 1987, the IUCN Botanic Gardens Conservation Secretariat was formed to promote and coordinate plant conservation and environmental education in botanic gardens (IUCN, 1987). Subsequently, the Botanic Gardens Conservation International (BGCI) was founded as a charity to support a global network of botanical gardens. BGCI has a membership of 650 botanic gardens in 118 countries, whose combined work forms the world's largest plant conservation network.

4.2　SPICE CULTIVATION IN PENANG AND THE FIRST BOTANIC GARDEN

After the Dutch took control of the Moluccas Islands from the Portuguese in the seventeenth century, they set a monopoly over the production and trade of nutmegs, mace, and cloves. With the British attempting to break the Dutch monopoly, this rivalry led to war between the Dutch and British. A subsequent treaty between the British and Dutch to divide the Far East (East India) into control between the two empires allowed the British to break the Dutch monopoly. This started the cultivation of nutmegs and cloves beyond the Moluccas Islands.

The person primarily responsible for establishment of the first botanical garden in Penang was the Irish botanist Christopher Smith (Langdon, 2015). Smith was trained in Kew Gardens, and by the time he arrived in Penang in 1795, he had already engaged in several successful expeditions in the region, collecting live plants and transplanting them to Kew Gardens and to other gardens in the colonies, especially Calcutta (Langdon, 2015). In 1796, Smith arrived in Ambonya or Ambon (Moluccas Islands) and collected "64,052 clove, nutmeg and other valuable plants" and sent shipments to the botanic gardens at Kew, the Cape Colony, St. Helena, Calcutta, Bencoolen (Bengkulu, Sumatera), and Penang (Langdon, 2015) for the British East India Company.

He subsequently made several more expeditions to the Moluccas Islands and transplanted many thousands of plants to the colonies with the British East India Company. This included further shipments of clove and nutmeg plants to Penang, and a botanic garden was set up there. The first plantings were in Ayer Itam and Sungai Keluang. The plants grew very well; other plants and spices were also planted there, including pepper, canary nuts, and sugar palms (*Arenga pinnata*). Christopher Smith was particularly keen also to grow sugar palms, not so much for the extraction of palm sugar but rather for the fibers from the leaf sheaths that can be made into cordage that was far superior to any of the existing ropes made from other fibers (Langdon, 2015).

Smith's introduction to Penang included possibly sago palm (*Metroxylon sagu*), gomuti or sugar palm (*Arenga pinnata*), canary nut (*Canarium commune*), cajeput (*Melaleuca cajuputi*), cinnamon tree producing *lawang* oil (*Cinnamomum culilawan*), langsat (*Lansium domesticum*), gandaria (*Bouea macrophylla*), gendarussa (*Justicia gendarussa*), the red lovi-lovi or batako palm (*Flacourtia inermis*), and *nam-nam* (*Cynometra cauliflora*). Smith also sent Moluccan varieties of *jambu* or rose apple, mango, mangosteen, banana, variegated pineapple, cinnamon, coffee, cacao, areca nut, and a host of others (Langdon, 2015).

Unfortunately, due to the poor state of the nutmeg and clove plantations in the botanical garden, the East India Company decided to abandon the plantings, and the land was auctioned off in 1805 (Langdon, 2015). It was also that the production of spices from the East India Company's plantation in Bencoolen (Bengkulu) was much better and was able to supply the trade requirement for the company. By this time, private growers in Penang had started planting nutmegs and cloves, and they thrived much better and started to supply their spices to Britain. The private growers at the height of the spice trade in Penang covered some 400 plantations in Penang and Province Wellesley, ranging in sizes from 1 to 1,000 acres (Langdon, 2015). It must be noted that Christopher Smith's efforts in transplanting the spice plants from the Moluccas paved the way for success of these private growers.

In 1823, Nathaniel Wallich proposed the establishment of the second botanical garden for Penang (Langdon, 2015) with George Porter appointed as the superintendent. It was unfortunate this too did not last; by 1830, following the abolition of Penang's status as the fourth presidency of India, the garden land was once again sold off. In 1884, Nathaniel Cantley, superintendent of the Singapore Botanic Gardens, established the present botanic garden and appointed Charles Curtis as the assistant superintendent. The garden was part of the Gardens and Forests Department, Straits Settlement and was engaged in the cultivation of essential commercial plants, inspecting crops, and advising the planting community generally (Banfield, 1947). Curtis subsequently became the curator of the garden and remained so until 1903 when he retired on account of ill-health. The garden, under the able hands of Curtis, was designed and landscaped, and the main plan remains relatively the same till the present day, although the garden went through a few master plans and improvements.

The initial site was only 29.14 ha, of which only 11.33 ha was developed for lawns and other related activities. The rest remained as natural forest environment. In 2003 and 2004, the Penang state government expanded the boundary to 242.07 ha to include the surrounding lowland and hill dipterocarp forests, making Penang Botanic Gardens the largest botanic garden in the country (Anon, 2019).

4.3 ARBORETUM AT THE FOREST RESEARCH INSTITUTE

The establishment of a Forest Department for the Straits Settlements and Federated Malay States of Malaya came with the appointment of Alfred M. Burn-Murdoch in 1901 as the Chief Forest Officer (Wong, 1987). Burn-Murdoch set the first herbarium for the department in Kuala Lumpur, with the aim of producing an account of the commercially important trees species of Malaya (Wong, 1987). The specimens were collected as reference specimens, and duplicates were submitted to H. N. Ridley for identification. The small herbarium was at the office of the Conservator of Forests, Strait

Settlements and Federated Malay States. His successor, G. E. S. Cubitt, continued with the collection although at a slower rate. In 1916, the Wray Herbarium of the Agriculture Department was transferred to the Forest Department in Kuala Lumpur (Cubitt, 1919). In 1918, Cubitt secured the services of Dr. F. W. Foxworthy, an American, as the first Forest Research Officer of the Federated Malay States and Straits Settlements. Under Foxworthy, the herbarium grew quickly. By the end of 1920, the herbarium contained more than 6,000 numbers (Wong, 1987). With the decision to form a Forest Research Institute (FRI), an area of about 324 hectares was acquired at Kepong in 1926, and the main office building was constructed in 1929 with the herbarium moving into the east wing of the building. C. F. Symington joined FRI in 1929 and began to assist the running of the herbarium. He contributed a large collection to the Kepong herbarium focusing on the important timber family Dipterocarpaceae, for which he was preparing a foresters' manual. By the time World War II broke out in Malaya, with J. G. Watson having succeeded F. W. Foxworthy, the herbarium had about 43,000 specimens. The herbarium holding in Kepong has since accumulated to more than 400,000 specimens.

Another significant development in the Forest Research Institute was the setting up of an arboretum in the late 1920s under J. G. Watson (Wong, 1987). Two main collections were initially established: the dipterocarp and non-dipterocarp collections. The Dipterocarp Arboretum is the finest in the world (Wong, 1987), and the Non-Dipterocarp Arboretum is among the finest in the country. Subsequent additions included a monocotyledon collection with particular emphasis on bamboos and rattans, Bambusetum, medicinal plant collection, and fruit tree collection in the main campus in the Forest Research Institute Malaysia (FRIM). The idea of starting a botanic garden in the FRIM campus was mooted in 1996 by Abdul Razak Mohd. Ali, the Director-General of FRIM at that time, assigned Saw Leng Guan to develop the framework for the garden. Saw formed a team, and in 1997, work for the Kepong Botanic Gardens (Plate 4.1) started. The

PLATE 4.1 Kepong Botanic Gardens, view across the lake.

site selected was in a valley of an ex-tin mining area with remnants of ponds and marsh area. Today Kepong Botanic Gardens covers an area of about 80 ha. The collection contains mainly native species of trees, shrubs, herbs, palms, gingers, aroids, and orchids, among others. The cleared-up lakes remain one of the main features and attraction of the garden. The function of Kepong Botanic Gardens is for education, research, and conservation of the rare, endemic, and threatened plant species.

4.4 SABAH

In 1916, D. M. Matthews set up the herbarium in North Borneo (Sabah) as he initiated his forest surveys and inventories around Sandakan and the east coast of Sabah (Sugau et al., 2016). The history of the Sandakan herbarium and its collections are well summarized by Sugau et al. (2016) for its centenary celebration. Here we highlight some of the important botanical events and the setting up of its botanical garden and arboretum in Sepilok. Among the early botanical exploration and collections from North Borneo before World War II included E. D. Merrill, D. D. Wood, F. W. Foxworthy, J. A. Agama, A. Villamil, M. Castillo, A. D. E. Elmer, and H. G. Keith. The collection was, however, destroyed during the war. After the war, José Agama and other forestry staff who had stayed on in North Borneo reported for duty in December 1945. Agama, then Assistant Conservator, was tasked with restoring the Forestry Department (Ibbotson, 2014), including the Sandakan herbarium.

Geoffrey Wood was appointed as the first dedicated forest botanist in 1954. He was an energetic collector and collected a vast number of specimens. He unfortunately died of burns from an accident at a camp during a trip in Brunei in 1957. Just a year before his death, a checklist of the forest flora of North Borneo by Wood and Agama (1956) was published. In May 1959, in the employ of the Forest Department North Borneo, Willem Meijer succeeded Wood as forest botanist in North Borneo. He served until 1968. Meijer continued and completed a detailed botanical documentation of the dipterocarps begun by Wood (Meijer and Wood, 1964).

On the night of 31 January 1961, fire spread from an adjacent veneer factory belonging to the British Borneo Timber Co. and razed the Herbarium to the ground. Some 15,000 herbarium specimens and several hundred wood specimens were lost in that fire (Sugau et al., 2016). Meijer, with much fervor, organized a collecting drive for Sabah (Wong, 2004) and effectively set up a new herbarium. By 1964, Meijer was able to report a new collection of 13,600 specimens, bringing the total to about 17,200. The herbarium has since grown, and by the centenary celebration, the herbarium in 2016 has some 200,800 accessions (Sugau et al., 2016).

The arboretum was set up by the Rainforest Interpretation Centre, and Rain Forest Walk (Rintis Belantara) was established at the Arboretum in Sepilok, Sandakan, in May 1994, with funding support through the *Deutsche Gesellschaft Fuer Technische Zusammenarbeit* (GTZ), Germany. After 2007, this was transformed into the present-day Rainforest Discovery Centre.

4.5 SARAWAK

A proposal to establish a herbarium in Sarawak was initiated by the Forestry Department in 1947, but a combined Museum-Forestry Department herbarium came to light in 1959. However, it was only in March 1961 that the herbarium was established and located in a building at Jalan Badruddin, Kuching. With the assistance of C. X. Furtado, valuable collections of G. H. Haviland, C. Hose, and J. and M. S. Clemens kept in Sarawak Museum were acquired. Subsequently, botanical collections were intensified, and the number of specimens grew to more than 60,000 in 1966 and 85,000 in 1974; by 1990, the number had grown to more than 130,000 specimens (Mokhtar, 1991). The Sarawak Forestry Herbarium is associated with both the Botanical Research Centre Semenggoh, Kuching and Sarawak Botanical Garden, though it is managed independently.

4.6 BOTANICAL GARDENS IN MALAYSIA

There are currently 19 public botanical gardens in Malaysia (Table 4.1). They range in size from 1.4 ha to 817 ha. The oldest gardens were set up by the British colonial administration (Penang Botanic Gardens and FRIM arboretum); the other gardens are mostly more recent in establishment, and they are of mixed origin, ranging from forest and agriculture departments, city and town municipalities, and universities.

Brief descriptions and functions of a few of the botanic gardens are given here.

4.6.1 Johor Botanic Gardens

This small botanic garden was established by the state government of Johor and the Department of National Landscape at Seri Medan, Batu Pahat, Johor. The area was formerly mined, and the debris was left unattended and unmanaged for some time, surrounding a pond that was created by tin mining. The areas around these areas were planted with trees, shrubs, and herbs and contain more than 100 plant species, mostly common ornamentals. In concept, its purpose is to introduce the rare, endemic, and other threatened plant species from the southern states of Johor, Melaka, and Negeri Sembilan. In 2009, there was an effort by both parties, namely the state government and Department of National Landscape, to improve it with more planting of plants and landscaping; however, it is left as it was till now. Currently it functions as a recreational area.

TABLE 4.1
Botanical Gardens in the Wide Sense in Malaysia

Botanical Gardens	City/Town	State	Size (ha)	Date/Year Established
Johor Botanic Gardens	Batu Pahat	Johor	NA	NA
Royal Palace Botanical Garden, Johor Bahru	Johor Bahru	Johor	NA	NA
Perdana Botanical Gardens, Kuala Lumpur	Kuala Lumpur	Kuala Lumpur	25.9	28 June 2011
Melaka Botanical Garden	Ayer Keroh, Melaka	Melaka	92.5	1 June 2006
Penang Botanic Gardens	Penang	Penang	242	1884
Perak Botanical Gardens	Taiping	Perak	22.6	Nov. 2018
Putrajaya Botanic Garden	Putrajaya	Putrajaya	93	31 August 2001
Rainforest Discovery Centre, Forest Research Centre, Sepilok, Sandakan	Sepilok, Sandakan	Sabah	4,300*	1997
Mount Kinabalu Botanical Garden	Gunung Kinabalu	Sabah	1.4	
Sabah Agriculture Park	Tenom	Sabah	200	
Botanical Research Centre Semenggoh	Semenggoh, Kuching	Sarawak	22	1976
Sarawak Botanical Garden	Kuching	Sarawak	83	2016
Sarawak Biodiversity Centre	Semenggoh, Kuching	Sarawak		2004
DBKU Orchid Garden	Kuching	Sarawak	6.2	2009
TUMBINA	Bintulu	Sarawak	57	1991
Kepong Botanic Garden, Forest Research Institute Malaysia, Kepong	Kepong	Selangor	70	1929, 1996
National Botanic Gardens	Shah Alam	Selangor	817	24 April 1986
Rimba Ilmu Botanic Gardens, University of Malaya	Kuala Lumpur	Kuala Lumpur	80	1997
Fernarium, Universiti Kebangsaan Malaysia	Bangi	Selangor	50	1986

* The Rainforest Discovery Centre is part of the Sepilok-Kabili Forest Reserve. NA – Not available.

4.6.2 ROYAL PALACE BOTANICAL GARDEN

The Royal Palace Botanical Garden is in Johor Bahru, Johor, and is otherwise known as Zaharah Botanic Gardens. It was established as a garden within the palace compound of Sultan of Johor. Presently, it functions as display gardens.

4.6.3 PERAK BOTANICAL GARDENS

This 50-ha botanic garden was established jointly by the state government of Perak and the Department of National Landscape. The area was formerly a nine-hole golf course with lakes or ponds in the middle. Located at the base of Larut Hills (Bukit Larut), the garden is in an ideal natural setting with its old mining ponds and lakes, part of the iconic Taiping Lake Gardens. The Taiping Zoo is located within the garden, and both the lake and the zoo are prime tourist attractions to Perak state. The garden aims to specialize its collection of rare, endemic, and other threatened plant species from the northern states of Perlis, Kedah, and Perak. Of late, development has progressed with a few thematic gardens, such as Perak endemics, Perakensis Gardens (many species named after the state of Perak and other places in the state are introduced), rare fruit trees, and economic or ethnobotanical plants. There is a future plan to extend it to about 110 ha.

4.6.4 RIMBA ILMU BOTANIC GARDENS, UNIVERSITY OF MALAYA

This 80-ha university botanic garden was established in 1974 for educational purposes and is modeled as a rain forest garden with the introduction of plant species that are considered rare, endemic, vulnerable, or otherwise threatened in the wild. Jones (1987) reported the garden's function has been extended for research and conservation, and more than 1,600 species have been planted. The core collection in the garden includes medicinal plants, palms, citruses, bamboos, and ferns, which reflect the interests of its founders. It is associated with the Rare Plants and Orchids Conservatory and the herbarium (KLU). Its displays include a broad scope of tropical rain forest, covering topics such as types of rainforests, biodiversity, and species richness as well as the values of forest conservation.

4.6.5 FERNARIUM, UNIVERSITI KEBANGSAAN MALAYSIA

The Fernarium or fern garden at this university is one of its kind in the country; its establishment was aimed at introducing species of ferns and lycophytes that are considered rare, vulnerable, and threatened in the wild. Malaysia has more than 1,167 species of pteridophytes, and more than 648 species occur in Peninsular Malaysia (Parris and Latiff, 1997). Like other seed plant species, many of the pteridophytes suffer similar fates as the country strives for socio-economic development. A 50-ha area of Bangi Forest Reserve was carved out for this purpose by Dr. Abdul Aziz Bidin, a pteridologist, and assisted by an able para-taxonomist Mr. Razali Jaman. More than 200 species of pteridophytes are currently present in this fern garden, including the existing species.

4.6.6 PERDANA BOTANICAL GARDENS, KUALA LUMPUR

Originally known as Perdana Lake Gardens and Public Gardens, Perdana Botanical Gardens was established by the British as part of the Lake Club and recreation park. Many tree species planted by the British are still thriving in the garden, and they give it a tropical ambience. Now it is taken over by the City Council of Kuala Lumpur and is being developed as a botanic garden. The garden contains numerous collections, including Forest Trees, Zingiberales and Heliconia, Lesser Known Fruit Trees, Laman Perdana, Cycad Island, Sunken Garden, Name of Places Tree Collection, Herb and Spice Collection, Brownea Street, Deer Park, Hibiscus Garden, and Orchid Garden. It spans an area of about 91.6 ha.

4.6.7 Putrajaya Botanic Garden

The garden is divided into eight botanical themes: Malaysian Ulam and Medicinal, Bambusetum, Zingiberales, Edible Fruit Arboretum, Lawn and Gramineae, Forest Fringe and Aboriginal Medicinal Plants, Conservatory, and Ecological Pond. The beauty of this park is in a green area overlooking the lake, which makes visitors fascinated by cultural elements incorporated in designs such as traditional wakaf, flower-theme park lighting, and plant-based scientific labels. It also has nurseries and plant propagation facilities. Its floral attractions bring together more than 750 plant species from 13 collections of plant themes coming from different continents.

4.6.8 National Botanic Gardens, Shah Alam

Originally this botanic garden was called Bukit Cherakah Botanic Gardens, but it was later renamed as National Botanic Gardens. As the name implies, it is supposed to be a national institution to showcase the plant species of Malaysia. Covering an area of 817 ha, it is located within one of the best-managed lowland dipterocarp forests. The garden specializes in providing opportunities for botanical garden displays and at the same time promoting agrotourism displays. Among the themes and collections developed in the garden include an animal park, ornamental garden, four-season temperate house, spice and beverages garden, paddy field, and herbs and medicinal garden.

4.6.9 Melaka Botanic Garden

Initially this state botanic garden was established as Ayer Keroh Recreational Forest in 1964; spanning about 359 ha, it functions as an arboretum managed by the Forestry Department of Melaka. It is situated at Ayer Keroh. Currently, it is jointly managed by the Forestry Department and the Melaka Tengah Municipality. It has changed from an arboretum to a garden with the introduction of ornamental shrubs and herbs, while leaving the original tree species intact. It functions as a display garden for education, recreation, ecotourism, and possibly also for education.

4.6.10 Penang Botanic Gardens

The original area of 29.14 ha when Penang Botanic Gardens (Plates 4.2, 4.3) was first founded in 1884 has now been extended to 242 ha. The garden is located about 6 km from George Town Heritage Site. Originally known as the Waterfall Gardens, it served as a catalyst for a period of intense discovery of Malaya's rich plant diversity. The accessibility of the forested hills adjacent to the gardens provided an excellent opportunity for the trained botanists stationed here, including Charles Curtis as well as his successors, to search for plant species new to science. Many plant species described for Malaya were discovered from this botanic garden. Many of its historical trees planted in the garden by the first curator, Charles Curtis, are still found growing on the grounds. The garden is now being purposed toward research, conservation, and environmental education (Anon, 2019).

4.6.11 Suriana Botanic Conservation Gardens

This garden is situated at Balik Pulau, Penang, and was established in 2000. It is a private botanic garden dedicated to plant research and conservation of rare and endangered native species, with a special interest in palms (Arecaceae) and gingers (Zingiberaceae). It is funded by Heritage Research Limited Company.

4.6.12 Rainforest Discovery Centre

Rainforest Discovery Centre (RDC; Plate 4.4) was established within a tropical rain forest and is part of the Sepilok-Kabili Forest Reserve by the Forest Research Centre, Sepilok, Sandakan. RDC

PLATE 4.2 Penang Botanic Gardens.

is situated next to the famous Sepilok Orang Hutan Centre. It functions as a botanical research and conservation center with a strong education emphasis on the tropical rain forest of Sabah. It has a good collection of native plants from Sabah in a very well-designed botanic garden. The garden and forest surrounding the center is strongly association with bird watching. Its interpretive center attracts a large number of students from in and around Sandakan. The free-standing canopy walk or Rainforest Skywalk spanning 620 m is the longest free-standing canopy walk in Malaysia. It is another great attraction for visitors. It takes visitors into the canopy of a lowland forest; the canopy walk enables bird lovers to see many lowland forest birds. RDC has also been identified as an Important Bird Area.

4.6.13 Mount Kinabalu Botanical Garden

Mount Kinabalu Botanical Garden is a small garden to display the unique flora of Mount Kinabalu, especially the orchids, rhododendrons, pitcher plants, and other shrubs and herbs. It is managed by Kinabalu Park, which is one of the parks under Sabah Parks.

4.6.14 Sabah Agriculture Park

Sabah Agriculture Park (Plate 4.5) was established at Tenom by the Department of Agriculture Sabah to introduce and display the various economic plants of Sabah, particularly the fruit tree species, orchids, medicinal plants, and others. It contains many local species of ethnobotanical importance and functions as a display garden as well as a place for the introduction of rare, endangered, or otherwise threatened plant species of agricultural importance.

4.6.15 Botanical Research Centre (BRC)

Botanical Research Centre (BRC) was established by Sarawak Forest Department in early 1976. It is located within Semenggoh Forest Reserve constituted in 1920, and in 2000, it was gazetted

PLATE 4.3 Waterfall at Penang Botanic Gardens.

PLATE 4.4 Rainforest Discovery Centre, Sepilok, Sabah. Rock garden with *Alocasia cuprea*.

PLATE 4.5 Sabah Agriculture Park, Tenom, Sabah.

as one of the Totally Protected Areas under the National Parks and Nature Reserves Ordinance, 1998, and known as Semenggoh Nature Reserve. The Centre provides a comprehensive collection of living plants collected from all regions of Sarawak for *ex situ* conservation. It also functions as a research and educational center on Sarawak's plants for scientists, students, and the public. Seven specialized gardens have been established between 1976 to 2007. These include the following: (1) Wild Orchids Garden, also known as Rena George Memorial Orchid Garden (named after Rena George, officer of Sarawak Forest Department, who died due to malaria after one of her field trips in 1994); (2) Nepenthes Garden; (3) Ethnobotanical Garden; (4) Wild Fruits Garden; (5) Bamboo and Ficus Garden; (6) Fernarium and Aroids Garden; and (7) Mixed Planting Garden, which comprises mainly palms and gingers collections. The species were collected from various parts of Sarawak by the staff of Sarawak Forest Department and, more recently, by the staff of Sarawak Forestry Corporation, as well as the visiting local and foreign researchers as early as 1977 as a result of various field trips in Sarawak. In 2009–2010 and 2012–2014, the gardens also added collection of plant species rescued from the forest areas affected by the development of Bakun Dam and Murum Dam. New species continue to be described based on the collections maintained at BRC Semenggoh. For example, two new species of *Areca* (*Areca gurita* and *Areca bakeri*) were described in 2011 (Heatubun, 2011) as well as two new species of *Licuala* (*Licuala maculata* and *L. rheophytica*) described by Saw (2012). Within the garden areas, 594 tree species from 84 families mostly native to Sarawak were recorded. Adjacent to the gardens is the Arboretum, which is home to more than 500 lowland tree species in 60 families and has 160 genera represented within its 4-ha plot (Ling and Julia, 2012). Currently, the management of BRC Semenggoh is under Sarawak Forestry Corporation since 2003. In recent years, BRC Semenggoh and the Arboretum also provided eco-tourism products for visitors to experience while visiting Semenggoh Wildlife Centre. For example, the visit to selected specialized gardens became part of the Semenggoh Ecotour Packages offered by Sarawak Forestry Corporation.

4.6.16 SARAWAK BOTANICAL GARDEN

This is a small garden in Petra Jaya, Kuching, that functions as a recreational garden, including education and conservation. Sarawak Botanical Garden was established in 2016 by Kuching North City Commission (DBKU) and supported by the Federal Urban Wellbeing, Housing and Local Government Ministry through its National Landscape Department. The development of hardscape and softscape at the garden are on-going.

4.6.17 SARAWAK BIODIVERSITY CENTRE

Sarawak Biodiversity Centre, Kuching (Plate 4.6), is the state-leading center for research and development and is dedicated to botanical drug development and biodiversity conservation. It has a small Laila Taib Etnobotanical Garden. The Laila Taib Ethnobotanical Garden was established by Sarawak Biodiversity Centre in 2004. The garden is located within the SBC office compound at Semenggoh, about 20 km from Kuching City Centre. It exhibits more than 400 species of ethnobotanical plants, representing 92 families, that were contributed by local communities in Sarawak that participate in SBC's Traditional Knowledge Documentation Programme (Sarawak Biodiversity Centre, 2018).

4.7 ROLE OF BOTANICAL GARDENS IN MALAYSIA

Malaysia launched its national strategy for plant conservation in 2009 (Saw et al., 2009). The national strategy was developed following a national workshop in 2004 involving experts from institutions with botanical interests in Malaysia. The national strategy was a response to the Global Strategy for Plant Conservation, which Parties of the Convention on Biological Diversity adopted in 2002. The national strategy forms a framework toward plant conservation in Malaysia in which

Botanical Gardens in Malaysia and Their Role in Conservation 77

PLATE 4.6 Sarawak Biodiversity Centre, Sarawak.

botanical institutions in Malaysia can participate. Botanical gardens will find the national strategy an important instrument toward their role to conserve and protect plant resources in the country. The national strategy is given in Table 4.2.

The rationale in each target is explained including the situation in Malaysia and how the target can be achieved. All the seventeen targets are still pertinent to the present botanical climate but could be reviewed and updated.

4.8 PLANT CONSERVATION IN MALAYSIA

Systematic plant conservation efforts are currently headed mainly by the forest institutions in Malaysia, i.e., Forest Research Institute Malaysia (FRIM), Forest Research Centre Sepilok, and Sabah and Botanical Research Centre Semenggoh, Sarawak. In 2005, the FRIM was awarded a grant to conduct conservation assessment and monitoring of threatened plants for Peninsular Malaysia. The project was awarded together with the Flora of Peninsular Malaysia project. The projects run in tandem – as families are revised for the Flora, conservation assessments are made (Saw et al., 2010). A similar conservation status assessment project is currently conducted in Sabah (Maycock et al., 2012). Yong et al. (2021a, 2021b) published the full listing of all species assessed for Peninsular Malaysia until 2020 with results from the publications of 11 volumes of the Flora of Peninsular Malaysia project. Some species were reassessed in this publication based on current available information (e.g., population size and distribution, trend on habitat loss and degradation, conservation measures in place). The summary of the assessments is provided in Table 4.3. Out of the 1,292 taxa assessed, 373 taxa or 28.87% were either extinct or in some category of conservation concern. They include 98 endemic species.

TABLE 4.2
National Strategy for Plant Conservation in Malaysia

Objective 1:	**Understanding and documenting plant diversity**
Target 1:	A widely accessible working list of known plant species as a step toward a complete national flora
Target 2:	A preliminary assessment of the conservation status of know plant species of the nation
Target 3:	Development of models with protocols for plant conservation and sustainable use, based on research and practical experience
Objective 2:	**Conserving plant diversity**
Target 4:	Put in place national policies and legislation that will meet the plant conservation needs of the nation
Target 5:	At least 10% of each of the nation's ecological habitats effectively conserved
Target 6:	Protection of 50% of the most important areas for plant diversity assured
Target 7:	At least 30% of production lands managed consistent with the conservation of plant diversity
Target 8:	60% of the nation's threatened species conserved *in situ*
Target 9:	60% of threatened plant species in *ex situ* collection within the country, and 10% of them included in recovery and restoration programs
Objective 3:	**Using plant diversity sustainably**
Target 10:	70% of the genetic diversity of crops and other major socio-economically valuable plant species conserved and associated indigenous and local knowledge protected
Target 11:	Management plans in place for major alien species that threaten plants, plant communities, and associated habitats and ecosystems
Target 12:	No species of wild flora endangered by national and international trade
Target 13:	30% of plant-based products derived from sources that are sustainably managed
Target 14:	The decline of plant resources, and associated indigenous and local knowledge, innovations, and practices that support sustainable livelihood, local food security, and health care halted
Objective 4:	**Promoting education and awareness about plant diversity**
Target 15:	The importance of plant diversity and the need for its conservation incorporated into communication, education, and public-awareness programs
Objective 5:	**Building capacity for the conservation of plant diversity**
Target 16:	The number of trained people working with appropriate facilities in plant conservation increased to achieve the targets of this Strategy
Target 17:	Networks for plant conservation activities established or strengthened at the national level

TABLE 4.3
Summary of the Red List of Plants of Peninsular Malaysia

Category	No. Taxa (%)	Endemic
Extinct (EX)	3 (0.15)	3
Regionally Extinct (RE)	1 (0.08)	0
Critically Endangered (CR)	77 (5.96)	29
Endangered (EN)	120 (9.29)	31
Vulnerable (VU)	173 (13.39)	36
Extinct or with Conservation Concern	**373 (28.87)**	**98**
Near Threatened (NT)	213 (16.49)	27
Least Concern	627 (48.53)	35
Data Deficient (DD)	79 (6.11)	16
Total	**1,292**	**176**

Conservation efforts for threatened species include detailed studies on some critically endangered and endangered species (Saw et al., 2010). These include population studies (location of extant populations, spatial mapping and enumeration of population, population structure and ecology), phenology, and reproductive biology. Output of the studies include *ex situ* conservation collection and planting of these threatened species in the Kepong Botanic Gardens. The majority of the species were Dipterocarpaceae. Findings, including conservation recommendations, were forwarded to the relevant stakeholders. In the case of Dipterocarps, the forest areas are under the jurisdiction and management of the Forest Department. The conservation team worked closely with the Forest Department to identify locations and populations of critically threatened species. These areas are then included in the high conservation value forest (HCVF) or annexed into the totally protected area network under Sustainable Forest Management (SFM).

4.9 MALAYSIAN BOTANICAL GARDENS NETWORK (MYBGNET)

The first meeting for the Malaysian Botanic Gardens Network (MYBGNet) was held in the Forest Research Institute Malaysia on 25 November 2016, with the second meeting on 23–24 April 2019 in Penang Botanic Gardens. The aim of the network is to foster cooperation among botanic gardens in Malaysia and support botanic garden activities beneficial to its members. The network is very much in line with Target 17 of the national plant conservation strategy. The network includes all botanical gardens found in Malaysia and government agencies that are important to the functioning of botanical gardens. Together with the botanical gardens listed in Table 4.1, the following agencies are included in the network: the Ministry of Energy and Natural Resources, Forest Department, Peninsular Malaysia, Department of Agriculture, and the Wildlife and National Parks Department.

The role of MYBGNet may become more important in coming years when more activities can be organized through the network to implement the national strategy for plant conservation and improve the curation of the garden collections found in the network.

4.10 CONCLUSION

Although Malaysia has among the richest flora in the world, its network of botanical gardens in the country is relatively poor. The botanical institutions that are involved in conducting botanical, plant ecological research and conservation are those linked with forestry research institutes and universities. Most of the other botanical gardens are mostly engaged in providing park facilities for public enjoyment with the display of plants and recreational and educational programs. There is a strong need to develop within Malaysia a dedicated network of botanical gardens for the country to deal with the challenges of plant diversity loss currently the country is facing. The national strategy for plant conservation that was passed in 2009 can be used as a framework for the network. This can lead the botanical gardens in the country toward a more proactive and dedicated function at a national level.

ACKNOWLEDGMENTS

The authors thank all those curators and officers who provided the information and data of various Botanic Gardens via the questionnaires, in particular Joan T. Pereira, Suhahida binti Mustafa, Narzielewati binti Salleh, Masnah binti Made, and Mohamad Nazrin bin Ahmad Azmi.

REFERENCES

Anon. 1980. The development of a Botanic Gardens. The Proceedings of a Conference held at Coffa Harbour Technical College, May 5–9, Department of Continuing Education, University of New England, Armidale 2353.

Anon. 2019. Special Area Plan for Penang Botanic Garden. Government of Penang Gazette Volume 63, No. 14, Additional No. 1.

Banfield, F.S. 1947. *Guide to the Botanic (Waterfall) Gardens Penang*. Georgetown Printers, p. 39.

Bramwell, D., O. Hamann, V.H. Heywood and H. Synge (eds.). 1987. *Botanic Gardens and the World Conservation Strategy*. Academic Press, London.

Cubitt, G.E.S. 1919. *Report on Forest Administration for the Year 1918*. Federated Malay States. F.M.S. Government Press, Kuala Lumpur, p. 15.

Heatubun, C.D. 2011. Seven new species of *Areca* (Arecaceae). *Phytotaxa* 28(1): 6–26.

Ibbotson, R. 2014. *The History of Logging in North Borneo*. Opus Publications, Kota Kinabalu.

IUCN. 1987. Botanic gardens and the world conservation strategy. Proceedings of an International Conference, 26–30 November 1985 held at Las Palmas de Gran Canaria.

Jones, D.T. 1987. Rimba Ilmu Universiti Malaya: Preserving the natural plant heritage of the world's oldest rain forest. In: Bramwell, D., O, Hamann, V.H. Heywood and H. Synge (eds.), *Botanic Gardens and the World Conservation Strategy*. Academic Press, London, pp. 345–346.

Langdon, M. 2015. *Penang: The Fourth Presidency of India, 1805–1830: Fire, Spice and Edifice*, Vol. 2. George Town World Heritage Incorporated, Marcus Landon, Penang p. 576.

Ling, C.Y. and S. Julia. 2012. Diversity of the tree flora in Semenggoh Arboretum, Sarawak, Borneo. *Gardens' Bulletin Singapore* 64(1): 139–169.

Maycock, C.R., C.J. Kettle, E. Khoo, J.T. Pereira, J.B. Sugau1, R. Nilus, R.C. Ong, N.A. Amaludin, M.F. Newman and D.F.R.P. Burslem. 2012. A revised conservation assessment of Dipterocarps in Sabah. *BIOTROPICA* 44: 649–657.

Meijer, W. and G.H.S. Wood. 1964. *Dipterocarps of Sabah (North Borneo)*. Sabah Forest Record, Sandakan, p. 5.

Mokhtar, A.M. 1991. Sarawak herbarium. In: Latiff, A. (ed.), *Status of Herbaria in Malaysia*. Malaysia National Committee on Plant Genetic Resources, Bangi.

Parris, B.S. and A. Latiff, 1997. Towards a pteridophyte flora of Malaysia: A provisional checklist of taxa. *Malayan Nature Journal* 50: 225–280.

Sarawak Biodiversity Centre. 2018. Sarawak Biodiversity Centre Annual Report 2018: 24.

Saw, L.G. (2012). *Licuala* (Arecaceae, Coryphoideae) of Borneo. *Kew Bulletin* 67: 577–654.

Saw, L.G., L.S.L. Chua and N. Abdul Rahim (eds.). 2009. *Malaysia: National Strategy for Plant Conservation*. Ministry of Natural Resources and Environment and Forest Research Institute Malaysia, Kuala Lumpur.

Saw, L.G., L.S.L. Chua, M. Suhaida, W.S.Y. Yong and M. Hamidah. 2010. Conservation of some rare and endangered plants from Peninsular Malaysia. *Kew Bulletin* 65: 681–689.

Saw, L.G. and R.C.K. Chung. 2015. The Flora of Malaysia projects. *Rodriguesia* 66 (4). DOI: 10.1590/2175-7860201566415

Sugau, J.B., J.T. Pereira, Y.F. Lee and K.M. Wong. 2016. The Sandakan herbarium turns a hundred. *Sandakania* 21: 1–20.

Wong, K.M. 1987. The herbarium and arboretum of the Forest Research Institute of Malaysia at Kepong: A historical perspective. *Gardens' Bulletin Singapore* 40: 15–30.

Wong, K.M. 2004. Meijer of Sandakan: A tribute to Willem Meijer, 1923–2003. *Sandakania* 15: 1–24.

Wood, G.H.S. and J. Agama. 1956. *Check List of the Forest Flora of North Borneo*. North Borneo Forest Records No. 6. Forest Department, Sandakan.

Yong, W.S.Y., L.S.L. Chua, K.H. Lau, K. Siti-Nur Fatinah, Y.H. Cheah, T.L. Yao, A.R. Rafidah, C.L. Lim, S. Syahida-Emiza, A.R. Ummul-Nazrah, A.T. Nor-Ezzawanis, M.Y. Chew, M.Y. Siti-Munirah, A. Julius, S.N. Phoon, Y.Y. Sam, I. Nadiah, P.T. Ong, R. Sarah-Nabila, M. Suhaida, A. Muhammad-Alif Azyraf, S. Siti-Eryani, J.W. Yap, M. Jutta, A. Syazwani, S. Norzielawati, R. Kiew and R.C.K. Chung. 2021a. Malaysia Red List: Plants of Peninsular Malaysia. Vol. 1, Part 1. Research Pamphlet No. 151, p. 107. Forest Research Institute Malaysia.

Yong, W.S.Y., L.S.L. Chua, K.H. Lau, K. Siti-Nur Fatinah, Y.H. Cheah, T.L. Yao, A.R. Rafidah, C.L. Lim, S. Syahida-Emiza, A.R. Ummul-Nazrah, A.T. Nor-Ezzawanis, M.Y. Chew, M.Y. Siti-Munirah, A. Julius, S.N. Phoon, Y.Y. Sam, I. Nadiah, P.T. Ong, R. Sarah-Nabila, M. Suhaida, A. Muhammad-Alif Azyraf, S. Siti-Eryani, J.W. Yap, M. Jutta, A. Syazwani, S. Norzielawati, R. Kiew and R.C.K. Chung. 2021b. Malaysia Red List: Plants of Peninsular Malaysia. Vol. 1, Part 2. Research Pamphlet No. 151, p. 753. Forest Research Institute Malaysia.

5 Ancillary Botanic Gardens
A Case Study of the American University of Beirut

Salma N. Talhouk, Ranim Abi Ali, Alan Forrest, and Yaser Abunnasr

CONTENTS

5.1 Botanic Gardens – Venues for Informal Botanical Learning and Nature Encounters 81
5.2 Challenges of Establishing Botanic Gardens 82
5.3 Institutions Housing Botanic Gardens 82
5.4 Ancillary Botanic Gardens (ABGs) 83
5.5 Case Study: ABG of the American University of Beirut 83
5.6 Retrofitting the AUB Campus into an ABG by Key University Stakeholders 85
5.7 Impact of the ABG on the Institution 85
    5.7.1 A Road Map for Conserving Critical Green Spaces Was Developed and Guided the University Master Plan 85
    5.7.2 Influence the Institution's Master Plan 87
    5.7.3 Introduce Sustainable Practices 87
5.8 Impact of the ABG on the University Community 87
    5.8.1 Train Community in Basic Plant Maintenance 87
    5.8.2 Engage Community in Mapping Environmentally Comfortable Paths 87
    5.8.3 Involve Students in Urban Conservation Research 87
    5.8.4 Label Plants and Provide Digital Links for Information 88
    5.8.5 Cater Plant Tours for Schools 88
    5.8.6 Promote Creativity by Organizing Nature-Themed Competitions 88
    5.8.7 Organize Social Events to Raise Environmental Awareness 88
    5.8.8 Launch a Volunteer Program 89
5.9 Impact of the ABG on Biodiversity 89
    5.9.1 Georeference the Living Plant Collection 89
    5.9.2 Define and Describe the Plant Collection 90
    5.9.3 Promote Bird Watching 90
5.10 Conclusions 90
References 92

5.1 BOTANIC GARDENS – VENUES FOR INFORMAL BOTANICAL LEARNING AND NATURE ENCOUNTERS

Botanic gardens open up possibilities for informal botanical learning and interactions with nature in the city, an important role considering that urban residents rarely have regular encounters with plants and nature (Cox et al., 2017). Depending on how and when it was established, a botanic garden offers unique experiences of nature telling a story about the culture or history of a community or geographic location (Paiva et al., 2020). Botanic gardens are also ideal venues for organizing citizen science projects, which provide strong datasets, particularly with the increasing need for knowledge on plants' responses to climate change (Primack et al., 2021). More importantly, botanic gardens indirectly contribute to long-term conservation actions by reversing people's detachment from nature, which leads to indifference to, and destruction of, the natural world (Slaughter, 1996; Williams et al., 2015; Wandersee and Schussler, 1999; Wilkins, 1988). By focusing on understanding, conserving, and sustainably using plant diversity, as well as raising awareness of environmental issues, botanic gardens are involved in many activities that contribute to the achievement of the Sustainable Development Goals (SDGs) (Sharrock, 2018). In fact, botanic gardens are uniquely situated to contribute to sustainability, education, and global conservation (Zelenika et al., 2018)

5.2　CHALLENGES OF ESTABLISHING BOTANIC GARDENS

Establishing a botanic garden, however, is a challenge in cities where real-estate value of land is prohibitive and where dense neighborhoods are rapidly expanding to accommodate high-population densities (Talhouk et al., 2014). In fact, there is evidence that the persistence of existing botanic gardens may be at stake. Cases where urban development projects have taken over gardens were reported in Kashmir and Iran, with Kashmir losing around 50% of its garden cover in the last decades (Kaul, 2021; Ershad, 2017). Maintaining botanic gardens is also a financial challenge because there are additional costs related to visitors, plant labels, and the need for trained personnel, such as gardeners, botanists, and tour guides, to develop, manage, and curate plant collections (Wong, 2005; Heywood and Iriondo, 2003; Wyse Jackson, 1999; Hohn, 2022; Trull et al., 2018).

5.3　INSTITUTIONS HOUSING BOTANIC GARDENS

Institutions that have no access to botanic gardens and are interested in offering their constituencies informal botanical experiences to enhance their programs have developed botanic gardens *in situ*, i.e., within their institutional grounds.

Schools have established informal botanic gardens for botanical learning within the school premises to promote dynamic and novel educational methods; for example, productive botanic gardens boost students' consumption of fruits and vegetables and provide an engaging environment for scientific learning (Berezowitz et al., 2015). Botanic gardens in schools were also shown to enhance student learning in the fields of plant and environmental sciences and contribute to the reduction of plant blindness (Sanders, 2007). For young learners, school gardens are places where they can discover and connect with nature, which may instill an early determination for environmental conservation (Sellmann and Bogner, 2013).

Universities have a long history with botanic gardens, which were integral to the advancement of plant-based medicinal knowledge (Forbes, 2008). However, interest in maintaining botanic education on campuses declined as the field of pharmacognosy became less popular and advanced botanical courses, such as plant identification and plant environmental/biological processes, were no longer offered (Bennett, 2014). Botanic gardens in university campuses are often described as sustainable; they include native plant collections and traditionally used plants that are displayed to promote natural history and the culture of the region where they are located. Examples of such gardens are found in the University of Missouri (declared a botanic garden in 1999), the University of Chicago (declared a botanic garden in 1997), the University of Maryland (designated an arboretum and botanical garden in 2008), Northwest Missouri State University (designated in 1993 by the Missouri State Legislature as the official Missouri Arboretum), the University of Nevada Las Vegas (established in 1985 as an arboretum and campus), the University of Nevada Reno (designated a state arboretum campus by the 1985 Nevada Legislature), and the University of Exeter (registered as a botanic garden campus at around 2010).

Hospitals, centers for mental well-being, and correctional facilities have also incorporated botanic gardens with a special focus; therapy gardens and edible gardens promote wellbeing and nurture interest in dietary changes. Such gardens are primarily maintained by patients or inmates

Ancillary Botanic Gardens

and supported by volunteers and staff of the host institutions (Stagg and Donkin, 2013). Examples include Horatio's Garden at Salisbury district hospital, the Chapel Garden in Norwich, the Morgan Stanley Garden in London, and the Prison Garden of Australia at Kirkconnell.

Civil society groups have developed public green spaces into botanic gardens to provide communities with a sense of ownership of place, to promote wellbeing, and to serve as a link to traditional knowledge (Graham et al., 2005; Qumsiyeh et al., 2017).

In line with this trend, the concept of ancillary botanic gardens was proposed by Talhouk et al. (2014) to further expand the scope and reach of informal botanical learning by exploring the transformation of urban institutional green spaces into botanic gardens.

5.4 ANCILLARY BOTANIC GARDENS (ABGS)

The concept of ancillary botanic gardens (ABGs) capitalizes on the fact that institutional green spaces in the city are many and widespread and can be used as venues for informal botanical learning (Talhouk et al., 2014). When many institutions establish ABGs, they will enhance the distribution of botanic gardens in the city, effectively expanding conservation areas within the urban context and reaching out to a larger number of citizens. It is worth noting that the current distribution of botanic gardens is skewed with the largest ratio of botanic gardens per person found in Europe and the USA, while the lowest ratio is found in Arab countries for which this concept was initially developed (Talhouk et al., 2014).

The term *ancillary botanic garden* was coined by Talhouk et al. (2014), who explained that ABGs are secondary or "ancillary" in space and function with respect to the institution's primary intent for the green space that it owns and manages. The concept was further elaborated in a subsequent study to encourage and facilitate the transformation of institutional green spaces into informal botanic gardens by developing guidelines that are useful to the retrofitting process (Melhem, 2019). Following a content analysis of 220 botanic gardens, an exhaustive list of botanic garden elements was produced and served as a basis for developing a checklist and guidelines, which were then tested on case study ABGs in Lebanon.

One of ABGs greatest strengths is in the potential capacity building and training among populations that lack opportunities for botanical education within the cultural context.

5.5 CASE STUDY: ABG OF THE AMERICAN UNIVERSITY OF BEIRUT

Following the development and publication of the ABG concept (Talhouk et al., 2014), we initiated the retrofitting of the campus of the American University of Beirut into an ABG. A university campus is not representative to all land situations and green spaces that can become ABGs, nor is the operational process described here applicable to all conditions. The details provided, however, may inspire or guide institutions or landowners interested in retrofitting their green spaces into ABGs to provide opportunities in informal botanical learning for their constituencies and to contribute to biodiversity conservation.

Located on the eastern shores of the Mediterranean Sea, the American University of Beirut (AUB) not only boasts extraordinary views of the sea and the mountains but also hosts a diverse flora and fauna, winding walkways, and secluded spaces, all of which were the result of historical and persistent attention to plants throughout the development of the institution (Figure 5.1).

As recorded by the university's first president, the campus was planted with trees to define the institution's boundaries; the university's first president Daniel Bliss states "Not far from twenty pounds sterling first given by Professor Dodge were expanded last spring in planting trees upon the property at Ras Beirut and building temporary walls for the purpose of defining boundaries" (Annual report, Syrian Protestant College issuing body, 1902–1903). Plants introduced to Campus were part of the AUB mission championed by administrators and faculty members, who contributed to the transformation of the campus from sparse rocky outcrop vegetation to dense trees and shrubs

FIGURE 5.1 A view of the campus of the American University of Beirut in the city of Beirut, Lebanon.
Source: Communication Department of AUB. 2021. American University of Beirut copyrighted.

(Bliss, 1902–1903; Syrian Protestant College issuing body, 1902–1903). According to the university records, the Arizona cypress trees were introduced in 1923. The planting of native trees and shrubs was the most significant contribution of West, who stated, "anywhere you see a Judas tree or a Viburnum shrub you can be sure that it came from a seedling which I dug with my own hands in the mountains and brought back to Beirut in my Knapsack" (West, 1931). Others also contributed to the importing and planting of tree species, such as the large eucalyptus tree, which was brought from Jerusalem in a biscuit tin in 1911. Californian and Australian seeds were transported in the early 1900s, and the large Ficus tree was planted in 1931. Between 1926 and 1931, seeds and hundreds of pine and cypress seedlings were brought from Palestine and Italy (West, 1931). In the 1960s, many areas on campus were planted with trees and shrub species imported from China and South Africa. With the commitment of all, the campus became a landmark and its trees a cosmopolitan living collection. In 1991, Abou-Chaar summarized the international "nature" of the campus:

> Of the big and beautiful trees on campus one may mention the stately Australian Grevillea and Queenslands's Brachychiton with its smooth green stems. Not forgetting of course, the beautiful Jacaranda of Brazil, or the Tipu tree of Bolivia, the Coral tree of South Africa, or the Lebbeck tree of tropical Asia. There are the Acacias and Eucalyptus of Australia, Dombeya and Poinciana from Madagascar, Bombax, Purging Cassia, Bauhinia and several species of Ficus including the Banyan tree, the Bo tree, the Rubber Tree, and the Benjamin Tree all from India. The Apple Blossom Cassia comes from Indonesia, the Redbud from West Asia, and South Europe. And of course, Lebanon's Cedar and the numerous Oaks, Cypresses, Palms and Pines are all represented.
>
> *(Abou-Chaar, 1991)*

Today, the ABG is benefiting from the species diversity on campus, which includes around 167 species of trees and shrubs, with an estimated 10,500 specimens. In addition, the campus hosts fauna: 28 migratory and resident bird species have been sighted, 8 reptile species, 3 small mammal species were observed, and 200 insect species are estimated to be present on campus (AUBotanic, 2019b).

5.6 RETROFITTING THE AUB CAMPUS INTO AN ABG BY KEY UNIVERSITY STAKEHOLDERS

The transformation of the AUB campus into an ABG was led by a university-wide committee, AUBotanic, that includes membership for various academic and administrative units. Members of the committee embarked on a challenging task to locate every tree on campus, to prepare labels and narratives for all trees and shrubs, to organize plant tours, to plan a design and management strategy for the campus, and to devise a fundraising strategy that supports the development and conservation of the campus for future generations. The ABG of the American University of Beirut was officially launched as part of the university's 150th celebration when the university declared its whole campus grounds an ABG, reaffirming its responsibility as a custodian of its natural environment and its commitment to informal botanical education.

Retrofitting the campus into an ABG began with the process of decision making on the administrative level, as follows:

1. **Identify members who champion the ABG concept.** Staff of the institution, including faculty and administrators, met informally on a regular basis to create a vision and the momentum to rally the necessary support for transforming the campus premises into an ancillary botanic garden.
2. **Form a committee to lead the ABG retrofitting process.** The grassroot efforts culminated in the formation of an official university committee (AUBotanic) with a mandate to lead the transformation of the campus into an ABG and ensure the university's commitment to the conservation of the campus natural environment.
3. **Elaborate a mission and vision of the ABG.** AUBotanic's mission and vision constitute a strong direction toward a biophilic and cost-efficient university campus that is resilient to local environmental issues and climate change. AUBotanic's mission is to conserve the living environment of the campus and increase the appreciation of the campus as a sanctuary for living things, to educate the university community about the natural environment with which we co-exist and the importance of preserving this environment for future generations, and to serve as a model for a sustainable relationship between people and nature. AUBotanic's vision is a series of simple to understand points: less water demand, naturally green, accessibility to the public, wide range of learning opportunities, all season garden, sustainable green campus, passages, increased green areas and green roofs, administrative regulations for campus growth and participatory involvement in any future development, and establishing the AUB campus as a botanic garden.
4. **Formalize the ABG Governance within the Institution.** Once the AUBotanic mandate, mission, and vision were set, AUBotanic was given official approval to proceed as an entity under the vice president for administration.

5.7 IMPACT OF THE ABG ON THE INSTITUTION

5.7.1 A Road Map for Conserving Critical Green Spaces Was Developed and Guided the University Master Plan

A typology of green spaces was developed by S.N. Talhouk and M. Fabian following the criteria described in Table 5.1 and was applied to each green space, which was delineated by buildings or built structures.

The total score for each area was calculated, and three typologies were defined based on the total score of the area as follows: The forest character (score 26–31) has a woodland feel and is defined by a vegetated soil, undisturbed micro ecological spaces with limited access. On the other hand, the urban character (score less than 20) included areas that are plaza-like areas consisting of paved

TABLE 5.1
Indicator Scoring System Defined by S.N. Talhouk and M. Fabian

Indicators	Types and Scores Assigned to Each				Impact
Understory *Low vegetation growing naturally*	Paved Area 1	Compacted Soil 2	Raised Bed 3	Natural Vegetation 4	Vitality of plants
Built Structure *Adjacent to building/wall/stairs*	Building 1	Stairs 2	Road 3	Footpath 4	Growth potential of roots and canopy
Woody Plant Cover *Grouping of canopy touching trees*	Zero 1	1–2 m 2	5–10 m 3	> 10 m 4	Micro-ecology
Light *Presence and type of artificial light*	Streetlight 1	Building Light 2	Natural Light 3		Flowering and fauna
Management *Maintenance intensity for each area*	Hardscape 1	Manicured 2	Productive 3	Natural 4	Sustainable management
Identity *Aesthetic and functional value*	Plaza 1	Park 2	Productive 3	Natural 4	Visual identity of space
Shade *Meters of shade provided by trees*	None 1	0–1 m 2	2–4 m 3	5–10 m 4	Environmental comfort
Institutional Value *Cultural, historic, and landmark value*	Green 1	Landmark 2	Notable 3	Historic 4	Cultural value
Accessibility *Accessible to pass through or sit in*	None 1	Partial 2	Full 3		Open space use

This system guided the masterplan in defining the aforementioned greenspace campus identities

Source: AUBotanic (2019a).

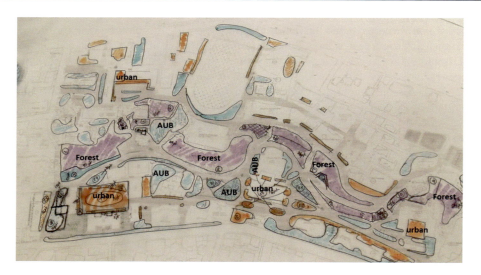

FIGURE 5.2 Typology of spaces in AUB ABG following an indicator scoring system developed by S.N. Talhouk and M. Fabian.

spaces where buildings are dominant, large trees are few, and where there is little or no shade. Areas with scores ranging between 20–29 were defined as an AUB character, which has a park feel and consists of spaces that are partially paved, containing large trees that balance the presence of buildings and provide shaded paths (Figure 5.2).

Using these newly defined typologies, AUBotanic championed the conservation of areas with a forest character and promoted the AUB character in future development projects.

5.7.2 Influence the Institution's Master Plan

AUBotanic influenced the campus master plan published in 2019. Working closely with architects, planners, and landscape architects, the committee members shared AUBotanic's mission and vision and provided support to guide planning decisions. As a result, the AUB campus master plan mentions the importance of conserving biodiversity, preserving green spaces, and transforming the campus into a botanic garden, and it stresses on the botanic garden initiative to stay active within the decision-making process of landscape management and campus development (Facilities Planning & Design Unit, Stage 3: Concept Master Plan, 2017).

5.7.3 Introduce Sustainable Practices

A new management strategy with a concise framework was developed to achieve a sustainable and affordable form of campus management (AUBotanic, 2020b). In addition, new projects were launched to promote composting of green garden waste for reuse as soil amendments and to collect air conditioning water for irrigation use during hot summer months. An internal campaign has also been launched to promote the installation of bio solar roofs instead of the current university trend to install solar panels on roofs.

5.8 IMPACT OF THE ABG ON THE UNIVERSITY COMMUNITY

Botanic gardens are ideal institutions to engage the community and boost its capacity (Chen and Sun, 2018; Pollock et al., 2015). With the decrease of botanical education, unique ways of providing environmental, botanic, and horticulture education in botanic gardens are applied at the AUB ABG. Today, the campus garden is used for academic purposes and general education; campus tours are regularly organized for young children from nearby schools and senior citizens enrolled in the University for Seniors Program. Following are some examples of how an ABG can engage its community constituency:

5.8.1 Train Community in Basic Plant Maintenance

Through regularly organized campus maintenance activities opened to the university community, alumni, students, faculty, and staff have received training in basic and safe gardening, including light pruning, weeding, raking, and planting.

5.8.2 Engage Community in Mapping Environmentally Comfortable Paths

A pathway comfort assessment to map the ideal walkways for AUB ABG visitors was conducted by volunteers and included measurements of impressions, sounds, smells, and temperatures during different seasons.

5.8.3 Involve Students in Urban Conservation Research

Students were involved in research projects as a means of offering experiential learning in urban biodiversity conservation. For example, they addressed the social challenges of promoting bird diversity in the presence of a high number of feral cats and a community that is strongly in favor of their protection and care.

5.8.4 Label Plants and Provide Digital Links for Information

Labels were placed on 200 species and approximately 2,000 specimen plants (Figure 5.3), including QR codes linked to a regional landscape plant database developed by the university (Talhouk et al., 2015).

5.8.5 Cater Plant Tours for Schools

AUB ABG has leveraged the support of its docents to offer plant tours to school children. For example, during spring 2019, 500 school children participated in plant tours.

5.8.6 Promote Creativity by Organizing Nature-Themed Competitions

Nature related competitions are ways to raise awareness and involve the community in exciting activities. For instance, AUB ABG launched the "ladybug hotel design competition" and received 55 entries from university student teams (Mounzer, 2019). The competition also included a children category with 30 school students, including students of displaced Syrian families, participating with their teachers in the design and development of insect hotels.

5.8.7 Organize Social Events to Raise Environmental Awareness

Every year, the AUBotanic hosts an open institutional event supported by the university event planning department (Figure 5.4). These events act as an informal educational introduction to a novel theme related to conservation and are often coupled with the results of a competition, an intricate

FIGURE 5.3 The IC labeling activity in 2018 involved docents setting up tree labels across the grounds of the International College in Lebanon, Beirut. The labels contain a QR code that students can use to access further knowledge of the respective tree using their phones.

Ancillary Botanic Gardens

FIGURE 5.4 AUBotanic holds an annual spring event every year in April or May. The first was for the announcement of AUB as a botanic garden; it was held as an exhibition of the many different species found on campus, displaying the rich biodiversity of the new botanic garden.

video presentation, or a list of speeches provided by influencers related to the overarching theme of the event. Themes covered by the annual events include celebrating plant diversity on campus, a grand opening of a forest trail, emphasizing the AUB campus as a bird sanctuary, and, the most recent, supporting the beneficial insects on campus.

5.8.8 Launch a Volunteer Program

AUB ABG has benefited from the diverse backgrounds represented in the AUBotanic committee to develop an interesting botanical docent training program that takes place over a semester and includes volunteering hours (Abi Ali, 2020). In addition to education and plant tours, the program covers the institution's history and natural history, ongoing sustainability and vegetation management practices, standard staff safety training, landscape character and identity, wellbeing, and fundraising.

5.9 IMPACT OF THE ABG ON BIODIVERSITY

The long-term goal of the ABG at AUB is to contribute to a vital, resilient, and diverse urban nature that is supported by sustainable practices. Following are listed impacts that the ABG may have on the biodiversity of the campus.

5.9.1 Georeference the Living Plant Collection

All woody plants were surveyed and georeferenced using ArcGIS software (Figure 5.5). This database is used for research and maintenance purposes. It is often shared with university students for their course projects and is used bi-annually to assess what aged trees must be removed, what

FIGURE 5.5 A screen capture of the georeferenced living plant collection of the American University of Beirut's campus.

species to add to the collection, and for development projects such as the recent space allocation for a Mediterranean collection.

5.9.2 Define and Describe the Plant Collection

A plant list was created and includes existing species and number of specimens per species, other native or ecologically adapted plant species, and invasive species. This list is used to guide new planting designs and replacements with the purpose to enhance the floristic diversity of the AUB campus, increase its native plant and international plant species collection, encourage the selection of native and drought tolerant species, and ensure a healthy resilient shade-providing campus canopy (AUB Botanic Garden Manual by the AUBotanic, 2019a).

5.9.3 Promote Bird Watching

Considering the lack of green spaces, AUB ABG, which falls within an important bird migratory route, also seeks to promote the value of urban nature afforded by the campus by organizing bird-watching tours for university students in partnership with a local bird conservation nongovernmental organization as a means to introduce this nature activity that is hardly practiced in the country (Figure 5.6).

5.10 CONCLUSIONS

Luck (2007) mentioned that it is the "appropriate management of people, not plants and animals, which determines the future state of our planet." Whether the effect is positive or negative depends on several variables and socio-economic factors, but one of them is the presence of professionally managed green spaces, which play a role in maintaining the relationship between people and nature (Luck, 2007). These spaces can serve any purpose, ranging from parks to botanic gardens, to less conventional urban green spaces like cemeteries and school playgrounds.

In this chapter, we explained how the transformation of an urban institutional green space into an ancillary botanic garden increased the value of plants in terms of their impact on the institution's

FIGURE 5.6 A bird-watching workshop that took place on 15 October 2019. It was organized by Mr. Fouad Itani, a professional in bird photography and a member of the NGO "Birds of Lebanon." Mr. Itani was also the instructor of the event.

commitment to conservation and mobilized people from the community to engage in nature-based activities.

Institutions establishing ABGs can help alleviate the absence of botanical gardens, especially in economically disadvantaged and biodiversity-rich countries. They may promote biodiversity conservation and provide a gateway of environmental education and a way for urban residents to connect to nature.

Institutional green spaces transformed into ABGs and enriched with a diverse plant collection may establish a new and different relation with their constituency or clients; they can serve as spaces for promoting biodiversity with emphasis on ecosystem services and the supportive role of nature. According to Wang et al. (2019), residents will value the uniqueness of ABGs by responding to its landscape features, vegetation structure and density, the surrounding urban context, and the level of biodiversity it contains. Institutions that own urban green spaces with tall trees and a higher diversity of flora are impactful; they affect users positively and promote nature connectedness, which benefits biodiversity as people are incentivized to conserve a place they give meaning to (Samus, 2020). In a recent interview with an undergraduate student during the Covid-19 lockdown, he expressed the enrichment the AUB campus has given him as the only green space he could visit during the lockdown, referring to it as nature therapy (AUBotanic, 2020a).

In increasingly urbanized societies, botanic gardens provide urban green space that may be the only access many people have to plants and nature. A more regular exposure to urban nature through the widespread presence of ABGs in cities may inspire youth to learn more about nature and the environment. Sharrock and Chavez (2013) explain botanical learning is on the decline, leading to an increasing gap in botanical knowledge and, as a result, patchy and inconsistent conservation activities around the world. This is further aggravated by the fact that there is a considerable imbalance in both the presence of botanic gardens and global conservation efforts, where biodiversity-rich countries have few botanic gardens or institutions dedicated to conservation activities. Heywood (2017) mentions this imbalance, stating that there is a call for changing old foundations to adapt to a changing climate and a changing society (Heywood, 2017). Westwood et al. (2020)

call for a massive scale up of garden-led conservation efforts to address a challenge at the scale of the global plant extinction crisis and to create new gardens in biodiversity hotspots and low-income economy countries of greatest conservation priority. The establishment of ABGs responds to this call, especially in economically disadvantaged countries. The establishment of ABGs is also in line with Reid and Gable (2020), who indicated that a small percentage of urban residents will ever visit formal botanic gardens; instead, the authors suggested that small-scale horticultural oases such as localized community, school, and demonstration gardens can play the same educational role and can have multiplied impacts that also include building a sense of community in underserved areas.

Botanic gardens serve as urban venues for information about plants to public audiences. Similarly, institutions that leverage their green spaces and transform them to ABGs can offer their constituencies an aesthetic experience and contact with a more natural environment, and they can organize plant-based exhibits that are the main vehicle for educational content (Krishnan et al., 2019). Institutions have many ways to engage their constituencies in their ABGs by incorporating plants that offer sensory experiences that tap into other senses, such as taste, smell, touch, and sound, providing fun and meaningful connections to plants and even raising deeper and more complex questions around food system values, inequalities, poverty, justice, and sustainability (Krishnan et al., 2019). In agricultural biodiversity-rich areas where botanic gardens are absent, ABGs can engage local constituencies with traditional botanical knowledge to showcase, conserve, and promote regionally important foods.

Considering the important ecosystem services provided by plants and ecosystems, a large number of local institutions retrofitting their green spaces into ABGs can open up local opportunities to revisit the relationship between plants and people and plants and local cultures (Sanders et al., 2018).

ABGs enriched with a diverse plant collection can attract non-plant organisms, offering a glimpse into ecological dynamics. ABGs are similar to botanical gardens, which have been described as artificial ecosystems, especially in urban areas, that allow the study of ecosystem dynamics relating to climatic factors (Faraji and Karimi, 2020). In fact, a widespread distribution of ABGs in ecologically diverse areas would provide local and regional data collection sites comprising a large number of native, imported, ecologically adapted, and invasive species introduced and maintained through the landscape industry.

Institutions establishing ABGs can promote their businesses by further engaging with their constituencies and clients beyond the institutional walls by offering various botanic garden–based activities, such as family programs, school programs, teacher training, adult education, and student internships, and influencing their environmental attitudes by promoting sustainable development (Faraji and Karimi, 2020).

REFERENCES

Abi Ali, R. 2020. End of docent training program 2019. *American University of AUB Botanic Garden* [Beirut]. www.aub.edu.lb/botanicgarden/Pages/2019-docent-program.aspx

Abou-Chaar, C.I. 1991. The green world of the campus: AUB campus: Perennial garden of our youth. *AUB Bulletin* 34: 3–6. https://aub.edu.lb.libguides.com/ld.php?content_id=18528964

AUBotanic. 2019a. *AUB Botanic Garden Manual*. AUB Botanic Garden, pp. 10–12. www.aub.edu.lb/botanicgarden/Documents/AUBotanic%20Policies%20July%202019.pdf

AUBotanic. 2019b. *AUBotanic Presentation to Board of Trustees*. American University of Beirut Botanic Garden. www.aub.edu.lb/botanicgarden/Documents/2019%20January%20-%20Aubotanic%20presentation%20to%20Board%20of%20Trustees.pdf

AUBotanic. 2020a. Nature therapy, a reflection. *AUBotanic Annual Newsletter* 1: 5. https://cms2.aub.edu.lb/botanicgarden/Documents/AUBotanic%20Newsletter%20Issue%201%20-%202020.pdf

AUBotanic. 2020b. Urban landscape management framework for the American university of Beirut. *AUB Botanic Garden*. American University of Beirut. www.aub.edu.lb/botanicgarden/Documents/AUBotanic%20Landscape%20Management%20Framework%20for%20the%20American%20University%20of%20Beirut%20-%20DRAFT%20Nov%2016,%202020.pdf

Bennett, B.C. 2014. Learning in paradise: The role of botanic gardens in university education. In: Quave, C.L. (ed.), *Innovative Strategies for Teaching in the Plant Sciences*. Springer, New York, NY, pp. 213–229. https://doi.org/10.1007/978-1-4939-0422-8_13.

Berezowitz, C.K., A.B. Bontrager Yoder and D.A. Schoeller. 2015. School gardens enhance academic performance and dietary outcomes in children. *J. School Health* 85(8): 508–518. https://doi.org/10.1111/josh.12278

Bliss, D. 1902–1903. *The Reminiscences of Daniel Bliss*. Fifth Annual Report of the Board of Managers, Fleming H. Revell Co., New York. 27 June 1871, p. 168. https://libcat.aub.edu.lb/record=b1198625~S7

Chen, G. and W. Sun. 2018. The role of botanical gardens in scientific research, conservation, and citizen science. *Plant Diversity* 40(4): 181–188. https://doi.org/10.1016/j.pld.2018.07.006

Cox, D.T.C., D.F. Shanahan, H.L. Hudson, R.A. Fuller, K. Anderson, S. Hancock, K.J. Gaston. 2017. Doses of nearby nature simultaneously associated with multiple health benefits. *Int. J. Environ. Res. Public Health* 14: 172. https://doi.org/10.3390/ijerph14020172

Ershad, A. 2017. By night, Iran's urban gardens are disappearing. *The Observers: France 24*. https://observers.france24.com/en/20170315-night-iran%E2%80%99-urban-gardens-are-disappearing

Facilities Planning & Design Unit, issuing body. 2017. Stage 3: Concept Master Plan. American University of Beirut MASTER PLAN. July, p. 36. AUB. www.aub.edu.lb/botanicgarden/Documents/Concept%20Master%20Plan%202017.pdf

Faraji, L. and M. Karimi. 2020. Botanical gardens as valuable resources in plant sciences. *Biodivers. Conserv.* https://doi.org/10.1007/s10531-019-01926-1

Forbes, S. 2008. How botanic gardens changed the world. Proceedings of the History and Future of Social Innovation Conference, Hawke Research Institute for Sustainable Societies, University of South Australia, Unisa, pp. 1–6. www.unisa.edu.au/siteassets/episerver-6-files/documents/eass/hri/social-innovation-conference/forbes.pdf

Graham, H., D.L. Beall, M. Lussier, P. McLaughlin and S. Zidenberg-Cherr. 2005. Use of school gardens in academic instruction. *J. Nutrition Education and behavior* 37(3): 147–151. https://doi.org/10.1016/s1499-4046(06)60269-8

Heywood, V.H. 2017. The future of plant conservation and the role of botanic gardens. *Plant Divers.* 39(6): 309–313. https://doi.org/10.1016/j.pld.2017.12.002

Heywood, V.H. and J.M. Iriondo. 2003. Plant conservation: Old problems, new perspectives. *Biological Conservation* 113(3): 321–335. https://doi.org/10.1016/S0006-3207(03)00121-6

Hohn, T.C. 2022. *Curatorial Practices for Botanical Gardens*. Rowman & Littlefield. ARBNET. http://arbnet.org/sites/arbnet/files/Hohn%202004%20Curatorial%20Practices%20for%20Botanical%20Gardens.pdf

Kaul, G. 2021. Disappearing gardens. *Greater Kashmir*. www.greaterkashmir.com/todays-paper/disappearing-gardens (Accessed 13 May 2022).

Krishnan, S., T. Moreau, J. Kuehny, A. Novy, S.L. Greene and C.K. Khoury. 2019. Resetting the Table for people and plants: Botanic gardens and research organizations collaborate to address food and agricultural plant blindness. *Plants, People, Planet* 1: 157–163. https://doi.org/10.1002/ppp3.34

Luck, G.W. 2007. A review of the relationships between human population density and biodiversity. *Biological Reviews* 82: 607–645. https://doi.org/10.1111/j.1469-185X.2007.00028.x

Melhem, M.G. 2019. *Guidelines for Establishing Ancillary Botanic Gardens*. Thesis. M.S.E.S. American University of Beirut. Interfaculty Graduate Environmental Sciences Program. http://hdl.handle.net/10938/21806

Mounzer, S. 2019. Buildings for Bugs? AUBotanic Event Celebrates Biodiversity. American University of AUB Office of Communications [Beirut]. www.aub.edu.lb/articles/Pages/AUBBotanicLadybug.aspx

Paiva, P.D.O., R.B. Sousa and N. Carcaud. 2020. Flowers and gardens on the context and tourism potential. *Ornamental Horticulture* 26(1): 121–133. https://doi.org/10.1590/2447-536X.v26i1.2144

Pollock, N.B., N. Howe, I. Irizarry, N. Lorusso, A. Kruger, K. Himmler and L. Struwe. 2015. Personal BioBlitz: A new way to encourage biodiversity discovery and knowledge in k – 99 education and outreach. *BioScience* 65(12): 1154–1164. https://doi.org/10.1093/biosci/biv140

Primack, R.B., E.R. Ellwood, A.S. Gallinat and A.J. Miller-Rushing. 2021. The growing and vital role of botanical gardens in climate change research. *New Phytol* 231: 917–932. https://doi.org/10.1111/nph.17410

Qumsiyeh, M., E. Handal, J. Chang, K. Abualia, M. Najajreh and M. Abusarhan. 2017. Role of museums and botanical gardens in ecosystem services in developing countries: Case study and outlook. *Intern. J. Environ. Studies* 74(2): 340–350. https://doi.org/10.1080/00207233.2017.1284383

Reid, K. and M. Gable. 2020. University-trained volunteers use demonstration gardens as tools for effective and transformative community education. *Acta Hortic.* 1298: 85–90. https://doi.org/10.17660/ActaHortic.2020.1298.13

Samus, A., C. Freeman, K.J.M. Dickinsom and Y. van Heezik. 2020. Relationships between nature connectedness, biodiversity of private gardens, and mental well-being during the Covid-19 lockdown. *Urban Forestry & Urban Greening* 69: 127519. https://doi.org/10.1016/j.ufug.2022.127519

Sanders, D.L. 2007. Making public the private life of plants: The contribution of informal learning environments. *Intern. J. Sci. Edu.* 29(10): 1209–1228. https://doi.org/10.1080/09500690600951549

Sanders, D.L., A.E. Ryken and K. Stewart. 2018. Navigating nature, culture and education in contemporary botanic gardens. *Environ. Edu. Res.* 24(8): 1077–1084. https://doi.org/10.1080/13504622.2018.1477122

Sellmann, D. and F.X. Bogner. 2013. Climate change education: Quantitatively assessing the impact of a botanical garden as an informal learning environment. *Environ. Edu. Res.* 19(4): 415–429. https://doi.org/10.1080/13504622.2012.700696

Sharrock, S. 2018. Botanic gardens and the 2030 sustainable development agenda. *B.G. Journal* 15(1): 14–17. www.jstor.org/stable/26596996

Sharrock, S. and M. Chavez. 2013. The role of botanic gardens in building capacity for plant conservation. *B.G. Journal* 10(1): 3–7. www.jstor.org/stable/24811260

Slaughter, R. (ed.). 1996. *New Thinking for a New Millennium* (Futures and Education Series). Routledge, Abingdon, UK.

Stagg, B.C. and M. Donkin. 2013. Teaching botanical identification to adults: Experiences of the UK participatory science project 'open air laboratories'. *J. Biol. Edu.* 47(2): 104–110. https://doi.org/10.1080/00219266.2013.764341

Syrian Protestant College, issuing body. President's annual Report, Annual Report of the Board of Managers of the Syrian Protestant College, American University of Beirut, 1902–1903, p. 22. https://libcat.aub.edu.lb/record=b1387201~S7

Talhouk, S.N., Y. Abunnasr, M. Hall, T. Miller and A. Seif. 2014. Ancillary botanic gardens in lebanon. *Sibbaldia: The International Journal of Botanic Garden Horticulture* (12): 111–128. https://doi.org/10.24823/Sibbaldia.2014.27

Talhouk, S.N., M. Fabian and R. Dagher. 2015. *Landscape Plant Database*. Department of Landscape Design & Ecosystem Management, American University of Beirut. https://landscapeplants.aub.edu.lb/

Trull, N., J. Penn and W. Hu. 2018. Visitor support for growth and funding in public built environments: The case of an arboretum. *J Hous and the Built Environ* 33: 829–841. https://doi.org/10.1007/s10901-018-9592-7

Wandersee, J.H. and E.E. Schussler. 1999. Preventing plant blindness. *The American Biology Teacher* 61(2): 82–86. https://doi.org/10.2307/4450624

Wang, R., J. Zhao, M.J. Meitner, Y. Hu and X. Xu. 2019. Characteristics of urban green spaces in relation to aesthetic preference and stress recovery. *Urban Forestry & Urban Greening* 41: 6–13. https://doi.org/10.1016/j.ufug.2019.03.005

West, W.A. 1931. History of AUB campus trees. *Al-Kulliyah, American University of Beirut Alumni Association* 28(1): 6. https://libcat.aub.edu.lb/record=b1198625~S7

Westwood, M., N. Cavender, A. Meyer and P. Smith. 2020. Botanic garden solutions to the plant extinction crisis. *Plants, People, Planet*. 3: 22–32. https://doi.org/10.1002/ppp3.10134

Wilkins, M.B. 1988. *Plant Watching: How Plants Remember, Tell Time, form Relationships and More*. Facts on File Publications, New York, USA.

Williams, S.J., J.P.G. Jones, J.M. Gibbons, et al. 2015. Botanic gardens can positively influence visitors' environmental attitudes. *Biodivers. Conserv.* 24: 1609–1620. https://doi.org/10.1007/s10531-015-0879-7

Wong, W.H. 2005. *Cost Recovery Strategies for Public Sector Botanic Gardens*. Doctoral dissertation, Nanyang Technological University, Nanyang Business School, Singapore.

Wyse Jackson, P.S. 1999. Experimentation on a large scale: An analysis of the holdings and resources of botanic gardens. *Botanic Gardens Conservation News* 3(3): 27–30. www.jstor.org/stable/24753880

Zelenika, I., T. Moreau, O. Lane and J. Zhao. 2018. Sustainability education in a botanical garden promotes environmental knowledge, attitudes and willingness to act. *Environmental Education Research* 24(11): 1581–1596. https://doi.org/10.1080/13504622.2018.1492705

6 Conservation of Threatened Plant Species and Protected Areas in Korean Botanical Gardens and Arboreta

Yong-Shik Kim, Hyun-Tak Shin, and Sungwon Son

CONTENTS

6.1 Plant Conservation in Korean Botanical Gardens and Arboreta 95
6.2 Single Species Recovery Projects in Botanical Gardens and Arboreta in Korea 96
6.3 Evaluation of Red List 96
    6.3.1 Background 96
    6.3.2 Evaluation of Korean Red List 97
6.4 Designation and Conservation of Protected Areas 98
    6.4.1 Background 98
    6.4.2 Status of Forest Genetic Resources Reserve 99
    6.4.3 Application of Key Diversity Areas in the Demilitarized Zone (DMZ) 100
6.5 Conclusion 100
References 101

6.1 PLANT CONSERVATION IN KOREAN BOTANICAL GARDENS AND ARBORETA

Compared to the functions of arboreta and botanical gardens that greatly developed in the West, the beginning of Korean arboretum was research on timber production during the Japanese colonial era with Hongnung and Gwangnung Arboretum at the center. Therefore, considering the general concept of arboreta and botanical gardens, Korean arboreta were deficient in their roles. In the 1970s, the Seoul National University Kwanak Arboretum and the Chollipo Arboretum, private arboreta, were vital in the development of botanical gardens and arboreta in Korea. Since Korea joined the Convention on Biological Diversity, a number of public, including provincial and private arboreta and botanical gardens, have been established. Specifically, 3 national, 33 public, 23 private, and 3 university arboreta are committed to the conservation of Korea plant biodiversity. Arboreta and botanical gardens in Korea have evolved into venues visited by 16 million people annually, owing to the rapid surge in users after the 1988 Seoul Olympics, and are used in numerous forms thanks to the popularity of concepts such as recreation and healing. However, a wide range of tasks should be done in terms of collection, display, education, research, and conservation, which are the distinctive functions of botanical gardens and arboreta, and outreach is needed in consideration of the local community and the GSPC 2020 as well.

    As of 2021, 96 botanical gardens and arboreta in Korea are open. Not all botanical gardens and arboreta actively participate in the conservation of species categorized as threatened, but most agreed on the need for an awareness campaign of conservation activities on the categorized threatened plant species. In particular, the Korea National Arboretum has been conserving rare plants in *ex situ* by growing, since 2010, a National Collection of Rare Plants Network jointly with nine local public arboreta.

As of 2020, there are 137 plant species in the Threatened category. In addition, the Korea National Arboretum and the National Baekdudaegan Arboretum are focusing an *ex situ* conservation of threatened plants through seed bank. In particular, the seed bank at the Korea National Arboretum reported that 345 species (90%) of 385 rare plants designated as threatened in Korea have already been secured.

This chapter intends to introduce the evaluation of the Red List conducted by Korea botanical gardens and arboreta and the cases of conservation of protected areas.

6.2 SINGLE SPECIES RECOVERY PROJECTS IN BOTANICAL GARDENS AND ARBORETA IN KOREA

Botanical gardens and arboreta in Korea have been contributing greatly to the restoration of threatened plants. Restoration projects targeting 195 threatened plant species have been executed so far, of which 54 were carried out by the botanical gardens and arboreta in Korea (KNA, 2019). Since the restoration project of *Berchemia racemosa* Siebold & Zucc., which belongs to the Rhamnaceae, was carried out by Hongnung Arboretum in 1990, the Korea National Arboretum, Halla Arboretum, Daea Arboretum, Chollipo Arboretum, Key-chungsan Botanical Garden, Yeomiji Botanic Gardens, Korea Expressway Corporation Arboretum, Hantaek Botanical Garden, Gohwun Garden, and Shingu Botanic Garden are carrying out single-species restoration projects. In particular, the Korea National Arboretum is mainly carrying out the single-species restoration project jointly with the public arboretum. Representatively, since 2015, it has been carrying out breeding restoration *of Thrixspermum japonicum* (Miq.) Rchb.f., collaborating with the Halla Arboretum. Several restoration projects are also being promoted. On the other hand, despite many constraints such as budget, the restoration activities by private arboreta are greatly encouraging. As a representative example, Yeomiji Botanic Gardens has been promoting the species restoration project of *Cymbidium lancifolium* Hook. In addition, the Chollipo Arboretum is promoting the restoration of *Ranunculus kadzusensis* Makino and *Iris dichotoma* Pall., and the Hantaek Botanical Garden, *Iris dichotoma* Pall. Of course, there are limitations. The most disappointing thing is the lack of documentation for each project and also the lack of systematic follow-up monitoring. Recently, the Korea National Arboretum has developed various capacity-building programs to improve the single-species restoration project in Korea. In particular, both in 2015 and 2018, scientists from the Center for Plant Conservation in the United States and IUCN Reintroduction Experts Group were invited to hold a workshop on the restoration of threatened plant species, and through this, practitioners were trained on the process of restoration of threatened plants. Both recovery works of *Aster altaicus* var. *uchiyamae* Kitam. (Kim et al., 2021) and *Traixspermum japonicum* (Miquel) Rchb. f. (Son et al., 2018) are good examples followed by IUCN Guidelines for re-introduction in Korea.

6.3 EVALUATION OF RED LIST

6.3.1 BACKGROUND

The International Union for Conservation of Nature (IUCN) Red List is a list that provides the most comprehensive global information on the threat of extinction of species, and it is estimated that more than 37,400 species are at risk of extinction (IUCN, 2021). The Global Strategy for Plant Conservation (GSPC) sets specific goals (16 targets) for plant conservation and recommends implementation at the national, regional, and global level.

Specifically, target 2 of the 16 targets is "evaluation of the conservation status for all known plants to support conservation activities." Therefore, numerous efforts are made across the globe, by governments and conservation-related NGOs, to contribute to target 2 of the global plant conservation strategy.

In Korea, a number of research studies have been conducted in recent years, such as the National Red List and global evaluation of species endemic to the Korean Peninsula. Particularly, in the field of plants, various ways to designate protected plants have been applied before the use of the Red List evaluation criteria. Plants subject to protection in Korea were mainly proposed or designated as "rare plants," "special wild plants," and "endangered plants." The term *rare plants* was first coined by Lee in 1959 in his research paper to disclose the new distribution sites of two species, *Cypripedium japonicum* and *Berchemia berchemiifolia*. Park Man-gyu (1975), Lee Young-no (1981), Lee (1987), and Hyun Jin-oh (2002) also presented a list of rare plants as a result of their academic research. Furthermore, the Ministry of Environment (1989, 1998) and the Korean Forest Service (1996) have legally designated and announced the list. The National Red List, which implements the criteria and categories of native plants in Korea introduced by the IUCN, was initially presented in the *Collections of Rare Plants in Korea* (2008, Korea National Arboretum). *Rare Plants Data Book Korea* evaluated 571 species, focusing on the taxa designated as protected plants and threatened plants by China, Japan, and Mongolia, and classified them into five categories [Extinct in the wild (EW), Critically endangered (CR), Endangered (EN), Vulnerable (VU), and Data Deficient (DD)}]. Subsequently, in 2012, the National Institute of Biological Resources introduced seven categories (CR, EN, VU, NT, LC, DD, and NE) following the evaluation of 534 taxa in the publication *Red Data Book of Republic of Korea Volume 5. Vascular Plants* (National Institute of Biological Resources, 2012). However, as for all the species that have been evaluated on the two national red lists published so far, a pool of target species was created based on the taxa mentioned in previous research, and only these groups were appraised. There are clear limits to the implementation of "assessment of all known plants" as stated by the GSPC.

The Korea National Arboretum announced the national red list for the second time after 2008. The newly released red list contains the assessment of all observable native plants found in Korea as well as re-evaluation of plants in the previous list so that it provides an important foundation for policy decision-making processes in terms of national plant conservation and management.

6.3.2 Evaluation of Korean Red List

Global Strategy for Plants Conservation (GSPC) target 2 recommends an assessment of the conservation status of all known plants. However, most of the "rare plants," "endangered plants," and species subject to national red list designation evaluation in Korea have been designated or evaluated at least once in the past literature. A checklist of native plants, which was the first effort in Korea to evaluate all native plants, was based on the Checklist of Vascular Plants in Korea by the Korea Forest Service (2021). According to the checklist, there are 3,849 taxa, 1,181 genus, and 211 families of Korean native plants (Korea National Arboretum, 2020).

The IUCN collected and utilized all data applicable to the five criteria for categorization of the Red List (A. population decline; B. geographic range; C. small or limited population size and decline; D. very small or limited population; E. quantitative analysis). Particularly, information on the distribution range corresponding to Criterion B is used the most in the evaluation of plant conservation status. For the distribution information of the species to be evaluated, the species occurrence information data were collected through the specimens owned by the Korea National Arboretum Herbarium (KH), the distribution maps of vascular plants in Korea (Korea National Arboretum, 2016), as well as the literature and reports containing the distribution information of each target species. For observation data, the rare plant distribution information database of the Korea National Arboretum was used. Based on the established distribution information, the extent of occurrence (EOO) and area of occupancy (AOO) were calculated using the Geospatial Conservation Assessment Tool (Bachman et al., 2011), respectively (IUCN, 2019).

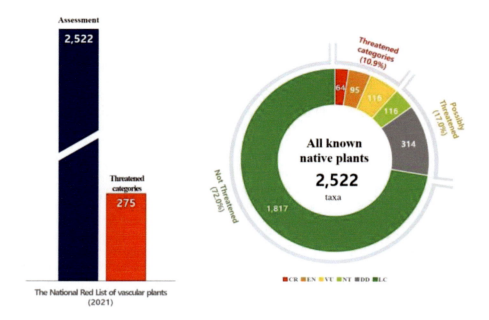

FIGURE 6.1 Rare plants in threatened categories among evaluated plants of Korea.

Among the native plants recorded in the Checklist of Vascular Plants in Korea, a total of 2,522 taxa were evaluated, and 275 taxa were found to belong to the threatened categories. Among the threatened categories, 64 taxa are classified as Critically Endangered (CR), 95 taxa as Endangered, and 116 taxa as Vulnerable' (VU). Near Threatened (NT), which is akin to the threat category, consisted of 116 taxa, and 'Data Deficient' (DD), which indicates a lack of sufficient information to directly or indirectly assess extinction risk, included 314 taxa.

Another 1,817 taxa were evaluated as a relatively threatened Least Concern (LC) (Figure 6.1). When 286 taxa belonging to the threat category are classified by families, Orchidaceae accounted for the largest number of 42 taxa, followed by 12 taxa in Asteraceae, and 11 each in Ranunculaceae and Scrophulariaceae. Rosaceae and Liliaceae each included 10 taxa.

6.4 DESIGNATION AND CONSERVATION OF PROTECTED AREAS

6.4.1 Background

The "Act on Creation and Promotion of Arboretums" enacted in 2001 played a decisive role for biodiversity conservation in arboreta. This law made it possible to create arboretums on general land, supported the operation costs of private arboretums by introducing an arboretum registration system, and focused on the preservation of forest genetic resources. By enacting this law, it was possible to systematically support the securing and management of domestic plant genetic resources.

At the moment, the Republic of Korea has designated and protected rare plants according to this law. Among them, 441 seeds of rare plants or nutrients, such as branches and roots, are collected and propagated, and they are stably managed in *ex situ* conservation facilities such as seed banks and conservation centers at the Korea National Arboretum. This laid the foundation for preparing in advance to prevent the extinction of rare plants in Korea, and it achieved the international standard for the conservation of rare plants presented by the Convention on Biological Diversity (CBD), an international convention for the conservation of living organisms on the planet.

6.4.2 Status of Forest Genetic Resources Reserve

Korea has designated 412 forest genetic resource protection zones of 1,720 km² to preserve the genes and species of plants distributed in forests or forest ecosystems (Figure 6.2). The protected areas are divided into seven types, including primitive, alpine plants, and rare plant habitats (Table 6.1). So far, the Korea National Arboretum has confirmed the habitats of 2,144 taxa through the forest genetic resource protection area research, and rare and endemic plants of 469 taxa have been identified.

TABLE 6.1
Status of Korean Forest Genetic Resources Reserve

Type	Alpine Floral Zone	Forest Marsh and River	Nature Reserve	Primitive Forest	Rare Korean Forest	Rare Plant Habitat	Useful Plant Habitat	Total
Number	4	92	61	59	38	119	39	412

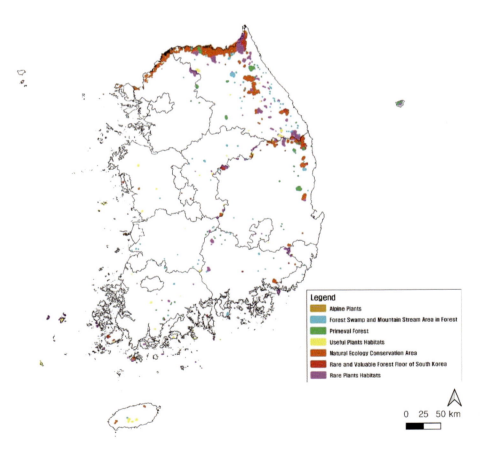

FIGURE 6.2 Designation type of forest genetic resources.

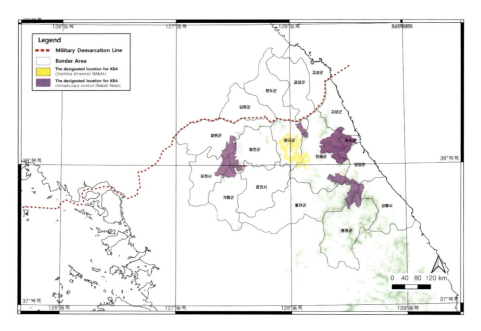

FIGURE 6.3 Proposal of KBAs on the DMZ.

6.4.3 Application of Key Diversity Areas in the Demilitarized Zone (DMZ)

First, by using the information of the identified rare plants, potential habitats were distributed, and the species that met international standards were analyzed. As a result, five species were selected, including *Echinosophora koreensis*, *Hanabusaya asiatica*, *Iris odaesanensis*, *Forsythia ovata*, and *Magaleranthis saniculijolia*. Key Biodiversity Areas were selected for *Echinosophora koreensis* and *Hanabusaya asiatica* as their habitats are limited to the DMZ area (Figure 6.3).

Key Biodiversity Areas for and *Echinosophora koreensis* and *Hanabusaya asiatica* were designated based on river basin zone standards and habitable areas. Finally, national forests that could be managed by the government were finally determined.

Additionally, the Korea National Arboretum has done a variety of research, including the establishment of Korea National Arboretum Plant Conservation Strategy 2030 (KSPC 2030 Korea Strategy for Plant), creation of a flora and habitat information system, conservation of plant species including genetic diversity, restoration using native plant species, discovery and protection of urban biodiversity, and designation of OECM (Other Effective Area-Based Conservation Measures). The Arboretum is also conducting research on new types of protected areas, including Punghyeol-ji, a place where cool and warm air occur in summer and winter, respectively. As a result, all northern lineage plants were confirmed to exist in Korea Punghyeol-ji areas, and thus a plan to designate them as legally protected areas is underway.

6.5 CONCLUSION

Botanical gardens and arboreta in Korea are making various efforts to protect not only species but also all habitats, conserving both by *in situ* and *ex situ* conservation, and are seeking ways to integrate protected areas toward the more effective way of management for conservation.

As mentioned earlier, botanical gardens and arboreta in Korea make a significant contribution to the conservation of threatened plants. The red listing work, which can be said to be the beginning of rare plant conservation activities, is led by the Korea National Arboretum. In particular, botanical gardens and arboreta in Korea play a basic role of *ex situ* conservation through field survey, collection, and propagation of target plants, and several botanical gardens and arboreta are more actively contributing to *ex situ* conservation of threatened plants through seed banking. Although it is still in its infancy, many botanical gardens and arboreta in Korea are interested in the single-species restoration of threatened plants.

REFERENCES

Bachman, S., M. Justin, W.H. Andrew, J. Torre and B. Scott. 2011. Supporting red list threat assessments with GeoCAT: Geospatial conservation assessment tool. *Zookeys* 150: 117–126.

Hyun, J.O. 2002. *Categorization of the Threatened Plant Species in Korea*. Ph.D. Thesis. Soonchunhyang University, Chungnam-do, South Korea.

IUCN. 2021. The IUCN Red List of Threatened Species. Version 2021–3.

IUCN Standards and Petitions Committee. 2019. Guidelines for Using the IUCN Red List Categories and Criteria. Version 15. Prepared by the Standards and Petitions Committee.

Kim, J.G., Y.Y. Lee, E.J. Cheong, Y.S. Kim and S.W. Son. 2021. Reintroduction planning and implementation of rare species Danyang aster, Yeoju, Republic of Korea. In: Soore, P.S. (ed.), *Global Conservation Translocation Perspectives: 2021: Case Studies from around the Globe*, IUCN SSC Conservation Translocation Specialist Group, Gland, Switzerland, pp. 346–349.

Korean Forest Service. 1996. *Rare and Endangered Species (Conservation Guidelines and Target Plant)*. Korean Forest Service Forestry Research Institute, Pocheon, p. 140.

Korean Forest Service. 2021. *Statistical Yearbook of Forestry*. Daejeon, p. 460.

Korea National Arboretum. 2008. *Rare Plant Data Book of Korea*. Korea National Arboretum, Pocheon, p. 332.

Korea National Arboretum. 2016. *Distribution Maps of Vascular Plants in Korea*. Korea National Arboretum, Pocheon, p. 809.

Korea National Arboretum. 2019. *100 Years History of the Arboretum & Botanical Garden in Korea*. Korean National Arboretum, Pocheon, p. 225.

Korea National Arboretum. 2020. *Checklist of Vascular Plants in Korea (Native plants)*. Korea National Arboretum, Pocheon, p. 1,006.

Lee, D.B. 1959. Newly found localities of *Cypripedium* and *Berchemia*. *Journal of Plant Biology* 2(2): 27–28.

Lee, T.B. 1987. Distribution status and conservation strategies of rare plants in Korea. *Nature Conservation* 59: 15–21.

Lee, Y.N. 1981. *Rare an Endangered Flora and Fauna of Korea*. Korea Nature Conservation Association, Seoul, pp. 153–271.

Ministry of Environment. 1989, 1998. Wildlife Protection and Management ACT.

National Institute of Biological Resources. 2012. *Red Data Book of Endangered Vascular Plants in Korea*. National Institute of Biological Resources, Incheon, p. 391.

Park, M.K. 1975. Research on rare and threatened plant in Korea. *Nature Conservation* 8(special issue): 3–24.

Son, S.W., J.Y. Jung, J.H. Pi, H.H. Yang, G.U. Suh, C.H. Lee, H.C. Kim and Y.S. Kim. 2018. First conservation translocation project of the East Asian *Trixspermum* on Jeju Island, South Korea. In: Soore, P.S. (ed.), *Global Reintroduction Perspectives: 2018: Case Studies from around the Globe*. IUCN, Gland, Switzerland, pp. 272–275.

7 Mongolian Botanical Gardens – Modern Plant Biodiversity Conservation Resources in Mongolia

Nanjidsuren Ochgerel, Luvsanbaldan Enkhtuyaa, and Victor Ya. Kuzevanov

CONTENTS

7.1	Introduction and a Historical Background	103
7.2	Objectives, Methods, and Resources	105
7.3	Results and Discussion	106
7.4	Conclusions	119
	7.4.1 Cross-Disciplinarity of Plant Conservation	119
	7.4.2 Leading Mission	125
	7.4.3 Conservation and the Pyramid of the Nature Management System	128
Acknowledgments		130
References		130

7.1 INTRODUCTION AND A HISTORICAL BACKGROUND

This work was conceived as an overview study of the history of the formation of the Mongolian botanical gardens, mainly the Botanical Garden of the Mongolian Academy of Sciences (MAS), as a science-intensive environmental toolkit for science and education, nature management, conservation and environmental management, as well as promoting the socio-economic development of modern society in Mongolia.

The unique nomadic civilization of the Mongols according to its richness of events and greatness of history influenced the development of civilizations in Asia and Europe (Karpini, 1957; Terguun, 2014). Many aspects of the cultural heritage of Mongolia continue to attract the attention of interdisciplinary researchers. However, many aspects of another important heritage – "natural heritage" – are often on the periphery of attention and can even be considered a secondary factor, although in recent decades, the role of the environmental factor, issues of rational environmental management, and environmental and economic development have become a priority in the competitiveness of regions and countries (Kalyuzhnova and Kuzevanov, 2010). The diversity of plants and the natural landscapes of pastures with steppe plants, which are at the base of the ecological pyramid, have traditionally had a great influence on the entire population of Mongolia, who at all times have depended on numerous farm animals with a dominant traditional nomadic way of life. Knowledge about plants, in addition to looking for productive pastures with the best breeds for livestock, was mainly focused around the properties of medicinal plants, the medical use of which became widely known traditions of Mongolian–Tibetan medicine (Bashkuev, 2014). Population growth accelerating over the past half century, the rapid demographic changes of Mongolian society, the transformation of socio-economic development, the industrial revolution, and urbanization have resulted in a

DOI: 10.1201/9781003281252-7

significant decrease of the predominantly nomadic way of life, and the settled way of life in rural and urban environments has increased sharply (Ochbadrakh and Ochirzhav, 2015). The transition from predominantly extensive economic management to intensive nature management led to a corresponding increase in the state's social demand for fundamental and applied science, for modern education, as well as for innovative scientific and technical tools for rational nature management and restoration of the environment. At the same time, accelerating urbanization and the urban way of life, undoubtedly, should have stimulated the revival of interest and the need of urban residents (about 70% of the country's population at present) to return to traditional contacts with nature, with animals and plants. And among rural residents (about 30% of the country's population) keeping a semi-nomadic or sedentary lifestyle, a natural interest in education and knowledge about new opportunities for the effective and successful use of shrinking and degrading pasture sites and natural complexes began to grow. It is happening in the context of global anthropogenic impacts and climate change. The processes of industrial development, the rapid changes of the economy and culture, and the construction and improvement of settlements have introduced an urgent need for large-scale use and enrichment of plant resources in Mongolia (Kubrikova, 2015).

Mongolia belongs to the vast ecological and geographical transboundary region of Baikalian Siberia around Lake Baikal practically in the center of Asia. The country, first, is inhabited by ethnic groups of Mongols, genetically and culturally closely related to the aboriginal ethnic groups inhabiting the territory around Lake Baikal (Buryats and others), and, second, it is connected with Baikal through the Selenga River, the largest among more than 300 rivers flowing into the lake through the unity of a giant drainage basin in North Asia (Vetrov, 1995; Dulov, 1995; Elokhina and Oleinikov, 2012). The population of Mongolia is historically characterized by the main involvement in animal husbandry for subsistence. Crop production and gardening has traditionally been regarded as a secondary business that was brought to the region rather late by settlers from Siberia and the European part of Russia and from adjacent regions of China. It is believed that the sharp continental climate of Mongolia, in terms of the possibilities of creating botanical gardens, has no analogues on the globe and is a vivid example of extreme living conditions for plants, animals, and humans (Banzragch et al., 1978; Ochirbat, 1994, 1996, 1999, 2006). Therefore, attempts are being made to stimulate the development of crop production and horticulture; among the nomadic ethnic groups in the area of Lake Baikal, these efforts have been perceived as a kind of burden and not an urgent need for survival. There are more than 3,191 species of vascular plants in Mongolia (Nyambayar et al., 2011; Urgamal et al., 2014; Gombobaatar et al., 2018; Urgamal, 2017, 2018a, 2018b, 2018c), many of which are traditionally used in natural habitats for grazing. Some of them are also used as well-known Mongolian–Tibetan medicines since only a few edible wild plant species are eaten by humans. It is believed that in Mongolia the first-ever successful attempts to develop crop production in the first state farms of grain crops in aimags (provinces) Kobdo and Ub date back to 1921 since agriculture was considered only as a secondary appendage of cattle breeding. Gardening as part of an independent way of life for a small percentage of the population took shape only by 1954 after the appearance of the first fruit and vegetable garden station in aimag Shamaar (Davaazhav, 2017). Understanding the need to introduce crop production, agriculture, and horticulture into everyday life as an independent, large type of social production in Mongolia was clearly identified by the 1960s, apparently due to the accelerating processes of urban development and increasing urbanization, problems with public health, desertification of land, and depletion of natural biological resources (Sanzheev et al., 2013). The transition of many Mongolian families from a predominantly nomadic lifestyle while residing in mobile nomadic yurts to gradual settlement for long-term settled living in stone buildings or stationary yurts in cities and large settlements was caused by the deployment of complex demographic processes new to Mongolia, the division of labor in the course of industrialization, and settlement and development of various natural resources of the country (Vinokurov and Alimaa, 2012). This undoubtedly caused gradual changes in the mentality and habits of the growing number of citizens and urban intelligentsia, who felt the need to return to traditional contacts with wildlife. It became especially in demand when the area of steppe natural landscapes was reduced, along with increasing environmental pollution and the removal of trees and shrubs in small towns and large cities.

It is believed now in Mongolia that one of the most significant inventions of mankind in the field of using and conserving plant biodiversity for the benefit of human well-being in European and Asian civilizations was the "idea of a botanical garden," which inspired the creation of about 3,000 botanical gardens worldwide that have become multifunctional environmental resources for rational use of biodiversity and the development of civilization (Prokhorov, 2004; Kuzevanov, 2010; Dodd and Jones, 2010). The modern "idea of a botanical garden" first came to Mongolian society in the middle of the twentieth century under the influence of scientists and educators, scientific and educational traditions of the Soviet Union (Banzragch et al., 1978; Erdenejav, 2009), as well as neighboring countries. It is known from history, for example, that the Mongolian Kublai Khan, a descendant of Genghis Khan and the founder of the Yuan State, created a unique botanical object – the Beihai Garden – which over time was recognized as an object of natural and cultural heritage of our time in China (Terguun, 2014).

The Botanical Garden of the MAS has become in recent decades the botanical resource contributing both in science and in certain aspects of the socio-economic development of Mongolia (Ochirbat, 1994, 1996, 1999, 2006, 2011; Ochirbat and Dorzhsuren, 2008; Enkhtuyaa and Ochgerel, 2008, 2015). Despite the fact that for a long time administrators and entrepreneurs ignored and overlooked the biodiversity importance, the economic significance of the conservation and rational use of plant biodiversity has become especially obvious over time. In the last two decades, the "idea of a botanical garden" has become particularly popular and caused some boom in the creation of new private and municipal botanical gardens predominantly in the format of zoo–botanical gardens in Mongolia.

Therefore, this chapter focuses on a relatively new phenomenon in Mongolian history, namely the "idea of a botanical garden" as a tool for biodiversity conservation that can help to identify solutions and answers in the right directions to the above environmental and social needs of the state and the growing needs of the population in the field of rational management of plant resources.

7.2 OBJECTIVES, METHODS, AND RESOURCES

At present, the attention of the population and the government of Mongolia to rational nature management, or to natural museums (zoos and botanical gardens and their analogues), has begun to noticeably increase due to the large-scale goals of environmental and socio-economic progress, that is, the development of the country's human capital (Bazaar, 2008; Badaraev, 2013; Potaev, 2015; Belozertseva et al., 2015). Therefore, the main objective of our study was to analyze, using the Botanical Garden of the MAS as an example, how and why the attitude toward the "idea of a botanical garden" and the modern understanding of the proper functions of a botanical garden has been developed for plant biodiversity conservation in Mongolia. The positioning of gardens in nature and society has changed in the Mongolia over the past decades due to great ecological and demographic transformations in the course of social and economic development of Mongolia. This study builds on the authors' many years of international experience and focuses on the past, present, and future role of Mongolia's botanical gardens as ecological and social resources for human well-being and environmental restoration. In our opinion, this work should help scientific research, as well as modern Mongolian society (administrative decision makers, scientists and teachers, entrepreneurs, farmers, adults and children, schoolchildren, students, environmentalists, etc.), to better understand the environmental and social importance of botanical gardens for their effective plant biodiversity conservation and use for the sustainable development of Mongolia.

The following terminological definitions were used in this work:

1. The botanical garden is a strictly protected, landscaped area of socio-ecological significance, containing documented plant collections and landscape gardens, where the managing organization creates resources for scientific research, education, and enlightenment;

public expositions of plants and technologies for biodiversity conservation; plant propagation; and the provision of services on the basis of knowledge about plants and their derivatives (Kuzevanov and Gubiy, 2014; Kuzevanov et al., 2021a). It is an extended definition from Peter Wyse Jackson's classic concise definition (1999).
2. Environmental resources are environment-forming components that integrate natural and biological resources, including habitats, living organisms and their links to each other and the environment, tangible products, and intangible results of human activity, into a set of factors that provide ecological balance in nature and the human environment (Kuzevanov and Nikulina, 2016).

The study included five main methodological approaches to the collection and analysis of materials: (1) collections of data about the history of formation and land use of the Botanical Garden of the MAS in Ulaanbaatar, including descriptions, maps, photographs, the current status and composition of plant collections, a variety of resources and functions, as well as directions and developmental trends; (2) collections of comparative data through special personal works and study tours to more than 200 botanical gardens and their analogues in 34 countries of the world, as well as in the transboundary geographical region of Baikalian Siberia; (3) collections and verification of plant names, materials, and archival documents of various botanical gardens, their networks (Consortium World Flora Online, 2013, 2020; BGCI, 2022), and relevant international and national organizations, including statistics from Mongolian sources, UN, UNDP, etc.; (4) compiling a bibliography, as well as collecting data through personal correspondence and private conversations about the features of the development of botanical, systematic, taxonomic studies of the Mongolian flora; and (5) statistical processing and methods of graphical presentation of data using standard statistical software in MS Office for analysis of plant collections, taxonomic data, and demographics of Mongolia. Population growth and human development index data for Mongolia were used from sources (World Bank, 2021; Barrientos and Soria, 2021; Worldometer, 2021; UNDP, 2020).

The source of data on the composition of the collections is the database and the archive of the Botanical Garden of the MAS dated 30 October 2021.

Modern synonyms for the names of vascular plants (flowering, coniferous, etc.) are given according to the current Plant List (Urgamal et al., 2014; 2019a, 2019b; Consortium World Flora Online, 2013, 2020), and family indexing by system (Urgamal et al., 2014; 2019b, Urgamal and Bayarkhuu, 2019) is based on the APG IV and LAPG IV classification systems (Angiosperm Phylogeny Group IV, 2016).

7.3 RESULTS AND DISCUSSION

A distinctive feature of the creation of new botanical gardens and their analogues in Mongolia at the beginning of the twenty-first century, as we see from Figure 7.1, is the almost parallel creation of urban zoos and their convergence in the format of zoo–botanical gardens as part of a system of science-intensive museums of nature.

The priority development of the "idea of a zoological park" at the beginning of the twentieth century, and then the "idea of a botanical garden" that came relatively later, is obviously associated with the absolute dominance of animal husbandry in the traditional nomadic way of life in Mongolia since the interests of local residents were mainly focused not on plants but on animal organisms (horses, sheep, cattle, goats, camels, etc.) on which daily life depended. From Figure 7.1, it is also visible that the increasing attention in the country to the establishment of zoos and botanical gardens has been intensified in recent decades, which is evidently associated with the growth of the socio-economic indicators of the country's development in the course of accelerating demographic growth (especially the increase in the urban population) (Worldometer, 2021; World Bank, 2021) and improving human well-being (UNDP, 2020). One of the "epochal" demographic periods is undoubtedly the 1970s, when the total number and percentage of urban residents in Mongolia first equalized with the rural population, and then the urban lifestyle began to dominate (see the

Mongolian Botanical Gardens – Modern Plant Biodiversity

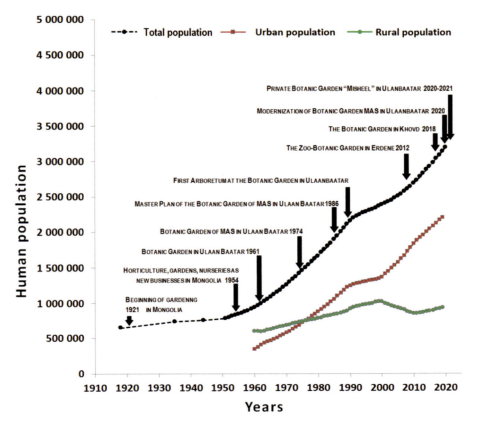

FIGURE 7.1 Comparison of key moments in the history of the formation of botanical gardens (and zoos) with the dynamics of population growth in Mongolia. The moments of creation of botanical gardens and zoos are indicated by black arrows. It shows the "demographic cross" – a graph of the change in the dynamics of the urban (red dots) and rural population (green dots).

"demographic cross" in Figure 7.1). This dominance of the urban sedentary way of life explains the growing social demand of the local population and the interest of the state in the creation of such museums of nature as botanical gardens and zoos.

Despite the fact that the scientific idea of creating a "botanical garden" appeared in the circles of the scientific intelligentsia of Mongolia first in the 1960s with the development of the Mongolian Academy of Sciences (MAS), it was not fully implemented. But it revived again in the early 1970s when cooperation between the USSR and Mongolia was especially active within the framework of the joint Soviet–Mongolian complex scientific expedition of the universities and academies of sciences of both countries (Dorofeyuk and Gunin, 2000; Pavlov et al., 2004). However, the "idea of a botanical garden" was then underestimated, and its full implementation was resumed only in 1974 in order to preserve the biological diversity and genetic resources of endemic, rare, and useful plants when a plot of land was specially allocated for the first Mongolian botanical garden (botanikiin tsetserleg) as an independent academic institution in the eastern part of Ulaanbaatar – the capital of Mongolia. In 1974, an international Mongolian–Soviet team of scientists, created through the initiative of the government and the Mongolia Academy of Sciences, developed a plan and main directions of fundamental and applied research to create the Botanical Garden of the MAS of the Mongolian People's Republic for the development of the national economy (Banzragch et al., 1978; Ochirbat, 1996, 1999, 2006; Enkhtuyaa and Ochgerel, 2008).

FIGURE 7.2 Location of botanical gardens and their analogues in the settlements of Mongolia and adjacent administrative regions of Russia and China, located in similar severe, sharply continental climatic conditions.

Source: Cartographic basis of the map is from Wikimedia https://commons.wikimedia.org/w/index.php?curid=69773

Further, thanks to local ideas of the late twentieth and early twenty-first centuries, conditions were also formed for initiating new projects to create four more botanical gardens: (1) the Regional Botanical Garden at the university in the city of Khovd in the western part of the country; (2) the State National Zoo–Botanical Garden in the city of Erdene, 76 km east of the capital; (3) the private corporate "Misheel Botanical Garden-Park" of the group of private companies "Misheel City" on the island of the Tuul River in Ulaanbaatar; and (4) the revival of the former zoo in the format of a mini-zoo–botanical park in Ulaanbaatar (Figure 7.2).

In the immediate vicinity of Mongolia in the adjacent administrative areas of Russia (Republic of Buryatia, regions of Irkutsk, Altai, Krasnoyarsk, and Zabaikalsky) and China (Ganxi Province and administrative regions Xinjiang Uygur and Inner Mongolia), where the population is about 84 million people, there are about 20 botanical gardens and their analogues (Figure 7.2), which undoubtedly contribute to the socio-ecological and economic development and enrichment of the botanical resources in similar severe, sharply continental climatic conditions. On the interstate transboundary natural Baikalian Siberia territory of Russia and Mongolia, united by the drainage basin of Lake Baikal and Lake Khuvsgul, where about 7.7 million people live, seven botanical gardens and their analogues are currently operating, and three more gardens are being created.

As part of joint developments under the leadership of the Main Botanical Garden of the USSR Academy of Sciences and the deep participation of the Institute of Botany of MAS, the State Design and Construction Institute and other institutions, local horticultural organizations, as well as specialists from the joint Soviet–Mongolian biological expedition and other botanical gardens of the USSR, the successful construction of the first academic Mongolian botanical garden started on an area of about 32 ha with an additional 38 ha of vacant land on the southeastern outskirts of Ulaanbaatar (Banzragch et al., 1978). The free landscaped plantations and collection plots were designed according to the principles of taxonomic and geographic grouping as well as decorativeness of plants. It was planned to create an arboretum on an area of 20 hectares as a windproof and build a basic collection of 574 species of trees and shrubs of 146 genera and 46 families. For the harsh climatic conditions of Ulaanbaatar, the following most promising priority genera and species of trees and shrubs were recommended, most of which, including at the level of groups of genera, have been successfully introduced in the Botanical Garden of the MAS at the present time (Table 7.1). Along the perimeter of the garden and in the alleys, decorative plantings and windbreaks

TABLE 7.1

Trees and Shrubs Recommended in the 1970s as the Most Promising Species for Primary Introduction in the Botanical Garden of the MAS

Modern Name	Synonyms Mongolian/English	Names of Available Representatives of the Genus in the Collection in 2021
Betula pubescens var. *pumila* (L.) Goverts.	Odo khus/White birch	*B. microphylla* Bunge *B. pendula* Roth *B. nana* subsp. *rotundifolia* (Spach) Malyshev *B. platyphylla* Sukacheva
Caragana arborescens Lam. *Caragana spinosa* (L) DC.	Udlyg hargana/Siberian pear Orgyest khargana/Prickly caragana	*C. arborescens* Lam. *C. bungei* (Ledeb.) Kuntze *C. fruiticosa* (Pall.) Besser *C. laeta* Kom. *C. microphylla* Lam. *C. pygmaea* (L.) DC. *C. spinosa* (L.) DC. *C. tibetica* Kom. *C. ussuriensis* (Regel) Pojark.
Cotoneaster lucidus Schltdl.	Gyalgar chargay/Brilliant cotoneaster	None
Cotoneaster melanocarpus Fisch. former A. Blitt	Khar zhimst chargay, Tom Urt/Black Cotoneaster	
Crataegus dahurica Koehne ex C.K.Schneid	Daguur doloogono/Dahurian hawthorn	*C. almaatensis* Pojark. *C. maximowiczii* CKSchneid. *C. sanguinea* Pall.
Hippophae rhamnoides L. (= *Elaeagnus rhamnoides* (L.) A. Nelson)	Yashilduu vhatsargana/Sea buckthorn	*H. rhamnoides* subsp. *mongolian-rus* L.
Larix sibirica Ledeb.	Sibir shines/Siberian larch	*L. sibirica* Ledeb.
Larix dahurica Turcz. ex Trautv (= *Larix gmelinii* (Rupr.) Kuzen.)	Gmeliniy shines/Dahurian larch	
Lonicera caerulea subsp. *altaica* (Pall.) Gladkov.	Khekh dalankhals/Blue honeysuckle	*L. tatarica* L.
Lonicera hispida Pall. ex Shult.	Arzgar Dalanhals/Honeysuckle	
Picea obovata Ledeb.	Sibir gatsuur/Siberian spruce	*P. obovata* Ledeb.
Pinus sylvestris L. _	Oin nars/Scotch pine	*P. silvestris* L.
Populus alba L.	Tsagaan ulias/White aspen	*P. balsamifera* var. *altaica* P.b.
Populus balsamifera L.	Gavar ulias/Balsam poplar, Hackmatack	*P. laurifolia* Ledeb. *P. tremula* L.
Populus laurifolia Ledeb.	Lavar navchit ulias/Laurel poplar	
Populus simonii Carrière	Simony ulias/Simonov poplar	
Ribes alpinum L.	Alpiin ukhriin nude/Alpine currant	*R. alpinum* L.
Ribes nigrum L.	Khar ulaagana/Black currant	*P. diacanthum* Pall. *R. nigrum* L. *R. rubrum* L.

(Continued)

TABLE 7.1
(Continued)

Modern Name	Synonyms Mongolian/English	Names of Available Representatives of the Genus in the Collection in 2021
Rosa rugosa Thunb.	Urchger sarnay/Rugosa Rose	*R. acicularis* Lindl. *R. baitagensis* Kamelin & Gubanov *R. davurica* Pall. *R. laxa* Retz. *R. oxyacantha* M. Bieb. *R. platyacantha* Schrenk *R. rugosa* Thunb. *R. spinosissima* L. *R. sherardi* Davis *R. xanthina* Lindl.
Salix alba L.	Tsagaan burgas/White willow	No
Spiraea media Schmidt	Dund tavilgana/English spirea	*S. alpina* Pall. *S. aquilegifolia* Pall. (= *Spiraea x vanhouttei* (Briot) CarriŠre.) *S. dahurica* (Rupr.) Maxim. *S. chamaedryfolia* L. *S. japonica* L.f. "Nana" *S. japonica* L.f. "Goldenflower" *S. japonica* L.f. "Genpei" ("Shirobana") *S. media* Schmidt *S. salicifolia* L. *S. vangutta* (= *Spiraea cantoniensis* Lour. x *S. trilobata* L.) *S. pubescens* Turcz.
Syringa vulgaris L.	Egel goltbor/Common lilac	*S. emodi* Wall. former Royle *S. josikaea* J.Jacq. *ex* Rchb.f. *S. komarowii* C.K.Schneid. *S. oblata* Lindl. *S. oblata* var. *dilatata* (Nakai) P.S.Green & M.C.Chang *S. pubescens* Turcz. *S. sweginzowii* Koehne & Lingelsh. *S. villosa* Vahl *S. vulgaris* L *S. wolfii* S. K. Shneid.
Ulmus pumila L. _	Odoi hailas/Siberian elm	*U. pumila* L.
Ulmus davidiana var. *japonica* (Rehder) Nakai.	Yapon hailas/Japanese elm	*U. davidiana* var. *japonica* (Rehder) Nakai.

Source: Banzragch et al. (1978).

made out of ornamental trees have been designed. Studies of the green fund of Ulaanbaatar and other settlements were carried out and proposals were made to improve the system of urban greening, including the use of grass coverings and wider use of ornamental and tolerant native plants of the genus *Carex* L., such as *C. duriuscula* C.A. May.

Based on the analysis of the natural vegetation of the Mongolian People's Republic and adjacent regions, it was proposed to plant at least 400 plant species of 40 genera in specialized collections on open areas of about 15 hectares: "Khangai and Altai area"; "Steppe area"; "Gobi Desert area"; "Endemic, relic, and rare species"; "Useful plants"; and "Introduced plants from other regions." To create displays, demonstrations, and experimental expositions of cultivated herbaceous flower-decorative, lawn, and ground cover plants, a plot of 4.2 hectares was allocated, and it was also planned to build a greenhouse complex and an exposition greenhouse with a usable area of at least 2,000 sq. m for the maintenance of exotic plants from subtropics. At the same time, in order to overcome significant natural and climatic restrictions, a program of long-term introduction studies was composed to make the gene pool of cereals, vegetables, fruits, and other cultivated plants tolerant to Mongolian conditions.

In accordance with the existing scientific paradigm and the joint Mongolian–Soviet R&D program, the Botanical Garden of the MAS was established as an innovative experimental scientific institution for Mongolia to solve the following scientific and practical problems (Banzragch et al., 1978; Ochirbat, 1999), many of which are still relevant today:

1. Study of the biology and ecology of Mongolian plants under stationary *ex situ* conditions, including genetic analysis of the country's flora and the history of its formation under an extremely unfavorable environment of a sharp continental climate
2. Preliminary analysis of the floras of adjacent and other botanical and geographical zones of the Earth in order to select plants that are prospects for mobilization and introduction in culture in Mongolia as well as those of specific scientific interest
3. Creation of extensive collections of local and foreign flora as a basis for scientific research and as a source of enrichment of the cultural flora of the country
4. Development of introduction methods and identification of plants useful for agriculture, industry, and landscaping and characterized by high frost and cold resistance, drought resistance, longevity, and productivity
5. Development of scientific and practical issues of crop production in cooperation with agricultural and scientific institutions
6. Development of approaches for protection of gene pools of valuable botanical objects and reproduction of plant genetic resources of the Mongolian People's Republic
7. Research and development studies of ornamental horticulture issues, the scientific basis of landscaping, and the accelerated introduction of promising ornamental plants into the practice of urban greening
8. Development of recommendations, together with departmental and sectoral scientific institutions, on the assortment of plantings for windbreaks and protective forest belts as well as for soil erosion control

Further development was planned with the prospect of placing the Botanical Garden of the MAS on an expanded area of about 147 hectares. In fact, it became possible to return to the performance of the key functions of the main Mongolian center of introduction, acclimatization, selection, and conservation of plants only by the end of the 1990s, when the Institute of Botany of MAS was able to resume its research activities with the development of economic growth and prosperity of the country. At the same time, the creation of the arboretum collection was continued on the lands of the botanical garden, where work was resumed on the introduction and cultivation of dozens of species of woody plants. The arrangement of the territory of the botanical garden was conceived as a platform for the development of a scientific complex in the interests of local residents on the basis of the resources of the Mongolian Academy of Sciences and with the involvement of universities and educational institutions of Ulaanbaatar-city.

In 1986, a new joint Mongolian–Soviet project was prepared in ten volumes – the first Master Plan for the development of the Botanical Garden of the MAS just for a total cultivated area of about 32 hectares. However, over the course of a complex transformation process in the context of the socio-economic crisis in the country during the late twentieth century and the transition of Mongolia to a free market economy, the first botanical garden of Mongolia and the Institute of Botany of the Academy of Sciences in Ulaanbaatar were sorely lacking in human resources to maintain living collections (there were not enough gardeners for planting and field work and a lack of tools and materials necessary for scientific research and for conventional horticultural techniques). As a result, many scientific and applied works with plants were suspended and could not be processed as expected. Unfortunately, as of 2017, the activities of the academic botanical garden experienced particular difficulties and began to decline, not only due to insufficient funding but also due to a lack of qualified personnel. Due to the difficulties of the transition to a market society, there was an acute shortage of skilled workers, and it was not easy to simultaneously conduct scientific research and maintain living collections that required a lot of daily care. Nevertheless, it was possible to establish plantings of various local species of medicinal, ornamental, rare, and endangered plants listed in the Mongolian Red Book on almost 20 hectares of developed land in the open space of the steppe type. In addition, the peculiarities of the mentality of local residents, who are not used to appreciating and preserving urban green spaces and maintaining a healthy ecological environment, did not prevent the discharge of wastewater and littering in the surrounding area, including on garden lands. Due to insufficient order in urban planning policy in the previous period, as well as uncontrolled fires and cutting down trees for firewood, many large trees and the entire citywide green fund were severely damaged. The scientists were helping to plant trees and shrubs in groups of groves (birch, pine, poplar, aspen, cedar, spruce, larch, etc.) to recreate the main green frame of the urban forest inside the city of Ulaanbaatar. During this period, due to the incessant "seizures" of expensive land in the growing and modernizing capital, a significant part of the land previously allotted for the botanical garden was lost for other purposes and for other urban development. For the remaining 18 hectares of land, a new Master Plan for the Botanical Garden of the MAS was drawn up in 2020–2021 and is being implemented in stages currently (Figure 7.3). The territory was conditionally subdivided into two large zones:

1. A green public and educational area in the northwestern part (8.02 ha), containing demonstration and exhibition areas and displays as well as public service facilities (shop, canteen, toilets, etc.)
2. A research zone in the eastern part (10 ha), in which the main plantations of woody plants, nurseries, and beds and an arboretum are located to be used in the research work of scientists and for teaching students. There are specialized gardens and collections, such as medicinal, fruit and berry, and rare and endangered plants, as well as a unique scientific and experimental site "Forest for young researchers." The main attention is paid to the introduction, acclimatization, and selection of various hardwoods (poplar, birch, fruit, and berry crops are especially in demand) and conifers (spruces, pines, and various larches). Currently, about 23,000 seedlings are grown annually in the arboretum and nursery to meet the local needs of citizens in high-quality seedlings and planting material, and the total number of cultivated plants in this zone is more than 60,000 living plant specimens.

The creation of a public landscape area with open displays, greenhouses, and areas for visitors at the scientific botanical garden is not a whim. This is a new modern factor based on the need to strengthen the positive scientific, educational, environmental, and social impact on the population and on Mongolian and foreign tourists, including on people with special needs and people with disabilities. Here they can get directly acquainted with scientific achievements and get to know the plant world better, as well as use it for relaxation and recreational purposes, or active spending of

Mongolian Botanical Gardens – Modern Plant Biodiversity

FIGURE 7.3 Scheme of the Master Plan of the Botanical Garden of the MAS in Ulaanbaatar as updated in 2021.

The red dotted line is the current external border (land area is about 18 ha). In October 2020, at the site marked with a gray rectangle (number 18), a large-scale interdepartmental and interdisciplinary project "Innovation Center of the Mongolian Academy of Sciences" ("The Innovation Cluster") was laid. Some premises of the Center will also house offices, laboratories, classrooms, and a herbarium of the Botanical Garden. The source of the illustration is the archive of the Botanical Garden of the MAS.

Legend with designations of numbers of main plots and collections of the garden: (1) display gardens; (2) green fencing of coniferous trees and mixed forest; (3) car parking; (4) flower beds; (5) aromatic and sensory garden; (6) educational garden; (7) dwarf coniferous plants; (8) garden of propagation and plant breeding; (9) collection and nursery of herbaceous perennials; (10) garden of rare and endangered plants of Mongolia; (11) plant introduction and evaluation plot; (12) collection and nursery of medicinal plants; (13) groves and "Forest of Young Explorers"; (14) garden of ornamental herbaceous flowering plants; (15) experimental site for seed propagation; (16) greenhouses; (17) collection and nursery of Flora of Mongolia; (18) Innovation Center of MAS.

free time in a comfortable natural environment. As a comprehensive museum of nature, the academic Botanical Garden in Mongolia became a "unique object of the country."

A large Innovation Center of MAS ("The Innovation Cluster") is currently being built on part of the former land of the botanical garden (Figure 7.4). The Botanical Garden will use some of the premises and a room for scientific laboratories, offices, a library, a herbarium, a seed storage, seminar rooms and classrooms, etc.

Collections of the Botanical Garden of the MAS currently hold quite diverse cultivated vascular plants, i.e., those introduced and preserved for many years on the territory of 18 hectares (Figures 7.5 and 7.6). The spectrum of species diversity of the collections accumulated over its entire history is quite rich and currently well represents the main families of the Mongolian flora (Figures 7.7, 7.8, and 7.9).

Over the past five years, the plant collection of the Botanical Garden of the MAS has been steadily replenished (Figure 7.7), mainly by local Mongolian species, and it represents 369 plant

FIGURE 7.4 Aerial view of the territory of the Botanical Garden of the MAS with buildings and structures built into the "Innovation Cluster" project created by the Academic Scientific and Technical Innovation Center of MAS for the collaborative use and a public–private partnership facilitation.The red dotted line marks the border of the territory of the Botanical Garden with living collections and expositions. The land of the Botanical Garden can be recognized as a characteristic green park area in the flat part of the city of Ulaanbaatar, the capital of Mongolia.

species of 166 genera and 61 families, which is about 11.5% of the registered 3,191 vascular-types plants in Mongolian flora (Urgamal et al., 2014, 2016, 2019a, 2019b, 2020), as well as 24.2% of 684 genera and 56.5% of 108 families, respectively. The collection of coniferous and deciduous trees and shrubs includes more than 150 species and 6 forms. The collection of ornamental and medicinal herbaceous plants consists of 190 species of 132 genera of 49 families of natural perennials.

Particular attention is paid to experimental research in order to increase the number of specialized collections of the genera *Betula*, *Syringa*, *Lonicera*, *Spiraea*, *Rosa*, *Iris*, *Paeonia*, *Lilium*, *Hemerocallis*, *Chelidonium*, and *Allium*, as well as to improve the technology of their cultivation and introduction tests. More than 110 promising species of trees, shrubs, and herbs have been studied and put into production in order to start creating a large-scale garden seedling that will contribute to the growth and development of the "green fund" of urban gardening for the beautification of public spaces and city streets with green spaces and flowers. As a result of botanical research and analytical work (Enkhtuyaa and Ochgerel, 2015), promising groups of trees and shrubs from 16 families of 536 species, as well as 29 varieties of ornamental herbaceous plants for introduction and acclimatization in the near future, have been identified.

The collection "Rare and endangered plant species of Mongolia" was established from the very beginning of the foundation of the Botanical Garden of the MAS. The main objective of the collection is to preserve plant species listed in the *Red Book* of Mongolia, including endemics and relics. On the territory of the garden, environments have been created that mimic their natural *in situ* habitat. On the basis of this collection, the biology of species is studied using a variety of methods and assessing the prospects for their introduction in urban environments and the possibility of repatriation into nature. The genetic material of sustainable populations under the conditions of introduction, especially of small or endangered species, is of great importance for the prospects for ecological restoration and repatriation of these species in nature. Thus, the botanical garden will contribute to the reconstruction of natural populations of rare and endangered plant species. Collecting rare and endangered species is the basis of the technology for biodiversity conservation

Mongolian Botanical Gardens – Modern Plant Biodiversity 115

FIGURE 7.5 Examples of collections and plants at the territory of the Botanic Garden of the Mongolian Academy of Sciences in Ulaanbaatar: (A–B) Collections and displays of the steppe vegetation; (C–D) trees and shrubs in the arboretum; (E) *Paeonia lactiflora* Pall.; (F) *Adonis mongolica* Simonov; (G–H) general views of the territory with collections of the Botanical Garden in springtime.

and rational use of plant genetic resources in Mongolia. The collection of rare and endangered plants, as an important part of the Botanical Garden of the MAS resources, can also contribute to solving issues in various interdisciplinary areas.

Of the 95 rare and endangered species, 39 are included in the list of 135 rare and endangered plants of the *Red Book of the Mongolian People's Republic* (Nyambayar et al., 2011; Shiirevdamba, 2013; Urgamal, 2017, 2018a, 2018b, 2018c; Urgamal et al., 2019a, 2019b), that is, about 28.9%, and 36 species are included in the special "Red List of Mongolia under the Plant Conservation Action Plan" (Nyambayar et al., 2011; Urgamal et al., 2016). A list of "very rare plants" cultivated in the Botanical Garden of the MAS includes 31 species and 27 genera in 18 families, according to the rarity category of the Red List of Mongolia, and 9 species and 8 genera in 6 families according to the IUCN classification (Table 7.2).

FIGURE 7.6 Examples of the introduced drought-resistant plants from the Gobi desert at the Botanic Garden of the Mongolian Academy of Sciences: (A) *Iris tenuifolia* Pall.; (B) *Saposhnikovia divaricata* (Turcz.) Schischk.; (C) *Lycium ruthenicum* Murray; (D) *Salsola laricifolia* Turcz. ex Litv; (E) *Nitraria sibirica* Pall.; (F) *Nanophyton erinaceum* (Pall.) Bunge.

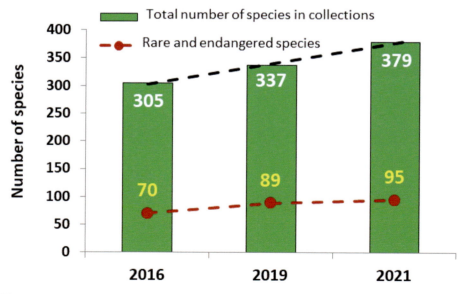

FIGURE 7.7 Dynamics of changes in living vascular plant collections of the Botanical Garden of the MAS in terms of the total number of species (green histogram and black dotted line) and the corresponding number of rare and endangered species (red dotted line) during 2016–2021.

Mongolian Botanical Gardens – Modern Plant Biodiversity

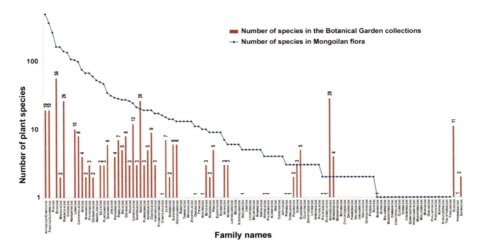

FIGURE 7.8 Species spectrum of vascular plant families in the entire collection of the Botanical Garden of the MAS (red bar chart) compared to the species composition of the Mongolian flora, ranked by frequency of occurrence (blue solid line).

The numbers on the histogram are the number of species in the corresponding families. The Y-axis is displayed in logarithmic coordinates for better visualization. The location of the bars above the solid line indicates the successful inclusion of additional alien species into the collection.

Sources: Urgamal et al. (2014, 2016, 2019b); Urgamal (2018a).

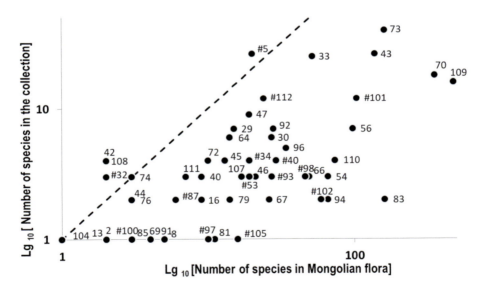

FIGURE 7.9 Comparison of the composition of the collection of living plants of the Botanical Garden and Research Institute of the Mongolian Academy of Sciences with the wild flora of Mongolia in terms of the number of species of vascular plants.

The numbers near the dots indicate the family index numbers according to the system (Urgamal et al., 2014, 2019b) based on the APG IV and LAPG IV classification systems (Angiosperm Phylogeny Group IV, 2016). The dashed line indicates the conditional boundary of the "ideal collection of aboriginal flora," when the number of native species in the collection is displayed to the right and below the dashed line, and in the limit along the dashed line, can coincide with the number of known species of Mongolian flora. The location of the dots to the left and above the dashed line indicates the successful inclusion of additional alien species introduced into the collection. Logarithmic scaling of the graph axes is used to better visualize the data.

TABLE 7.2
List of "Very Rare Plants" Cultivated in the Botanical Garden of the MAS

			Category of Rarity	
Species Accepted Name	**Family Name**	**Growth Form**	**Mongolia Red List Status**	**IUCN Status**
Acorus calamus L.	Acoraceae	H	VR	LC
Sambucus manshurica Kitag.	Adoxaceae	S	VR	-
Sambucus williamsii Hance	Adoxaceae	S	VR	-
Allium macrostemon Bunge	Amaryllidaceae	H	VR	LC
Allium obliquum L.	Amaryllidaceae	H	VR	-
Ferula ferulioides (Steud.) Korovin	Apiaceae	H	VR	
Cynanchum thesioides (Freyn) K. Schum.	Apocynaceae	H	VR	-
Anemarrhena asphodeloides Bunge	Asparagaceae	H	VR	
Convallaria keiskei Miq.	Asparagaceae	H	VR	
Saussurea involucrata Matsum. & Koidz.	Asteraceae	H	VR	-
Solidago virgaurea subsp. *dahurica* (Kitag.) Kitag.	Asteraceae	H	VR	LC
Sedum roseum (L.) Scop.	Asteraceae	H	VR	-
Rhododendron lapponicum (L.) Wahlenb.	Ericaceae	S	VR	
Caragana tibetica Kom.	Fabaceae	S	VR	
Sophora flavescens Aiton	Fabaceae	H	VR	LC
Lilium martagon L.	Liliaceae	H	VR	
Lilium pensylvanicum Ker Gawl.	Liliaceae	H	VR	
Tulipa uniflora (L.) Besser ex Baker	Liliaceae	H	VR	
Paris verticillata M. Bieb.	Melanthiaceae	H	VR	
Peganum harmala L.	Nitrariaceae	H	VR	
Cypripedium calceolus L.	Orchidaceae	H	VR	LC
Cypripedium macranthos Sw.	Orchidaceae	H	VR	LC
Neottianthe cucullata (L.) Schltr.	Orchidaceae	H	VR	EN
Orchis militaris L.	Orchidaceae	H	VR	LC
Paeonia lactiflora Pall.	Paeoniaceae	H	VR	-
Lancea tibetica Hook. f. & Thomson	Phrymaceae	H	VR	-
Aconitum kusnezoffii Rchb.	Ranunculaceae	H	VR	-
Adonis mongolica Simonov.	Ranunculaceae	H	VR	-
Trollius sajanensis Sipliv.	Ranunculaceae	H	VR	-
Rosa laxa Retz.	Rosaceae	S	VR	-
Saxifraga hirculus L.	Saxifragaceae	H	VR	LC

Abbreviations: H – Herbaceous; S – Shrub; VR – Very Rare; EN – Endangered; LC – Least Concern.

Sources: Table includes indication of rarity category according to the *Red Book* and the "Red List" of Mongolia (Nyambayar et al., 2011; Shiirevdamba, 2013; Urgamal, 2018a, 2018b, 2018c; Urgamal et al., 2019a) and according to the IUCN classification (IUCN, 2022).

In addition, the collection also contains 25 endemic species of Mongolia (6.8% of the total size of the garden collection), including 19 sub-endemic and 6 neo-endemic species.

As can be seen from Figure 7.7 and Figure 7.8, it is local Mongolian plants that make up the dominant basis of living collections, while alien species-introducers make up only about 4%. These indicators are in good agreement with the global goals of the international Convention on Biological Diversity of the United Nations (SCBD, 1992) and the goals of the international Global Strategy for

Plant Conservation (SCBD, 2002; BGCI, 2010), which sets the tasks for the professional community of the world's botanical gardens to conserve and manage a significantly greater diversity of plants (at least 70% of the local flora and rare plants) than any other branch of public production and nature management institutions.

7.4 CONCLUSIONS

7.4.1 Cross-Disciplinarity of Plant Conservation

Prominent botanists of the Botanical Garden of the Research Institute of MAS really make an important contribution to the fundamental and applied aspects of the conservation and rational use of plant biodiversity. For example, a solid result of this joint global work with the Institute of General and Experimental Biology of MAS, as well as with the National Mongolian University, the State Pedagogical University, was the investment of a lot of time and effort in fundamental research for the development and publication of three editions of the *Red Book of Mongolia* in 1987, 1997, and 2013 (Shiirevdamba, 2013) available for scientists, educators, and the general public. Particular merit in this achievement belongs primarily to those who can provide the most reliable information from their scrupulous field surveys and laboratory research and/or studies of *ex situ* cultivation of plants. However, such intense activity is usually not visible to ordinary citizens and the general public.

More recently, deputies of the Great People's Khural, municipal administrators, and corporations have become convinced of the importance and popularity of projects to create new, small botanical gardens, reconstruct a large academic garden, and establish a national plant gene bank in Mongolia (Orgil, 2019). An important role in this was played by review lectures and publications, which were convincing of the relevance and contingency of botanical gardens for the preservation of the national gene pool of plants.

About 600 species of rare and very rare useful plants are introduced in the Botanical Garden of the Institute of Botany of the Academy of Sciences; Gobi-Altai, Eastern, Bayankhongor, Khentii, Central aimags; and in the Khan-Ula region of Ulaanbaatar. Among them are *Astragalus borealimongolicus* Y.Z. Zhao, *Glycyrrhiza uralensis* Fisch., *Scutellaria baicalensis* Georgi, *Saposhnikovia divaricata* (Turcz.) Schischk., *Ephedra*, and other plants that are in demand for export and for which it is necessary to develop large-scale *ex situ* production to protect natural populations. Modern gene bank and cell culture technologies are also being developed for the propagation of rare and threatened plants of especially high demand, such as *Ferula ferulaeoides* (Steud.) Korov, *Zygophyllum potaninii* Maxim., and others (Suran et al., 2016; Bayarmaa et al., 2018).

Information on scientific and common Mongolian names, national distribution, legal status, and habitat information for a total of 148 rare species, as well as digital photos and distribution maps of endangered species, are available online. Based on such scientific materials, an action plan for the conservation of endangered species has also been developed, identifying the predominant threats and conservation measures to involve the public in biodiversity conservation. Experts involved in the development of Mongolia's Red List of Plants and Summary Action Plan helped to develop Mongolia's biodiversity database and legislation on protection of plant biodiversity during the last two decades. The online databases allow researchers to more easily collaborate and revise Mongolian taxa and will continue to improve the documentation of the Mongolian flora. To date, the GBIF database (Munkhnast et al., 2020; Shukherdorj et al., 2022) includes 2,249 taxa (about 73% of the Mongolian flora).

The staff of the Botanical Garden and the Research Institute of MAS, with their investigations in recent decades on plant identification, biodiversity dynamics, and distribution of rare and endangered species of Mongolia (Enkhtuyaa and Ochgerel, 2015; Urgamal and Bayarkhuu, 2019; Shukherdorj et al., 2022), make fundamental contributions to the Mongolian Biodiversity Strategy (Ministry of Environment and Tourism of Mongolia, 2019), to the implementation of the Global Strategy for

Plant Conservation (BGCI, 2010, 2019), and to Aichi Biodiversity Targets (SCBD, 2010). For these great purposes, the employees of the Botanical Garden and the Research Institute of MAS have also been involved in the creation of a strong legal framework and laws for the implementation of plans for the conservation of biodiversity and natural resources. As a result, more than 50 environmental laws and programs were developed in order to give legal provisions for biodiversity conservation and sustainable use under the leadership of the Ministry of Environment and Tourism and the Ministry of Education, Culture, Science, and Sports, and with the involvement of research materials from specialists from other institutions, universities, and non-governmental environmental organizations (Table 7.3). Laws such as the Law on Special Protected Areas, the Law on Environmental Protection, the Law on Natural Plants, the Law on Plant Protection, the Law on Forest, the Law on the Payment for the Use of Natural Resources, and the Law on Transparency of Public Information are especially important for the conservation of plant biodiversity, as are national programs, such as the Government Strategy to Protect, Restore *in situ*, and Cultivate Endangered Species of Animals and Plants 2012–2016 and the National Program on Natural Plant Conservation for 2013–2021. The Mongolian Parliament also approved the National Program for Special Protected Areas, which aims to increase the coverage of protected areas up to 30%, which is undoubtedly the most powerful way to create refuges for the sustainable support of plant biodiversity. Most of the country's zones have been classified as nature reserves and protected areas as a biodiversity conservation measure. The

TABLE 7.3

Some Examples of Key Environmental Laws and Programs in Mongolia

Years	Laws, Programs and Strategies
1994	Law on Special Protected Areas; National Programme on Special Protected Areas
1995	Law on Environmental Protection
1995	Law on Natural Plants
1995	Law on Payments for the use of Water, Forest, and Natural Resources
1995, 2002	Law on Land
1995, 2007, 2012	Law on Forest; State Policy on Forests
1996	National Biodiversity Strategic Action Plan; National Programme on Biodiversity
1995, 1996, 2007	Law on Plant Protection
1997	State Policy on Ecology
1997	Law on Special Protected Area Buffer Zones
2002	National Programme on Protection and Sustainable Use of Rare Plants in Mongolia
2003	National Programme for Combatting Desertification
2005	"Green Belt" National Programme
2012	Law on Wildlife
2012	Law on Environmental Impact Assessment
2012	Government Strategy to Protect, Restore *in situ*, and Cultivate Endangered Species of Animals and Plants (2012–2016)
2012	Law on Soil Protection and Prevention of Desertification
2012	Law on the Payment for the Use of Natural Resources
2013	National Programme on Natural Plant Conservation for 2013–2021
2015	National Biodiversity Strategic Action Plan for Mongolia 2015–2025
2017	Health and Environment National Program
2017	National Programme "Environmental Health"
2021	Law on Transparency of Public Information

Note: These laws and programs required the knowledge, expertise, and scientific results of staff members of the Mongolian Botanical Garden and Research Institute of Mongolian Academy of Sciences.

plants of Mongolia are experiencing natural and anthropogenic threats, such as spring fires after dry winters, overgrazing, desertification in the southern part of the country, and air pollution. Without the understanding and support of the population, it is impossible to overcome the threats to environmental security and the risk of biodiversity loss.

However, in the rapidly changing conditions of socio-economic development in the country, the laws and programs can no longer fully ensure the preservation of environmental conditions that ensure the sustainable reproduction of biological resources. Therefore, despite the theoretical qualification of all aspects and technologies, the work on the restoration of rare and endangered plants is not yet being carried out in full, although research and work is being carried out to restore steppe landscapes and forests, and specially protected natural areas are being created (Regdel et al., 2012).

Previous checklists of vascular plants in Mongolia have included both native and non-native taxa; however, there have been few dedicated studies on non-native taxa, especially invasive species. Recently, reviews and reports on invasive species of Siberia, the Russian Far East, and Mongolia have been published (Vinogradova et al., 2020), aimed at raising public awareness of invasive species and strengthening the rules for their biosafety.

Undoubtedly, the strongest role of the Mongolian Botanical Garden is precisely the scientific influence of the largest *ex situ* collection of various plants on the public ecological consciousness of the population and officials, and especially on young people (schoolchildren, university students, young entrepreneurs, etc.). About 2,000 schoolchildren from nearby schools began to visit the academic Botanical Garden every day during the growing season. An additional subject, "Ecology," has been introduced in schools' curriculum. On TV and radio, at least once every other week, special environmental programs in the Mongolian language are broadcast, including programs and videos about botanical gardens, plants, and the importance of biodiversity in people's daily lives. The Botanical Garden and Research Institute of MAS shares its publications with the city's libraries. Popular botanical and environmental literature has become widely available to rural residents in the format of regular books and also in digital format via the Internet. For example, scientists from the Botanical Garden and the MAS Research Institute have published many highly requested scientific-popular publications on *Red Book* and Red List of rare and endangered plants in recent years.

However, it is important to note that a certain gap in communication between scientists and the local population still exists due to some lack of scientific and popular science literature in the Mongolian language on the importance of biodiversity, as scientific publications in English are traditionally used to evaluate scientific productivity of scientists. Therefore, it is necessary to develop the right decisions in this matter, which must be consistent with the Aichi Targets, so that people are well-informed about the value of biodiversity and about the measures that they can take for its conservation and sustainable development, since use and values of biodiversity are incorporated into national and local development strategies for the improvement of human well-being (SCBD, 2010). As we can see at present, the Mongolian botanical scientists managed to "infect" some competent entrepreneurs with mutually beneficial ideas of the "botanical garden," or the idea of restoring natural ecosystems and reforestation. The townspeople used to litter and pollute the botanical garden land, and now they appreciate and care for it since the scientists of the garden, with their devotion to science and the conservation of plant biodiversity, set a worthy example and managed to competently convince and motivate a critical percentage of the population about the usefulness of the botanical garden for everyone.

The combination of tangible and intangible resources in the unity of the accumulated largest collections and demonstration objects of the botanical garden (Kuzevanov and Sizykh, 2006; Gorbunov and Kuzevanov, 2022), as well as the sharing of traditional indigenous and scientific knowledge, can have a strong ideological impact on the interests of people and decision makers. At the same time, intangible components of biodiversity conservation obviously have the same huge significance as material resources. Perhaps the garden's intangible resources may have a much more powerful impact on people and society in the long run. The staff of the Botanical Garden and Research Institute of MAS, being part of the prestigious national Mongolian Academy of Sciences, along with other scientific institutions and universities, have had a scientific influence and made an important contribution to the formation of environmental legislation in Mongolia through the change of public consciousness. The goals of the GSPC are used as the most relevant guiding principles for the participation of botanical gardens, including national and foreign collaborating gardens, in the tasks of studying, conserving, and managing the plant biodiversity of Mongolia together with the local population.

The work on biodiversity conservation "has a beginning, but there is no end, no boundaries," as humanity will have to constantly care about maintaining the ecological balance on Earth. We understand that biodiversity conservation can never be completed; it must be carried out continuously. The loss of each collection plant is an irretrievable loss of a valuable tangible resource that can only be replenished by bringing another plant from outside or growing from seeds, cuttings, and cell cultures. Cultivated plants are an expensive material resource, so the holding of each plant in the collection should have well-defined goals for conservation and rational use. The number of plants is actually limited by the boundary of the fence of the territory of the botanical garden, therefore, naturally, it cannot be unlimited since it is a "limited resource." The amount of available material resources and the size of *ex situ* plant collections are usually physically limited by the size of the botanical garden. That is, tangibles are limited resources, while intangibles (knowledge, technology, stories, etc.) have no limits in their growth and development. At the same time, intangibles can be associated with each plant specimen or species, directly related to their conservation and use both inside the garden and far beyond its borders. People are intellectual generators and carriers of intangibles. Unlike limited tangibles, the intangibles are actually "unlimited resources" that can be replicated and distributed without loss of quality and content (Kuzevanov and Sizykh, 2006). Therefore, within the framework of the activities of employees in botanical gardens, it is intangible resources based on scientific knowledge that are becoming more and more important principle factors influencing the ideology and processes of conservation and rational use of plant biodiversity, which go far beyond the borders of the garden.

It is well known that the global network of botanical gardens of the world contains in *ex situ* collections at least 30% of the 350,699 known species of living plants (including 41% of all endangered species and 2% of endangered species included in renewal and restoration), nature reserves, representing almost two-thirds of genera and more than 90% of plant families (Mounce et al., 2017; Urgamal, 2018a, 2018b, 2018c; Westwood et al., 2020).

Thus, the achieved indicators of the accumulation of plant biodiversity in the collection of the Botanical Garden of the MAS can be recognized as a pretty successful contribution fully consistent with the 5 goals and 16 tasks of national and international development set by the United Nations in the Global Strategy for Plant Conservation (BGCI, 2010, 2019).

The Botanical Academy of MAS in Ulaanbaatar, as a result of its fundamental scientific works in creating unique collections of plants and applied research, has obtained a well-deserved position in the community of botanical gardens of the world and joined the Botanic Gardens Conservation International (BGCI, 2020). This happened due to the fact that the attitude toward "the idea of a botanical garden" among the country's leadership, population, and scientific and business circles of Mongolia underwent cardinal positive changes during the last 50 years, particularly in the early twenty-first century, in the course of ongoing global shifts in the restructuring of the nomadic and sedentary lifestyle of communities and the acceleration of socio-economic development. To comprehend why botanical gardens are needed, people come from different fields of activity, from different disciplines, driven at first by pragmatic personal motivations and interests. A profound understanding that the botanical garden is not just about "botany" but a much larger cross-disciplinary resource than a simple garden of vegetables, fruits, and herbs appeared in Mongolia precisely after the acceleration of urban civilization development. It became clear even to the "root-and-grass" people on the streets that botanical gardens are useful ecological resources that preserve the natural and cultural heritage for humans and are potentially ways to bring the richness of plant diversity to the streets and courtyards. It also became clear that scientists of the academic Botanical Garden have the knowledge to study, advise, and teach how to preserve forests and pastures for the well-being of a country that is very dependent on livestock and natural resources. This explains exactly how the interest of the community made it possible to basically preserve the botanical garden and stop the "unstoppable seizures of land" and the reduction of the territory and collections. Despite the past serious backlog to the "demographic cross" of the 1970s (see Figure 7.1), at the beginning of the twenty-first century, intensive outstripping development began in the creation of botanical gardens, zoos, and nature museums of various departmental and regional subordination, especially in the last decade. In many ways, this is obviously due to the global trend of revising the role of botanical gardens in modern Mongolian society, namely, with the expansion of their functions from a predominantly narrow, scientific botanical role to a broader socio-ecological role as unique, specialized, multifunctional organizations in the national system of division of labor and sustainable development trends. Therefore, it is no coincidence that, in addition to the Academy of Sciences and the highest state structures of the central government of Mongolia, local authorities at the level of aimags, private corporations, and large industrial enterprises in Ulaanbaatar, Erdene, and Khovd (Figure 7.2) began to independently establish new botanical gardens and their combinations with zoos (Figure 7.1). For example, it is very symbolic that the local authorities in the city of Khovd, as an argument for the initiative to create a new botanical garden with the support of the Russian Botanical Garden of Altai State University, proclaimed its priority environmental role as a factor in maintaining a healthy environment and ecological balance (Batimeg and Lhagwa, 2018), which is exactly in line with the new the integral role of the modern concept of botanical gardens as "ecological resources" (Kuzevanov and Nikulina, 2016). For example, while creating the botanical garden "Michelle" by the private corporation Michel Ltd., as well as a private mini-zoo in Ulaanbaatar and the municipal zoo–botanical garden in Erdene, the main priority is the rational use of natural resources and introduction of high technologies and education for local residents, meeting the growing demands of the local population for "green" contacts with nature. In particular, the botanical and ecological resources of the gardens will be simultaneously used for recreational purposes of human development, raising public awareness and entertainment, landscaping and improvement of the ecological environment of urban areas in large settlements and cities, and restoration of disturbed populations and landscapes in nature. In fact, the key role of modern botanical garden turns out to be largely related to a science-based formation of a healthy and safe environment in the urban area through the performance of four main interdisciplinary functions (Figure 7.10): (1) scientific research and development (R&D), (2) nature protection, (3) education and enlightenment of the people, and (4) commercialization of innovations. In order to create and run a botanical garden, it is no longer enough to know botany and horticulture, but it is required to have sufficient knowledge in a wide range of scientific disciplines. This vector model of the main functions of the botanical garden as a cross-disciplinary intermediary between nature and society (Figure 7.10), describing the diversity of functions and flows of material, intangible resources, and areas of application, shows how the influence of the botanical garden can be organized to have a positive impact simultaneously on nature (biodiversity) and on a market economy society. It is proved that the main goal of the tangible and intangible resources of a modern botanical garden is the use of its scientific plant collections, demonstration plots, and nurseries in order to support the vital activity of natural systems necessary for the sustainable development of cities and for environmental and social security in the country and worldwide to improve the environment and well-being of people. In the context of the economic crisis and climate change, each botanical garden, as an ecological resource, is able to substantially contribute to socio-economic development and also act as an anti-crisis tool in the transition to the "sixth technological order" (Kuzevanov and Gubiy, 2014; Gorbunov and Kuzevanov, 2022):

1. Provide people with economically significant plant resources to overcome hunger and poverty.
2. Contribute to poverty reduction, and provide knowledge and survival skills through the introduction of environmental innovations, technologies, and new demanded plants.

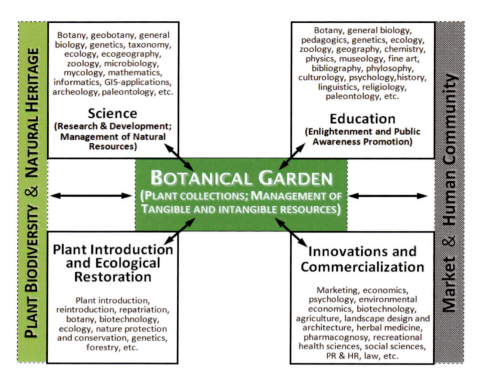

FIGURE 7.10 The multi-vector model of the modern botanical garden as a cross-disciplinary intermediary between nature and society. The botanical garden is a socio-ecological actor that brings together various scientific disciplines and humanitarian activities associated with four main interdisciplinary functions: (1) science, (2) education and enlightenment, (3) plant introduction and eco-restoration, and (4) innovations and commercialization. The arrows indicate direct and feedback links of the botanical garden as an intermediary in the circulation of material (tangible) and non-material (intangible) resources between nature and society.

Source: The scheme has been supplemented and modernized based on an original model (Kuzevanov and Sizykh, 2010; Kuzevanov et al., 2021a).

3. Contribute to overcoming environmental illiteracy in nature management through the dissemination of knowledge and best practices, training, and education.
4. Promote an environmentally friendly and healthy lifestyle in cities.
5. Serve as one of the most inexpensive tools of social adaptation and rehabilitation of people with special needs (the disabled, the infirm, children, the elderly) through the introduction of horticultural therapy into practice.
6. Promote the development of various forms of innovative "green business" and the creation of new jobs.
7. Create buffer systems based on plants of economic and ecological importance to neutralize the adverse effects of climate change, extreme conditions, and shortage of food resources.
8. Promote mutually beneficial international relations through the exchange of knowledge and genetic resources of economically valuable plants, seeds, and technologies, including support in the creation of botanical gardens.

Despite the fact that people who are historically accustomed to a nomadic lifestyle have to painfully change their worldview, in the same context of changes, there are also changes in the attitude toward the "botanical garden idea" in Mongolia on the part of the population and administrative decision makers. Therefore, we can positively assess the intentions of the local authorities of the

Khovd aimag, which began in 2018, to create another new Mongolian botanical garden in the Buyant River valley on an area of 6 hectares of land belonging to the Khovd State University in the city of Khovd, the oldest and fastest growing administrative, economic, and cultural center with a population of 34,100 people in the western part of the country (Batimeg and Lhagwa, 2018). Scientists of the Altai State University, headed by the director of the Botanical Garden, Professor A.I. Shmakov, have been laying the foundation of this new Mongolian Botanical Garden since 16 September 2018, and they are going to provide further assistance and scientific support for it (Kozerlyga, 2018).

During the development of the new botanical garden, about 100 species of plants were planted, including decorative woody and fruit and berry plants. The garden is expected to primarily feature introduced plants from arid habitats, with particular attention to the indigenous flora of western Mongolia, which is characterized by harsh climate, drought, and desertification. From the very beginning of the design, it was planned to build an infrastructural road and path network, an artificial water reservoir, two greenhouses, an alpine hill, an arboretum, landing sites for trees and shrubs, and collections of medicinal plants and herbaceous perennials. In the future, it is planned to conduct joint practical classes for Mongolian, Russian, and other foreign students, as well as the implementation of term papers and theses. It is expected that such public gardens will become an economically important and innovative technological tool for the market economy for the development of tourism through excursions, as well as for the large-scale production and plant sales of valuable plants, growing seedlings for landscaping the city of Khovd (Batimeg and Lhagwa, 2018). The City Council of Khovd also formulated the main botanical garden's role as a valuable instrument for maintaining the balance of a healthy ecological environment, which corresponds to the new definition of the contemporary role of the world's botanical gardens as "ecological resources" (Kuzevanov and Nikulina, 2016). The initiators explicitly affirm that they expect the establishment of the University Botanical Garden in the city of Khovd to help reduce desertification and maintain the ecological balance in the western part of Mongolia.

Similarly, in accordance with the multi-vector model in Figure 7.10, the main stated goal is to create the first private botanical garden "Misheel" (land area 11 hectares), which the company Misheel Group opened for visitors on the island site in Ulaanbaatar in 2021. The goal of this site is the conservation and restoration of rare species of native trees, shrubs, and herbaceous plants, as well as the development of an environmental education facility for schoolchildren (Anonymous, 2017, 2020; Orgil, 2019). This botanical garden is being implemented as part of a public–private partnership between the Ulaanbaatar Mayor's Office and the private company Misheel Group. While designing and planting a garden on a separate island territory, natural paths and trails were taken into consideration, and growing trees were carefully preserved. The garden was designed by Japanese landscape architect Dr. Ogata Motomu of Chiba University in Japan.

In the course of establishing the botanical gardens of Mongolia, their plant collections are mainly focused on utilitarian purposes (decorative, aesthetic, or economic). However, in the last decade, new large-scale ecological challenges have arisen for botanical gardens: (1) the introduction of various species for the purposes of ecological restoration, (2) participation in landscaping to improve the comfort and environmental safety of the urban environment, and (3) education, public awareness promotion, and enlightenment of the local population and tourists.

7.4.2 Leading Mission

An important leading mission of the Botanical Garden of the MAS is to assist other botanical gardens and plant nurseries in Mongolia in developing their knowledge and collections. Obviously, the Academy of Sciences and the university system are becoming increasingly important factors in the conservation of nature and the development of society in Mongolia. For example, Figure 7.11 shows that over the past 50 years in Mongolia, along with a fairly smooth demographic growth (compare with Figure 7.1), there has been a steady exponential increase in the interest of the public to the "botanical garden" theme (identified with Google Search Engine Trends). A similar increase in

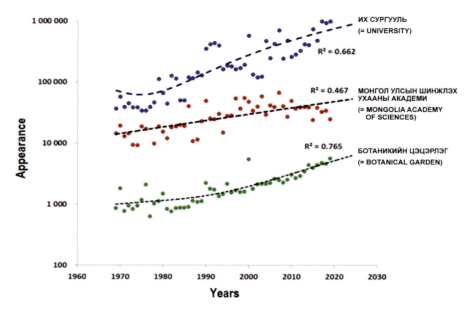

FIGURE 7.11 Long-term trends in the frequency of annual occurrence (mentions) of Mongolian words and phrases "БОТАНИКИЙН ЦЭЦЭРЛЭГ" (botanical garden), "МОНГОЛ УЛСЫН ШИНЖЛЭХ УХААНЫ АКАДЕМИ" (Mongolia Academy of Sciences), and "ИХ СУРГУУЛЬ" (university) on the Internet.

Source: Identified using the Google Search Engine Trends (https://trends.google.com/trends/).

general interest is characteristic of two other topics closely related to education and science-intensive subjects – "university" and "Mongolia Academy of Sciences" (Figure 7.11). This phenomenon in recent years also shows the growing demand of the Mongolian people to improve the education and scientific foundations of their knowledge of the natural world (Kuzevanov et al., 2021b).

The managers of the Botanical Garden of the MAS set the following specific goals, which must be achieved by solving a number of issues of material resources and intangible assets, including the problems of investment and financing to improve the functioning of the Botanical Garden in the field of plant biodiversity conservation:

1. Development and dissemination of technologies for cultivation and introduction of various economically significant and rare plants
2. Formation and creation of a plant gene bank in *ex situ* conditions for the purpose of technological and successful cultivation and *ex situ* conservation of at least 70% of potentially endangered plant species in accordance with the provisions of the Global Strategy for Plant Conservation
3. Improving the database and knowledge base about plants, including information on the conservation and protection of rare and endangered medicinal and ornamental plants
4. Expansion of the variety of demonstration collections and special exhibition areas, including a stock tropical greenhouse and an artificial lake with an area of about 1.2 hectares
5. Collecting herbarium specimens and seeds for exchange, as well as widely informing about the results of studies of introduced plants, including in printed or electronic form and in a series of journals
6. Carrying out regular cultural and educational events for various age groups and segments of the population in the field of botany, dendrology, ecology, nature conservation,

introduction, breeding, ecological restoration, ornamental plantings, and landscape architecture
7. Development and promotion of a system of continuous learning/education and cognitive testing

For example, in 2019, deputies of the People's Great Khural (Parliament) of Mongolia put forward the idea of creating an ecological research center and a special gene bank on the basis of the Botanical Garden of the MAS for the conservation and protection of plants of the Mongolian flora. At the same time, Mongolia turned to the Government of the Irkutsk Oblast with a request for assistance in the development of the academic botanical garden in Ulaanbaatar and in restoring the gene fund of endemic, rare, and useful plants of the country (Anonymous, 2017, 2020; Orgil, 2019). As the deputies of the People's Great Khural of Mongolia noted, the rich experience of the neighboring region in Baikal Siberia, where several botanical gardens and dendrological parks are located (e.g., Botanical Gardens of the Irkutsk State University, Botanical Gardens of the Agrarian State University, Dendrological Park of the Baikal Museum of the Russian Academy of Sciences, the municipal Tomson's Garden, etc.) could certainly breathe new life into the academic botanical garden in Ulaanbaatar and other similar organizations in the country. Since many botanical gardens are considered as special scientific and educational objects and nature conservation resources with high social and environmental potential (Andreev et al., 2005; Adonina et al., 2006; Golding et al., 2010; Prokhorov, 2018; Soltani and Annenkova, 2020), the Botanical Garden of MAS is considered a "unique object" in Mongolia, the mission of which, as a public environmental and scientific institution, is not only to study and grow economically valuable, rare, and endangered plant species but also to promote the ecological ideas of environmental restoration, sustainable development, and conservation of nature for the sake of the well-being of present and future generations, increasing the country's competitiveness. Therefore, it is quite logical that the Ministry of Education, Culture, and Science of Mongolia announced the beginning of the development of special measures for botanic gardens' development as urban "green zones," sites for recreational and educational activities for citizens, and the placement of a modern interdisciplinary "Innovation Cluster" (Figure 7.4) for a private–state partnership under the Mongolian Academy of Sciences sponsorship.

At present, a small staff of employees is putting all their enthusiasm and increasing efforts into the restoration, rehabilitation, and conservation of ornamental, food, medicinal, rare, and endangered plants based on fundamental and applied scientific research and innovations of different research institutes of the Mongolian Academy of Sciences. Mongolia's interest in such cooperation in the field of science and ecology is also related to the protection of the water resources of Lake Baikal as a UNESCO World Heritage Site linked through the river Selenga with the largest Mongolian lake Khuvsgul (Pavlov et al., 2004). Obviously, according to task No. 13 of the "Global Strategy for Plant Conservation" (BGCI, 2010, 2019), the national network of botanical gardens of Mongolia, under the auspices of the Botanical Garden of the Academy of Sciences and other academic institutions and universities of the country, undoubtedly, are able to make a decisive contribution to "stopping the depletion of plant resources and the loss of related knowledge of indigenous communities, maintaining sustainable livelihood practices, and improving human health and food security at the local level." It is assumed that such international cooperation and mutually beneficial scientific and technical exchange between neighboring territories and countries (Oyuungerel et al., 2011; Osodoev et al., 2014), as well as participation in the organization and management of fundamental and applied scientific research and experiments in close connection with production development, environmental management, and rational consumption, will contribute to the achievement of significant benefits in certain socio-economic areas and improve the well-being of the population of Mongolia in the process of scientific and technological revolution during the transition to the "sixth technological order."

7.4.3 CONSERVATION AND THE PYRAMID OF THE NATURE MANAGEMENT SYSTEM

The botanical gardens of Mongolia are becoming exceptional places for the accumulation and conservation of the plant genetic resources of the country and adjacent territories, as well as sites for the concentration of various environmental resources, including zoo–botanical collections, where potentially profitable marketing technologies are used. They will serve as multifunctional scientific, educational, and socio-ecological factors to protect the environment and improve human well-being in a region with extreme climatic conditions. It is also obvious that focusing exclusively on the rather narrow task of plant conservation is no longer enough for such a multifunctional institution as a botanical garden. Botanical gardens can greatly strengthen their conservation mission if they manage to integrate into the nature management pyramid (Figure 7.12), where their functions are in demand on all floors of this pyramid (Table 7.4). Historical experience shows that botanical gardens can fully realize their potential only with solid support from the local population, that is, when people feel the obvious and immediate usefulness of botanical gardens for them and for all strata of the society. The idea of participating in the observation and conservation of plant biodiversity so well motivated residents of all ages that volunteers provided more than 20,000 images of 1,780 taxa over the course of three years as part of the citizen science contribution to the "Flora of Mongolia" project on the iNaturalist platform (Ueda et al., 2019).

The forecast for the next 20–25 years, as well as the understanding of botanical gardens as interdisciplinary ecological resources in connection with demography, economics, and ecology, should undoubtedly contribute to their advanced innovative development and modernization in the first half of the twenty-first century (Kuzevanov and Gubiy, 2014; Kuzevanov, 2016). This will

TABLE 7.4

The Key Functions of Botanical Gardens in the Pyramid of Nature Management (see Figure 7.12)

Participation in the creation and preservation of the natural and cultural heritage of mankind

Education of students, school children, and the population in the field of botany and ecology and related disciplines

Dissemination of scientific knowledge and technologies of environmental management

Greening of cities and rural settlements

Participation in the design of parks and public spaces, a healthy and comfortable urban environment

Social adaptation, rehabilitation, and education by methods of horticultural therapy

Provision of services, consulting in the field of botany, horticulture, ecology, etc.

Experimental commercial production of high-quality planting material

Research and development for processing of plant residues and wastes (composting, etc.)

Research and development to create ecological and botanical innovations (new varieties, forms, methods, etc.) and technologies based on plants and their derivatives

Conservation of rare/endangered plants in *ex situ* living collections and seed banks

Creation and maintenance of large collections of living plants and artifacts

Research on the introduction of native and foreign plants in culture

Reintroduction and repatriation of plants *in situ*

Ecological restoration of disturbed natural/industrial landscapes

Ecological restoration of disturbed plant populations

Reforestation

Participation in conservation of biodiversity of plants, animals, fungi, and microorganisms

Participation on environmental monitoring and maintaining the regime of strictly protected natural areas

Participation in the development of decisions and legislation on environmental management

Research and development of databases and knowledge bases about plants and collections

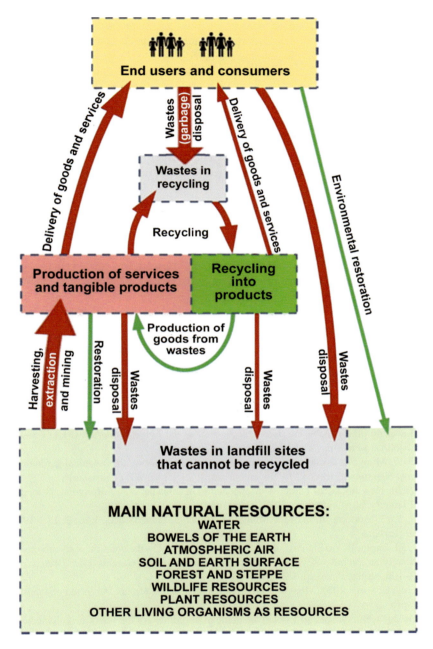

FIGURE 7.12 The pyramid of the nature management system, where botanical gardens are able to participate at various stages of the division of labor in the processes of production and implementation of services, material products/goods, as well as in the processing of wastes, the restoration of natural resources, and plant biodiversity conservation.

Source: Modified from Kuzevanov (2017). See Table 7.2 for details.

better unlock the potential of botanical gardens and increase their contribution to socio-economic development, division of labor, and environmental management (Figure 7.12) during the scientific and technological growth and development, simultaneously with the involvement of more countries and regions participating in the "race" to create multidisciplinary botanical gardens for balanced urban development.

ACKNOWLEDGMENTS

The authors would like to thank Dr. David A. Galbraith of the Royal Botanic Gardens (Hamilton, Ontario, Canada) and Prof. Pullaiah Thammineni of Sri Krishnadevaraya University (Anantapur, Andhra Pradesh, India) for their great contribution to the initiation and preparation of this chapter.

The authors express their cordial gratitude to colleagues from Russia, South Korea, and Mongolia for their support in carrying out the work, for their help in systematizing materials, and for valuable discussions on modern problems of the botanical gardens: Dr. A. A. Prokhorov, Dr. Yu. N. Gorbunov, Dr. E. N. Kuzevanova, Dr. E.V. Gubiy, Professor G. Ochirbat, Dr. E. Ochirbat, Dr. Ts. Tsedenbal, Professor N. A. Nikulina, and Professor Yong-Shik Kim. The work was carried out in accordance with the budget theme of the project for the development of the Botanical Garden of the MAS.

REFERENCES

Adonina, N.P., S.V. Aparin, M.N. Ber, K.N. Bochkareva, N.S. Danilova, et al. 2006. Botanical gardens and dendrological parks of higher educational institutions of the Federal Agency for Education of the Ministry of Education and Science of the Russian Federation. *Hortus Botanicus* 3: 28–104. https://clck.ru/rn4T3 [Russian]

Andreev, L.N., M.N. Ber, A.A. Egorov, R.V. Kamelin, E.A. Lurie et al. 2005. Botanical gardens and dendrological parks of higher educational institutions. *Hortus Botanicus* 3: 5–27. https://clck.ru/rn4nS [Russian]

Angiosperm Phylogeny Group IV. 2016. Angiosperm Phylogeny Group classification for the orders and families of flowering plants: APG IV. *Botanical Journal of the Linnean Society* 181 (1): 1–20. https://doi.org/10.15468/fzuaam https://clck.ru/sKGsP

Anonymous. 2017. Mongolia asks the Angara region for help in restoring the botanical garden in Ulaanbaatar. *Oblastnaya Newspaper*, Irkutsk, 14 November. https://clck.ru/rsWZY [Russian]

Anonymous. 2020. New botanical garden will appear in Ulaanbaatar. *News Mongolia*, Buryatia, Kalmykia, Tyva. 12 January. https://clck.ru/rnExC [Mongolian]

Badaraev, D.D. 2013. Modernization of the Mongolian society: Social aspects. *Vlast* 2: 188–191. https://clck.ru/rnAG9 [Russian]

Banzragch, D., P.I. Lapin, B. Ya. Sigalov and N. Ulziyhutag. 1978. On the creation of the first botanical garden in the Mongolian People's Republic. *Bulletin of Main Botanical Garden of USSR Academy of Science* 107: 102–106. https://clck.ru/rnBEJ [Russian]

Barrientos, M. and C. Soria. 2021. Mongolia-urban population. *Index Mundi*. Published on the Internet. https://clck.ru/rspJb

Bashkuev, V. Yu. 2014. Buddhism and Tibetan medicine in the context of the scientific and medical study of the Mongolian peoples (1920s–1930s). *Vestnik of Buryat Scientific Center of the SB RAS* 4 (16): 276–281. https://clck.ru/rnCrN [Russian]

Batimeg, N. and N. Lhagwa. 2018. The foundation of the botanical garden was laid on the basis of the Khovd University internship. *Khovd News* (KhovdNew.MN), 20 September. https://clck.ru/rnCKJ [Mongolian]

Bayarmaa, G.-A., N.N. Lee, H.D. Kang, B. Oyuntsetseg and H.K. Moon. 2018. Micropropagation of the Mongolian medicinal plant *Zygophyllum potaninii* via somatic embryogenesis. *Plant Biotechnology Reports* 12: 187–194. https://doi.org/10.1007/s11816-018-0484-9

Bazaar, B. 2008. The socio-economic development of Mongolia in the transition period. *Izvestiya of Baikal State University* 3(59): 102–104. https://clck.ru/rn9oc [Russian]

Belozertseva, I.A., D.D. Dorzhgotov, D.D. Enkhtaivan and A.A. Sorokova. 2015. Soil-ecological zoning of the transboundary territory of Russia and Mongolia. *Modern Problems of Science and Education* 2–2: 737. https://clck.ru/rnDmn

BGCI. 2010. Global Strategy for Plant Conservation 2011–2020. BGCI, Richmond, p. 40. https://clck.ru/rsmqG

BGCI. 2019. Declaration of the World Forum on the Global Strategy for Plant Conservation (GSPC). Dujiangyan, Sichuan Province, China. https://clck.ru/sJndb

BGCI. 2020. The Mongolia Garden Search. Botanic Gardens Conservation International, London. https://clck.ru/rsqW3

BGCI. 2022. The Garden Search of the Botanic Gardens Conservation International. London. https://tools.bgci.org/garden_search.php

Consortium World Flora Online. 2013. The Plant List. Version 1.1. Published on the Internet. www.theplantlist.org/

Consortium World Flora Online. 2020. Published online. Project of the Consortium World Flora Online. www.worldfloraonline.org

Davaazhav, B. 2017. The development of agriculture in Mongolia in the twentieth century. *Izvestiya of Altai State University* 5(97): 125–128. https://clck.ru/rnKUa [Russian]

Dodd, J. and C. Jones. 2010. Redefining the Role of Botanic Gardens: Towards a New Social Purpose. BGCI, Richmond, p. 143. https://figshare.com/ndownloader/files/18290612

Dorofeyuk, N.I. and P.D. Gunin. 2000. Bibliographic index of literature based on the research results of the Joint Russian-Mongolian complex biological expedition of the Russian Academy of Sciences and the Mongolian Academy of Sciences (1967–1995)]. Proceedings of RMKBE, Bioinformservice, Moscow, Vol. 41, p. 386. [Russian]

Dulov, A.V. 1995. Historical and geographical aspects of the use of the natural resources of Siberia in the 17th and early 20th centuries. Geography Research of Asian Russia: History and Modernity. ISU Publ, House, Irkutsk, 103–106. [Russian]

Elokhina, Yu. V. and I.B. Oleinikov. 2012. Baikalian Siberia in the policy of regional development: Resources and opportunities. *Izvestya of Irkutsk State University (Political Science and Religious Studies)* 1(8): 67–74. https://clck.ru/rnMLd [Russian]

Enkhtuyaa, L. and N. Ochgerel. 2008. Botanical gardens of the world. *Proceedings of the Mongolian Academy of Sciences* 1: 6–7. [Mongolian]

Enkhtuyaa, L. and N. Ochgerel. 2015. Collections of rare herbaceous plants of the Botanical Garden of the Institute of Botany of the Mongolia Academy of Sciences. Ecosystems of Central Asia in Modern Conditions of Socio-Economic Development: Proceedings of the International Conference, 8–11 September, Ulaanbaatar (Mongolia), Vol. 1, pp. 103–106. [Mongolian]

Erdenejav, G. 2009. *A Brief History of Botanical Science in Mongolia*. Mongolian Academy of Science, Ulaanbaatar, p. 232. [Mongolian]

Golding, J., S. Guzewell, H. Kreft, V. Ya. Kuzevanov, S. Lehvavirta, I. Parmentier and M. Poutasso. 2010. Species-richness patterns of the living collections of the world's botanic gardens: A matter of socio-economics? *Annals of Botany* 105(5): 689–696. DOI: 10.1093/aob/mcq043 https://clck.ru/rsnm3

Gombobaatar, S., K. Natan, M. Urgamal, K. Terbish and C. Gantigma. 2018. *Biodiversity of Mongolia: Chapter 11. Global Biodiversity: Selected Asian Countries*, 1st ed., Vol. 1. Apple Academic Press, Waretown, p. 44. DOI: 10.1201/9780429487743-11

Gorbunov, Yu. N. and V. Ya. Kuzevanov. 2023. Chapter 4: The role of Russian botanical gardens in plant biodiversity conservation. In: Pullaiah, T., and D. Galbraith (eds.), *Botanical Gardens and Their Role in Plant Conservation*, Vol. 3. CRC Press, Canada, pp. 63–89.

IUCN. 2022. The IUCN Red List of Threatened Species. Version 2021–3. https://clck.ru/sKbCK

Kalyuzhnova, N. Ya. and V. Ya. Kuzevanov. 2010. The role of the environmental factor in the competitiveness of the region. *Economics of the Region* 3(23): 54–62. https://clck.ru/rnMjo [Russian]

Karpini, D. 1957. History of the Mongols. State Publishing House of Geography Literature, Moscow, p. 125. https://clck.ru/rnNTY [Russian]

Kozerlyga, A.G. 2018. Altai State University scientists took part in the laying of the first botanical garden in Mongolia. *Altai State University News*, 10 May. https://clck.ru/rnSUp [Russian]

Kubrikova, Yu.A. 2015. Economic development strategy of Mongolia in the conditions of productive growth. *European Journal of Social Sciences* 1–2: 17–23. https://clck.ru/rnTcK [Russian]

Kuzevanov, V.Ya. 2010. Botanic gardens as ecological resources for the civilization development. *Botanical Gardens: Problems of Introductions (Series Biology)*. Tomsk State University Publishing House, Tomsk, pp. 218–220. https://clck.ru/rnvGb [Russian]

Kuzevanov, V.Ya. 2016. Long-term trends in the dynamics of the world's botanic gardens. *Bulletin of Irkutsk State Agricultural Academy* 72: 62–71. https://clck.ru/sLSMs

Kuzevanov, V.Ya. 2017. Contemporary botanic gardens as eco-social resources of urban development: From the idea of "garden in the city" to the "city in the garden". Modern Problems of Education and Science: Proceedings of the Intern, Scientific-Practical, Conference, 17 March, ISU Publishing House, Irkutsk, pp. 174–180. https://clck.ru/rujy6 [Russian]

Kuzevanov, V.Ya., L. Enkhtuya and N. Ochgerel. 2021a. Academic botanic garden as an ecological resource for the socio-economic and sustainable development of Mongolia. *Hortus Botanicus* 16: 3–29. DOI: 10.15393/j4.art.2021.7685. https://clck.ru/rnQyn

Kuzevanov, V.Ya. and E.V. Gubiy. 2014. Botanical gardens as world ecological resources for innovative and technological development. *Izvestiya of Irkutsk State University (Series Biology. Ecology)* 10: 73–81. https://clck.ru/rspu5

Kuzevanov, V.Ya. and N.A. Nikulina. 2016. On the definition of the term "ecological resources". *Bulletin of KrasGAU* 5(116): 77–83. https://clck.ru/rnv6o [Russian]

Kuzevanov, V.Ya., G. Ochirbat, L. Enkhtuya and N. Ochgerel. 2021b. Botanic gardens of Mongolia: New scientific, ecological and socially significant resources: History of establishment and prospects. *East Siberian Journal of Biosciences* 102: 64–74. DOI: 10.51215/1999-765-2020-102-64-74. https://clck.ru/rnRgW

Kuzevanov, V.Ya. and S.V. Sizykh. 2006. Botanic gardens resources: Tangible and intangible aspects of linking biodiversity and human well-being. *Hiroshima Peace Science* 28: 113–134. https://clck.ru/ge78B

Ministry of Environment and Tourism of Mongolia. 2019. Sixth National Report to the Convention on Biological Diversity (2015–2018). Ministry of Environment and Tourism of Mongolia, Ulaanbaatar, p. 168. https://clck.ru/sHTmg

Mounce, R., P. Smith and S. Brockington. 2017. *Ex situ* conservation of plant diversity in the world's botanical gardens. *Nature Plants*. DOI: 10.1038/s41477-017-0019-3

Munkhnast, D., C. Chuluunjav, M. Urgamal, L.J. Wong and S. Pagad. 2020. GRIIS Checklist of Introduced and Invasive Species-Mongolia. Version 2.7. Invasive Species Specialist Group ISSG. Checklist dataset. https://doi.org/10.15468/ https://clck.ru/sG2HC

Nyambayar, D., B. Oyuntsetseg, R. Tungalag, Ts. Jamsran, C. Sanchir et al. 2011. Mongolian Red List and Plant Conservation Action Plans. Regional Red Book Series, Vol. 9. Plants (Part 1). National University of Mongolia and Zoological Society of London, London, p. 183. https://clck.ru/rsroa [English and Mongolian]

Ochbadrakh, N. and O. Ochirzhav. 2015. Analysis of socio-economic development of the aimags of Mongolia. *Modern Problems of Science and Education* 1(1): 491. https://clck.ru/rsYHy [Russian]

Ochirbat, G. 1994. *Biological Resources of Melliferous and Perganate Plants and Ways of Their Rational Use*. PhD thesis (Biol. Sciences). Ulaanbaatar, p. 56. [Mongolian]

Ochirbat, G. 1996. Current status and results of the introduction work of the Botanical Garden of Mongolia. Central Siberian Botanical Garden Publ. House, Novosibirsk, pp. 96–98. [Russian]

Ochirbat, G. 1999. Botanical Garden. Mongolian Science in XX Century, Ulaanbaatar, pp. 211–212. [Mongolian]

Ochirbat, G. 2006. Present situation and some results of introduction works of Botanic garden of MAS, its perspective. The First Meeting of the East Asia Botanic Gardens Network (EABGN), Kunming, China, 19–20 August, pp. 172–173.

Ochirbat, G. 2011. The list of nectar and pollen plants of Mongolia. *Scientific works. Ulaanbaatar* 1: 76. [Mongolian]

Ochirbat, G. and Ch. Dorzhsuren. 2008. Botanical Garden of Mongolia. 2nd EABGN Seminar, Seoul, Korea, 9–13 June, pp. 169–171.

Orgil, H. 2019. There Is an Opportunity to Restore the Botanical Garden. Peak, Ulaanbaatar, 11 June, p. 2. https://clck.ru/sHGhz [Mongolian]

Osodoev, P.V., A.S. Mikheeva and G.C. Tsybekmitova. 2014. Ecological and geographical problems of nature management in the basins of transboundary rivers in the Asian part of Russia: River Selenga (Russia-Mongolia), river Argun (Russia-China). *Modern Problems of Science and Education* 5: 680. https://clck.ru/rsXWY [Russian]

Oyuungerel, B., B.M. Neronov and A.A. Lushchekina. 2011. Transboundary cooperation between Russia and Mongolia: Protected natural areas and ecotourism. *Space and Time* 2(4): 148–153. https://clck.ru/rsbiQ [Mongolian]

Pavlov, D.S., O. Shagdarsuren, R.V. Kamelin and N. Ulziyhutag. 2004. The 35 years of the Russian-Mongolian (Soviet-Mongolian) complex biological expedition. *Arid Ecosystems* 10(24–25): 8–16. [Russian]

Potaev, B.C. 2015. Some problems and ways of socio-economic development of Mongolia. *Vestnik of Buryat State University* 2–1: 88–92. https://clck.ru/rsd7k [Russian]

Prokhorov, A.A. 2004. *Ecological Problems of Biological Diversity Conservation on the Example of Genetic Resources of Russian Botanical Gardens*. PhD Thesis (Doctor of Biological Sciences). Petrozavodsk, p. 42. https://clck.ru/rsdya [Russian]

Prokhorov, A.A. 2018. The botanical garden is not a public garden, but a scientific research tool. *Hortus Botanicus* 13: 750–753. DOI: 10.15393/j4.art.2018.5764. https://clck.ru/rseYv [Russian]

Regdel, D., Ch. Dugarjav and P.D. Gunin. 2012. Ecological demands to social-economic development of Mongolia under climate aridization. *Arid Ecosystems* 18 (1–50): 5–17. https://clck.ru/sKsj5

Sanzheev, E.D., A.S. Mikheeva, B.S. Batomunkuev, D.A. Darbalaeva, D.T. Zhamyanov and P.B. Osodoev. 2013. The impact of desertification processes on the health of the population of Mongolia (according to sociological surveys in model territories). *Modern Problems of Science and Education* 5: 552. https://clck.ru/rsfF2 [Russian]

SCBD. 1992. Convention on Biological Diversity. Secretariat of the Convention on Biological Diversity. United Nations, New York, p. 23. https://clck.ru/NyqRL

SCBD. 2002. Global Strategy for Plant Conservation 2002–2010. Secretariat of the Convention on Biological Diversity. Montreal, Canada, p. 18. https://clck.ru/rnJwH

SCBD. 2010. Strategic Plan for Biodiversity 2011–2020, Including Aichi Biodiversity Targets. Secretariat of the Convention on Biological Diversity. Nagoya, Aichi Prefecture, Japan. https://clck.ru/sKqUe

Shiirevdamba, T. (ed.). 2013. Red Book of Mongolia (Mongol ulsyn Ulaan Nom). Admon Press, Ulaanbaatar, p. 535. https://clck.ru/sJqwq [Mongolian]

Shukherdorj, B., M. Urgamal, B. Oyuntsetseg, A.P. Sukhorukov, Z. Tsegmed et al. 2022. Flora of Mongolia: Annotated checklist of native vascular plants. *PhytoKeys* 192: 63–169. https://doi.org/10.3897/phytokeys.192.79702. https://clck.ru/sKJ84

Soltani, G.A. and I.V. Annenkova. 2020. Resource potential of botanic gardens and their use. Sustainable development of specially protected natural areas. The 7th All-Russia Scientific-Practical Conference, Sochi. Donskoy Publishing Center, Sochi, 1–3 October, pp. 304–311. https://clck.ru/rshBk [Russian]

Suran, D., T. Bolor and G.-A. Bayarmaa. 2016. In vitro seed germination and callus induction of Ferula ferulaeoides (Steud.) Korov. (Apiaceae). *Mongolian Journal of Biological Sciences* 14 (1–2): 53–58. http://dx.doi.org/10.22353/mjbs.2016.14.07

Terguun, D. 2014. Khubilin Khaan. Great Yuan Country Mongolia. Khot, Ulaanbaatar, p. 755. [Mongolian]

Ueda, K.-I., S. Loarie, P. Leary, A. Shepard, T. Iwane et al. 2019. "Flora of Mongolia" Project on the iNaturalist Platform. Published on the Internet. https://clck.ru/sLS3Z

UNDP. 2020. Human Development Report-2019. Beyond Income, Beyond the Average, Beyond Today: Inequalities in Human Development in the 2nd Century, Human Development Report, United Nations Development Program, p. 351. https://clck.ru/NAQdU

Urgamal, M. 2017. Review of endemic vascular plants of Mongolian flora. *Problems of Botany in South Siberia and Mongolia* 16: 96–100. https://clck.ru/rsiea [Russian]

Urgamal, M. 2018a. *Catalog of Species of Rare and Endangered Vascular Plants of Mongolia*. Bambi Sun Press, Ulaanbaatar, p. 195.

Urgamal, M. 2018b. Current Red List plants and the rare and threatened plants of vascular flora in Mongolia. EABCN Workshop, Yanji (China), 20 June, p. 46. https://clck.ru/rstYH

Urgamal, M. 2018c. Catalogue of rare and endangered vascular plants of Mongolia. *Problems of Botany in South Siberia and Mongolia* 17: 139–142. https://clck.ru/rsiEe [Russian]

Urgamal, M. 2020. New information on the distribution of some collections held in the Botanical Garden's Plant Bank (UBA). *Mongolian Journal of Botany* 2: 129–131. https://clck.ru/sKJXh [Mongolian]

Urgamal, M. and S. Bayarkhuu. 2019. Plant diversity of Mongolia and their quantitative changes in last 30 years (1989–2017). Global Biodiversity, pp. 2–3 https://clck.ru/sKHra

Urgamal, M., O. Enkhtuya, N. Herlenchimeg, E. Enkhzhargal, T. Bukhchuluun, G. Burenbaatar and S. Javkhlan. 2016. Current review of plant diversity in Mongolia. *Proceedings of the Mongolian Academy of Sciences* 1: 3(219) and 86–94. [Mongolian]

Urgamal, M., S. Gombobaatar and N. Herlenchimeg. 2019a. The current status of the regional red list of Mongolian plants. Proc. of the International Conference Dedicated to the 50th Anniversary of JRMKBE RAS & MAS. RAS, Moscow, 23–25 October, pp. 1–2. https://clck.ru/rsvpw [Russian]

Urgamal, M., V. Gundegmaa, Sh. Baasanmunkh, B. Oyuntsetseg, D. Darikhand and T. Munkh-Erdene. 2019b. Additions to the vascular flora of Mongolia. *Proceedings of the Mongolian Academy of Sciences* 58(1): 41–53. https://clck.ru/rsuh4 [Mongolian]

Urgamal, M., B. Oyuntsetseg, D. Nyambayar and C. Dulamsuren. 2014. *Conspectus of the Vascular plants of Mongolia* (eds. Sanchir, Ch. and Ts. Jamsran). Admon Press, Ulaanbaatar, p. 282. https://clck.ru/sG2Db

Vetrov, V.M. 1995. Baikal Siberia in antiquity. Interuniversity. Proc. Sci. Conf. ISU Publ. House, Irkutsk, p. 220. https://clck.ru/rnGy8 [Russian]

Vinogradova, Y.K., E.V. Aistova, L.A. Antonova, O.A. Chernyagina, E.A. Chubar et al. 2020. Invasive plants in flora of the Russian Far East: The checklist and comments. *Botanica Pacifica* 9(1): 103–129. https://doi.org/10.17581/bp.2020.09107

Vinokurov, M.A. and D. Alimaa. 2012. Development of human potential in Mongolia in the context of the demographic situation in the country. *Izv. Irk. State Economic Academy* 2: 115–122. https://clck.ru/rnJ6Q [Russian]

Westwood, M., N. Cavender, A. Meyer and P. Smith. 2020. Botanical garden solutions to the plant extinction crisis. *Plants, People, Planet* 00: 1–11. https://doi.org/10.1002/ppp3.10134

World Bank. 2021. Urban population – Mongolia. *The World Bank*. https://clck.ru/rsssj

Worldometer. 2021. Population of Mongolia. https://clck.ru/rswWJ

Wyse Jackson, P.S. 1999. Experimentation on a large scale: An analysis of the holdings and resources of botanic gardens. *BGCNews* 3(3): 53–72. https://clck.ru/eojVW

8 Jawaharlal Nehru Tropical Botanic Garden and Research Institute, a Treasure House of Tropical Plant Germplasm, Blends into the Western Ghats, the Biodiversity Hotspot in Indian Region

R. Prakashkumar and R. Raj Vikraman

CONTENTS

8.1 Introduction 135
8.2 Vision and Objectives 136
8.3 Theme Group Germplasm Conservatories and Landscapes 137
8.4 Conclusion 147
References 148

8.1 INTRODUCTION

Jawaharlal Nehru Tropical Botanic Garden and Research Institute (JNTBGRI) is a unique institution of its kind in India that conserves the tropical plant germplasm, comprising more than 4,000 species (Mohanan and Mathew, 2016), and is committed to multi-disciplinary research on the sustainable utilization of the tropical plant species. The prodigious efforts of Prof. A. Abraham, the founder director who attended and was inspired by the United Nations Conference on the Human Environment, held in Stockholm in 1972 that recommended the construction of botanic gardens all over the tropics, was accepted in principle by the government of Kerala in 1979 with the objective of conserving the plant resources in the tropics in general and of the country in particular (Abraham, 1987; Namboodiri, 1994). The 121-hectare land sanctioned for the garden is at Palode, 39 km east to the capital city Thiruvananthapuram on the state highway to Shenkottah, right at the foothills of Agasthyamalai hyper diversity ranges of the Western Ghats, one of the 36 biodiversity hotspots of the world., This area supports an enormous plant wealth that habitats nearly 5,800 species, including 56 endemic genera and 2,100 endemic species. The endemic flora of Western Ghats has the prevalence of more than a dozen monotypic genera and maximum taxa representation of more than 15 genera (Rao, 2013). It is the rich germplasm center of wild relatives of many crop plants, fruits, vegetables, spices, and condiments. But an alarming rate of loss of biodiversity happens in the region day after day due to anthropogenic interventions, such as cultivation encroachments; grazing; indiscriminate exploitation for timber, firewood, and for non-timber products, including medicinal; recurrent forest fires; collection for medicinal purposes; and spreading of invasive alien species. The area is geographically located at 8° 45' 15" N latitude and 77° 1' 29" E longitude and an average altitude of about 100 m (Nayar et al., 1986). Out of the 121-hectare forest land, nearly one-third is undisturbed reserve forest preserved as such, and the remaining portion, degraded in anthropogenic activities, is used for developing the garden.

Besides undertaking the committed responsibility of botanic gardens in general, developing an *ex situ* conservatory garden of indigenous tropical species, JNTBGRI has adopted as its mission an integrated research on the sustainable utilization of these plants. The Saraswathi Thangavelu Extension Centre of JNTBGRI at Puthenthoppu, in the north, coastal side of Thiruvananthapuram city, was also established in 1996 particularly for bioinformatics. The institution started as an autonomous R&D organization and was brought under the Kerala State Council for Science, Technology, and Environment (KSCSTE) when it was established in 2003. Accepting the call from the Global Strategy for Plant Conservation (GSPC) in 2002 for the conservation of threatened plant species in their country of origin, JNTBGRI is also focusing on the *ex situ* conservation of endemic species and on the restoration of degraded ecosystems through the reintroduction of saplings developed in the garden back to the degraded natural habitats.

Through the 42 years of determination and perseverance of the limited number of staff, JNTBGRI has flourished into one of the best tropical plant collections all over the world and a noticeable R&D organization doing research on sustainable utilization of plants with the support of state- and central-funded projects. Being a postgraduate and doctoral research center recognized by several universities in the country, the research output has supplemented the academic field also. The garden, which maintains more than 4,000 tropical plants in about 50,000 accessions, has been bestowed with many national and international honors recognizing its esteemed contributions in the field. The Ministry of Environment, Forest and Climate Change, Government of India, recognized it as a "National Centre of Excellence in *ex situ* conservation and sustainable utilization of tropical plant diversity." The Non-Aligned and Other Developing Countries Science and Technology Centre (NAM S&T Centre) recognized the Institute as a "Centre of Excellence for the Conservation of Biodiversity and Plant Science Research." JNTBGRI enjoys the membership of international organizations like BGCI for the conservation of botanic gardens and IUCN for the conservation of nature, as well as the Global Genome Biodiversity Network, and others.

8.2 VISION AND OBJECTIVES

The JNTBGRI garden is maintaining and enriching at every moment, through new accessions and multiplication, its documented collections of living plants with the objective of conservation, scientific research, public education, and aesthetic display. The garden had given emphasis on conserving rare and threatened plants in compliance with international policies, sustainability, and ethical initiatives even before the 2018 updating of BGCI criteria of defining botanic gardens in regard of it. The objectives of the garden and the institute are clearly and elaborately visualized and have been since the time of its inception. Those objectives were recorded as follows: (1) to make a comprehensive survey of the economic plant wealth of Kerala; (2) to conserve, preserve, and exploit the plant wealth of Kerala; (3) to introduce, cultivate, and culture plants of India/other countries with comparable climatic condition for the economic benefit of Kerala and of India; (4) to carry out botanical, horticultural, and chemical research for plant improvement and utilization; (5) to offer facilities for the improvement of ornamental plants and to propagate them in the larger context of establishing nursery and flower trade; (6) to organize germplasm collection of economic plants of interest to the state in the case of those species for which separate centers are not already in existence; (7) to establish a model production center for translating the fruits of research to public advantage, leading to plant-based industrial ventures; (8) to engage in activities conducive to helping botanical teaching and to creating public understanding of the value of plant research in general, and the need for preserving our plant wealth; (9) to establish an arboretum in approximately half the area of the garden, with representative specimens of trees of Kerala and India and trees of economic value introduced from other tropical areas of the world; (10) to establish a garden consisting of medicinal plants, ornamental plants, and various introduced plants of economic or aesthetic value; (11) to establish laboratories for botanical, horticultural, and chemical research, with the aim of improvement and utilization of plants of medicinal and ornamental value; (12) to prepare a flora of Kerala; (13) to establish a tissue culture facility with special reference to the improvement of seeds/fruits/flowers and quick and easy propagation; (14) to organize breeding for plant improvement and production of hybrid seeds, in the case of species for which such facilities are currently lacking or inadequate; (15) to be engaged in garden planning and research; (16) to serve as a source of supply of improved plants not readily available from other agencies; (17) to do chemical screening of plants of potential medicinal importance; (18) to work in collaboration with similar institutes in India and outside; and (19 to promote and establish modern scientific research and development studies relating to plants of importance to India, and to Kerala in particular (Abraham, 1987). Based on these objectives formulated by the founder director, the vision of the venture is briefed as follows.

- *Ex situ* conservation of indigenous, particularly endemic species of the country
- *In situ* conservation studies and eco-restoration of the IUCN Red Listed species
- Documentation and systematic studies of plant biodiversity of the country
- Bioprospecting of indigenous species, involving biotechnological, phytochemical, and pharmacological studies for their effective use in the long run
- Documentation of plants related to indigenous/traditional knowledge
- Education on conservation and sustainable utilization of the plant wealth through live displays, trainings, and extension programs

These basic objectives are accomplished through the two live plants divisions and six research divisions. The two live plants divisions, the Plant Genetic Resource division and the Garden Management, Education, Information, & Training division, maintain different theme-based conservatory gardens – their enrichment, through field explorations, multiplication, and proper display, is supported with effective information, education and communication (IEC). These live plants divisions are also involved with extension activities disseminating botanical knowledge and sales and distribution of plants to generate awareness of plant conservation in the general public. The divisions Biotechnology & Bioinformatics, Conservation Biology, Ethnomedicine & Ethnopharmacology, Microbiology, Phytochemistry & Phytopharmacology, and Plant Systematics & Evolutionary science are involved in research activities that will open new prospects for conservation and sustainable utilization of the plant resources.

8.3 THEME GROUP GERMPLASM CONSERVATORIES AND LANDSCAPES

The more than 4,000 species/infra-specific taxa and hundreds of cultivars of vascular plants are maintained as several theme groups (Anonymous, 2019) as well as in developing the aesthetic landscape. The plants that are well adapted to thrive in the edaphic and climatic conditions of the region are grown open, where as those that were collected from different environmental conditions, like high altitude, moist evergreen, desert and arid regions, forest underneath, etc., are grown in the microclimatic conditions created, such as greenhouses, shade houses, and mist houses with controlled light, temperature, and humidity. A basic informal design style is followed in developing the garden landscapes due to these reasons: (1) any species of conservation importance can be included irrespective of their physical diversity in form, color, or pattern; (2) plants could be deployed in a nature-simulating way of interacting associations to achieve the level of ecosystem conservation rather than being confined to individual plants; and (3) regarding the location of JNTBGRI landscape, the informal style could provide a proper blend to the natural forest background of the Western Ghats hillocks. Like any other botanic garden, the basic objectives for the displays are plant conservation, dissemination of botanic knowledge, and modulation of the environment aesthetics.

The indigenous species, especially the endemic and IUCN Red Listed species of the region, have been given the prime importance in selecting plants. The exotic species are included for botanical education and aesthetics with care in checking the chances of invasion into the natural flora.

The frontage garden (Plate 8.2C; Plate 8.3D) and the extending landscape (Plate 8.3C) on the sides of the road leading to the Core garden, comprising the Aquatic plant conservatory, Arboretum, Carnivorous plants collection, Central Nursery, Lake shore leisure trail, Medicinal plants display, Orchidarium, Palmetum, Visitors Management Centre, Zingiber garden, etc., are designed in a formal style with carpet beds, topiaries, clipped hedges, etc. for catching attraction of the general public. Displays of general interest aimed toward the school students and the general public, who prefer to spend only a few hours in the garden, are focused at the Frontage garden. Baobab tree (*Adansonia digitata*), the tree species so far recorded with the largest trunk size and longest life span, and the petrified form fossil of angiosperm tree trunk (Plate 8.1A) of the Tertiary period in the Cenozoic era, excavated from sandstones in Tiruvakkarai, South Arcot District, Tamil Nadu, are among them. There are prestigious tree plantings by great personalities in the frontage garden. A live display of vascular plant evolution sequence explained by cladogram and descriptive boards, the Vinery (Plate 8.3B) showing the morphological diversity in climbing adaptations, the Topiary garden, the Rose garden, and the Carpet beds (Plate 8.2B), etc., are other interesting components of the frontage garden. The Thai species *Saraca declinata* (Jack) Miq.; the most advanced Gymnosperm *Gnetum gnemon* L. and *Gnetum ula* Brongn.; the purported *Shimshibha* tree of Ramayana *Amherstia nobilis* Wall.; the Talipot palm *Corypha umbraculifera* L., bearing the largest inflorescence; the Indonesian Tiger Orchid, *Grammatophyllum speciosum* Blume (Plate 8.4B), with the record of the world's tallest orchid species; etc., are other curious sights in the frontage and the subsequent landscapes leading to the Core garden area. The information, education, and communication (IEC) in the frontage garden is well supported by direction boards, general explanatory boards, and individual plant labels with QR code leading to the link of detailed information.

Orchidaria and the associated carnivorous plants collection is one of the top priority areas for the general public. About 650 species and many numbers of cultivars, including about 20 cultivars hybridized in JNTBGRI, are in the collection. There are three greenhouses with controlled environments for growing plants of other climatic conditions and two open plots for plants that prefer bright light for flowering. *Paphiopedilum druryi* (Bedd.) Steinthe, a species endemic to the Agastyamalai Hills of the Western Ghats and the only southern Indian orchid species in the genus, rediscovered in 1972 after its original description in 1870, has been multiplied through tissue culture by the scientists of JNTBGRI, conserved *ex situ* as well as planted in large numbers at its degraded habitat in Agasthyamalai to restore the wild population. Besides *ex situ* conservation, horticultural value-added programs are carried out to develop quality cultivars so that nine new cultivars (Plate 8.4C) were registered in the Royal Horticultural Society's International Orchid Register, another ten are awaiting registration, and more are under experiment. The carnivorous plants collection of 35 species, mostly from high altitude and temperate region, are also kept in the same houses designed for orchids. It is a section of high education interest, especially for lower students and the general public. Groups differing in morphology and function of insect trapping mechanisms, such as pitcher plants, sundews, bladderworts, and butterworts, are represented Species of *Nepenthes* (Plate 8.4D), *Sarracenia*, and *Brocchinia reducta* with leaf -modified pitfall traps; Sundews (*Drosera*) trapping prey in sticky hairs on leaves; Bladderworts (Utricularia) with leaf tip bladders; and Butterworts (*Pinguicula*) with sticky, glandular leaves to trap and digest insects are in the collection. The active insect-trapping plant Venus flytrap (*Dionaea muscipula*) is also in the collection.

The largest conservatory garden, the arboretum (Plate 8.1D), spreads over 25 hectares of land and is designed in a natural style to blend with the continuing natural forest. Nearly 800 tree species, including more than 250 endemics and about 50 IUCN Red Listed species, are conserved here in multiple accessions. Bringing together different species of certain genera, such as *Ficus* (60 spp.), *Garcinia* (15 spp.), *Goniothalamus* (6 spp.), *Humboldtia* (5 spp.), *Madhuca* (6 spp.), *Memecylon* (25 spp.), *Syzygium* (35 spp.), etc., at the same place in multiple accessions facilitates students and

Jawaharlal Nehru Tropical Botanic Garden & Research Institute

PLATE 8.1 A. Fossil of Angiosperm tree trunk; B. Bambusetum; C. A N Namboodiri Cacti House interior; D. Arboretum

PLATE 8.2 A. Bromeliad Garden; B. Carpet bed (Ghandhi ji); C. frontage garden entrance; D. Palmetum

Jawaharlal Nehru Tropical Botanic Garden & Research Institute 141

PLATE 8.3 A. Fern garden; B. Vinery; C. Frontage extension landscape to the core garden area; D. Frontage garden; E. *Baccaurea courtallensis* (Wight) Müll. Arg., male tree

PLATE 8.4 A and E: *Itti Achuthan Vydhyan* traditional physician's garden; B. *Grammatophyllum speciosum* (Tiger orchid); C. *Paphiopedilum* 'M S Valiathan'(registered orchid hybrid); D. *Nepenthes khasiana* Hook.f.

researchers to conduct variability and evolutionary studies and also creates an atmosphere of new genetic combinations leading to variation. The Palmetum (Plate 8.2D), which harbors about 160 species of palms and rattans (canes) of both indigenous and exotic origin, is developed in the 5-hectare area underneath the evergreen forest canopy. The rivulet passing through is kept perennial by providing rain water–harvesting stratagems, like small weirs and ponds, so that a moist tropic atmosphere suitable for palms is generated. Species with conservation status, such as *Arenga wightii* Griff., *Bentinckia condapanna* Berry ex Roxb., *Calamus andamanicus* Kurz, *Calamus vattayila* Renuka, *Corypha umbraculifera* L. (flowering plant with the largest inflorescence in the world), *Korthalsia rogersii* Becc., *Pinanga dicksonii* (Roxb.) Blume, and many ornamental plants are in the collection. A small pond was constructed in a natural style to grow *Nypa fruticans* Wurmb., the only palm adapted to the mangrove biome.

The medicinal plant display garden that maintains more than 1,200 species in multiple accessions displays almost all the medicinal plants mentioned in Ayurveda and other medicinal plant systems and also the ones practiced by different tribal folks in the region. All the plants in the display carry labels with their scientific name, Sanskrit name, local name, usage in medicine, and also the floral distribution. Live display consist of traditional formulations and combinations, like *Chyavanaprasam*, *Dashamoolam*, *Dashapushpam*, *Nalpamara*, etc., and plants used in the modern homeopathic and allopathic systems are made here. The herbal garden concept of a traditional physician (Plate 8.4A and E), constructed in the *Chittar* river bank, is named after Sri Itty Achudan, distinguished herbalist of the seventeenth century associated with the Hortus Malabaricus, compiled by Hendrik van Rheede (1669 to 1676). Health-related dictums mentioned in the Sushruta Samhita (sixth century BCE) and Charaka Samhita (around 400–200 BCE) are written in indelible natural die on the early-style earthen compound wall of the Itty Achudan memorial garden. Another interesting component of the medicinal plants garden is the ginger conservatory, which maintains 82 species of the families Zingiberaceae and the closely related Costaceae. *Costus comosus* var. *bakeri* (K. Schum.) Maas (Red Tower Ginger), *Costus stenophyllus* Standl. & L.O. Williams (Bamboo Ginger), and *Etlingera elatior* (Jack) R.M. Sm. (Torch ginger) are the major attractions of the collection.

Other than the display garden, there is a Field Gene Bank garden for medicinal plants, where plantings are made in an informal pattern simulating the natural vegetation. Variants of same species from different localities are planted in adjacent beds for ensuring the consistency of characters, phytochemical studies for active principles, and subsequent pharmacological efficiency among the variants. The Andaman garden in the Field Gene Bank maintains about 120 species collected by exploring the islands, including 21 Red Listed and strictly endemic species, facing threat of being vanished away forever in tsunami-like natural disasters. The taxonomic display named "Systematic garden," which groups live plants as per the Bentham & Hooker (modified) system of classification, gives a practical familiarization of plant families to college students. The Bambusetum (Plate 8.1B), the field germplasm collection of 70 species and 35 hybrids of Bamboos (plants belong to the subfamily, Bambusoideae of family Poacae), is developed on about 5 hectares in the degraded portions of the existing deciduous forest, giving a perfect blend to the natural vegetation. The Climber-Bamboo (*Dinochloa andamanica* Kurz), Monastery Bamboo (*Thyrsostachys siamensis* Gamble), Giant Bamboo (*Dendrocalamus giganteus* Munro), Painted Bamboo (*Bambusa vulgaris* Schrad. "Vittata"), Dwarf Buddha Belly (*Bambusa vulgaris* "Wamin"), and the topiaries done on the Garden Bamboo (*Bambusa multiplex* (Lour.) Raeusch.ex Schult.) are specimens that attract the public. The garden is also involved with the eco-restoration programs of riverbank protection and bio-fencing with bamboos and also arranging trainings and workshops for local people on making handicrafts and household utensils from bamboo parts.

The Gymnosperm garden with 20 species of Cycads, the group dominated in the Jurassic era of about 200 million years back, with *Bowenia* (1 spp.), *Ceratozamia* (1sp.), *Cycas* (7 spp.) (Plate 8.5B), *Dioon* (2 spp.), *Zamia* (5 spp.), *Encephalartos* (3 spp.) (Plate 8.5C), and *Macrozamia* (1 sp.), stands as one of the rarest in the country. The rare indigenous conifer species *Nageia wallichiana* (Presl) Kuntze and the Andaman species *Podocarpus neriifolius* D.Don are also conserved

PLATE 8.5 A. *Victoria amazonica* (Poepp.) Klotzsch; B. *Cycas beddomei* Dyer; C. *Encephalartos gratus* Prain; D. *Tillandsia ionantha* Planch.; E. *Syzygium laetum* (Buch.-Ham.) Gandhi; F. *Humboldtia brunonis* Wall.

ex situ. The most advanced Gymnosperm genera *Gnetum* is represented with two species in the garden. The Southeast Asian species *Gnetum gnemon* L. is planted in the garden, while *Gnetum edule* (Willd.) Blume (Syn. *Gnetum ula* Brongn.) is a part of the natural riparian vegetation in the garden. The Fern garden (Plate 8.3A), maintained along with the Cycads collection, comprises nearly 230 species, including the primitive Lycophytes (Club mosses, Quillworts, and Spike mosses) and Euphyllophytic Polypodiophytes (Polypodiopsida, Marattiopsida, Equisetopsida, and Psilotopsida). Conservation demanding and educationally important species like *Angiopteris helferiana* C. Presl; *Bolbitis presiliana* (Fée) Ching.; *Huperzia* (5 species); *Psilotum nudum* (L.) P.Beauv.; *Ophioglossum reticulatum* L. with the highest recorded chromosome count of any organism (2n = 1,260); the "tree ferns," *Alsophila nilgirensis* (Holttum) R.M.Tryon; *Gymnosphaera gigantean* (Wall. ex Hook.); S.Y. Dong, "climbing ferns," *Lygodium* (4 species); and ornamentally important *Blechnum brasiliense* Desv., *Nephrolepis biserrata* (Sw.) Schott, *Platycerium bifurcatum* (Cav.) C.Chr., *Pteris argyraea* T. Moore, *Pteris ensiformis* Burm. f., *Selaginella erythropus* (Mart.) Spring, and *Selaginella willdenowii* (Desv.) Baker. are conserved in controlled, moist tropical microclimatic conditions generated by providing proper grade of shade nets and regulated misting.

The orchard comprises nearly 100 indigenous species of wild fruits that the tribal and local people living in the premises of forests once depended upon to meet their basic food needs or as an additional food supplement. The project established nearly 30 years ago has full-grown fruiting trees *Aporosa cardiosperma* (Gaertn.) Merr., *Baccaurea courtallensis* (Wight) Müll. Arg. (Plate 8.3E), *Elaeagnus conferta* Roxb., *Elaeocarpus serratus* L., *Flacourtia indica* (Burm.f.) Merr., *Syzygium cumini* (L.) Skeels, *Flacourtia montana* J.Graham, *Garcinia gummi-gutta* (L.) Roxb., and others, such as *Salacia beddomei* Gamble, *Salacia fruticosa* Wall. ex M.A. Lawson, *Syzygium caryophyllatum* (L.) Alston, *Syzygium zeylanicum* (L.) DC., etc. These trees that have been traditionally used for many generations are multiplied and distributed for public environment restoration initiatives as well as to individuals. High-quality variants of demanding indigenous fruit trees, such as *Artocarpus heterophyllus* Lam., *Mangifera indica* L., *Phyllanthus emblica* L., etc., and dioecious fruit trees, such as *Aporosa cardiosperma* and *Baccaurea courtallensis*, whose sex determination is not possible until flowering, were multiplied vegetatively to ensure the quality and sex of the mother plants in saplings for distribution to the public. Selected species are subjected to chemical studies for nutritional value and toxicity by the Phytochemistry and Phytopharmacology divisions of JNTBGRI.

The garden maintains a separate conservatory for the lesser exploited wild ornamental plants also. With the objective of promoting the use of indigenous plants in ornamental gardens that are now dominated with more than 80% of alien species, potential species were identified, conserved *ex situ*, multiplied, and distributed for planting in gardens and avenues. A few of the species thus conserved and popularized are shrubs: *Alstonia venenata* R.Br., *Dichrostachys cinerea* (L.) Wight &Arn., *Ixora brachiata* Roxb., *Ixora polyantha* Wight, *Murraya paniculata* (L.) Jack, and *Pinanga dicksonii* (Roxb.) Blume; under shrubs: *Azanza lampas* (Cav.) Alef., *Barleria courtallica* Nees, *Barleria cristata* L., *Barleria prionitis* L., *Barleria strigosa* Willd., *Eranthemum capense* L., *Melastoma malabathricum* L., *Osbeckia aspera* (Meerb. ex Walp.) Blume, *Osbeckia virgata* D.Don ex Wight & Arn., and *Phyllanthus gageanus* (Gamble) M.Mohanan; herbs and ground cover plants: *Rhynchospora corymbosa* (L.) Britton, *Acrotrema arnottianum* Wight, *Artanema longifolium* (L.) Vatke, *Begonia dipetala* Graham, *Stenosiphonium wightii* Bremek., *Strobilanthes barbatus* Nees, *Strobilanthes ciliates* Nees, *Strobilanthes lawsonii* Gamble, *Strobilanthes lupulina* Nees, and *Gymnostachyum febrifugum* Benth.; Climbers: *Thunbergia mysorensis* (Wight) T. Anderson, *Cissus discolor* Blume, *Thunbergia fragrans* Roxb., *Jasminum flexile* Vahl, and *Vanilla wightii* Lindl. ex Wight; Lianas: *Combretum malabaricum* (Bedd.) Sujana, Ratheesh, & Anil Kumar; Trees: *Alstonia scholaris* (L.) R. Br., *Butea monosperma* (Lam.) Kuntze, *Cochlospermum religiosum* (L.) Alston, *Diospyros buxifolia* (Blume) Hiern, *Diospyros ebenum* J.Koenig ex Retz., *Filicium decipiens* (Wight & Arn.) Thwaites, *Humboldtia decurrens* Bedd. ex Oliv., *Humboldtia vahliana*

Wight, *Mesua ferrea* L., *Pterocarpus marsupium* Roxb., and *Syzygium laetum* (Buch.-Ham.) Gandhi (Plate 8.5E).

Another significant initiative is the RET Species Park that was established with MoEF assistance to Lead Botanic Gardens. All three targets of botanic gardens, viz. conservation, education, and aesthetics, are tied together in this 2.5-ha area in an informal-style landscape design that blends perfectly in the forest surroundings. Following are some of the Red Listed and Rare status species out of the 142 endemic species conserved in this garden: *Abutilon ranadei* Woodrow & Stapf, *Aglaia malabarica* Sasidh., *Anacolosa densiflora* Bedd., *Arenga wightii* Griff., *Bentinckia condapanna* Berry ex Roxb., *Buchanania barberi* Gamble, *Calamus brandisii* Becc., *Calliandra cynometroides* Bedd., *Cinnamomum chemungianum* M. Mohanan & A.N. Henry, *Cinnamomum riparium* Gamble, *Cynometra beddomei* Prain, *Cynometra bourdillonii* Gamble, *Cynometra travancorica* Bedd., *Dimocarpus longan* Lour., *Dysoxylum beddomei* Hiern, *Garcinia travancorica* Bedd., *Garcinia wightii* T. Anderson, *Gluta travancorica* Bedd., *Goniothalamus rhynchantherus* Dunn, *Goniothalamus wynaadensis* (Bedd.) Bedd., *Gymnacranthera canarica* (Bedd. ex King) Warb., *Hopea ponga* (Dennst.) Mabb., *Humboldtia bourdillonii* Prain, *Humboldtia brunonis* Wall. (Plate 8.5F), *Humboldtia decurrens* Bedd. ex Oliv., *Hydnocarpus macrocarpus* (Bedd.) Warb., *Prioria pinnata* (Roxb. ex DC.) Breteler, *Knema attenuata* (Wall. ex Hook.f. & Thomson) Warb., *Madhuca bourdillonii* (Gamble) H.J. Lam, *Myristica andamanica* Hook.f., *Myristica malabarica* Lam., *Nothopegia aureofulva* Bedd. ex Hook.f., *Ochreinauclea missionis* (Wall. ex G. Don) Ridsdale, *Poeciloneuron pauciflorum* Bedd., *Rauvolfia hookeri* S.R. Sriniv. & Chithra, *Sageraea thwaitesii* Hook.f. & Thomson, *Saraca asoca* (Roxb.) W.J. de Wilde, *Syzygium palghatense* Gamble, *Syzygium stocksii* (Duthie) Gamble, *Tabernaemontana gamblei* Subr. & A.N. Henry, and *Vateria indica* L. The pathways that decide the skeletal structure of the landscape are marked with a trailing herb with Red status, *Strobilanthes wightii* (Bremek.) J.R.I. Wood.

The Aquatic plants conservatory maintains many conservational, educational, and aesthetically important species of fresh water angiosperms. The Giant water lily, *Victoria amazonica* (Poepp.) Klotzsch (Plate 8.5A), and the smallest flowering plant, *Wolffia globosa* (Roxb.) Hartog & Plas, grown in the same spot is an educational curiosity. Many species of plants with heterophylly and phenotypic plasticity in leaf form, a curious feature of the same plant with submerged and areal stem or aquatic and terrestrial phases in life, are represented with many examples. Students could have an encounter with *Limnophila indica* (L.) Druce, *Limnophila heterophylla* (Roxb.) Benth., *Hygrophila difformis* (L.f.) Blume, *Hygrophila balsamica* (L.f.) Raf., and *Cabomba furcate* Schult. & Schult.f., which develop surprisingly different leaves when grown underwater and emerged in air. Plants variously adapted for floating, such as *Trapa natans* L., *Pontederia crassipes* Mart., *Ludwigia adscendens* (L.) H. Hara, *Ceratophyllum demersum* L., and *Hydrilla verticillata* (L.f.) Royle, etc., and insectivorous species *Utricularia gibba* L. and *Utricularia aurea* Lour., etc., are all direct experience for students those who have only book knowledge of school lessons.

The rock garden and associated A. N. Namboodiri Cacti House (Plate 8.1C) accommodates nearly 300 species/variants of succulents and about 50 Bromeliads (Plate 8.2A). Mostly exotic species are grown in the interest of educating students rather than conservation. Some plants that adapted in different ways to survive scarcity of water, mainly of the families Cactaceae, Euphorbiaceae, Apocynaceae, Agavaceae, Liliaceae, Bromeliaceae, etc., are conserved by providing a suitable edaphic and environmental atmosphere. Plants that can survive the heavy rain are planted in the open rock garden, designed in a natural style, while the plants from higher altitudes or subtropical arid regions that cannot tolerate the heavy rain and intense heat are kept in the microclimatic glass house. Both in the open rock garden and inside the glass house, the terrine is made into natural style mounts to avoid water lagging, and plantings are made in rock pockets made in the slopes. Various forms of "phylloclades" (Cacti, Euphorbias, Dogbanes, Milkweeds, *Muehlenbeckia*, etc.) and "cladodes" (*Asparagus* and *Ruscus*), leaf succulents (*Aloe*, *Sedum*, *Haworthia*, *Dischidia*, *Hoya*, etc.), and stoloniferous succulents (Bulbous type Lilies and Amaryllis) are displayed with supporting education labels. *Boucerosia frerei* (G.D. Rowley) Meve & Liede and *Euphorbia vajravelui*

Binojk. & N.P. Balakr. are the two endemic succulent species with threatened status among the majority alien species grown in educational and aesthetic interests. The Bromeliads, which lead a unique kind of epiphytic or saxycolous life by absorbing water and minerals through the special epidermal cellular "trichomes," are grown fastened on tree branches and drift woods or fixed in crevices of artificially made rubble grounds. The two groups Cacti and Bromeliads, whose natural distribution is restricted to the two American continents only, are purported to have evolved after the Gondwana splitting. "Tank epiphytes," which store water in their leaf clefts, including *Aechmea bracteata* (Sw.) Griseb., *Aechmea gamosepala* Wittm., *Billbergia zebrina* (Herb.) Lindl., *Billbergia pyramidalis* (Sims) Lindl., *Lutheria splendens* (Brongn.) Barfuss & W. Till, *Neoregelia spectabilis* (Antoine) L.B.Sm., Spanish moss (*Tillandsia usneoides* (L.) L.), the Pink quill (*Wallisia cyanea* Barfuss & W.Till), the Earth Stars (*Cryptanthus* spp.), the Saw blade (*Dyckia brevifolia* Baker), and *Tillandsia ionantha* Planch. (Plate 8.5D) are among the more than 50 educationally and aesthetically interesting Bromeliad species in the collection.

While giving prime importance to conserving indigenous species, alien species of educational and aesthetic importance are also grown in the garden with strict care to limit the chances of escape from cultivation and subsequent naturalization that would affect the growth and development of the indigenous flora. The information, education, and communication (IEC) system is achieved through the visitor's management center and arranges guided tours for visitors and participates in exhibitions conducted in schools and other public involved programs to convey plant conservation awareness and also to invite their attention to the activities going on in JNTBGRI. In addition to Central Nursery activities of multiplication and distribution to promote plant conservation, there is a commercial tissue culture section for large-scale multiplication and distribution of ornamental and other useful plants of public interest as well as conservation importance. The research divisions of JNTBGRI are focusing on identification of yet-to-be-explored plant wealth of the nation and attempting to develop technologies for their utilization without compromising their availability for the future generation. *In situ* conservation programs of keystone and Red Listed species in the Western Ghats, emphasizing the reasons of population decline and their restoration with saplings developed from the *ex situ* collection, is carried out in the conservation biology division. Biotechnological research in JNTBGRI focuses mainly on developing plant conservation strategies and bio-production of plant-specific compounds to minimize the direct exploitation of plants. Pharmacological, biochemical, and biotechnological collaborative studies concentrate on bioprospecting of plants for medicinal drugs and other valuable bio-chemicals. Traditional use of indigenous plant materials by ethnic groups for livelihood is streamlined by analyzing the nutritive value and toxicity to protect the cultural identity of food practices of the Indigenous people. The Microbiology division of the institute, which studies the bioprospecting and sustainable utilization of the microbial wealth of the Western Ghats, focuses mainly on fungi, including the mycorrhizal species, which is one of the key influential factors in the forest environment. The research activities predominantly focusing on the sustainable utilization of the plant wealth of the country have resulted in the publication of more than 1,893 (until 2015) research publications, including papers in peer-reviewed journals, book chapters, and books; more than 100 species/infraspecific taxa are so far reported new to science by the scientists of JNTBGRI (Dan Mathew et al., 2016).

8.4 CONCLUSION

Besides fulfilling the mandatory objectives of every botanic garden – the *ex situ* conservation of indigenous species and dissemination of botanic knowledge – the rich germplasm collection of JNTBGRI is a resource for multidisciplinary research on sustainable utilization of indigenous plant wealth, conducted in the research divisions of the Institute. Through using the garden progenies of endemics for eco-restoration in their ruined habitats in the biodiversity hotspots, particularly the Western Ghats, the *ex situ* conservatories serve as a reservoir for the retrieval of lost vegetation from natural habitats. The on-going scientific search and evaluation of indigenous useful plants in the Institute is adding much to the traditional and modern medical and other systems of human life. At present, JNTBGRI maintains a very rich collection of indigenous plant species under the theme groups Aquatic plants, Arboretum, Bambusetum, Fernery, Medicinal plants, Orchids, Wild fruits, and Wild ornamentals, all of which are among the largest collections of their kind in India. To accomplish the committed mission to conserve *ex situ* the indigenous flora of the country, with the estimated record of about 20,000 indigenous vascular plant species (angiosperms, gymnosperms, and pteridophytes) including about 4,100 endemics, nearly 2,200 threatened species (Arisdason and Lakshminarasimhan, 2020), and many more unexplored species, particularly in the four world biodiversity hotspots in the country, germplasm enrichment eco-restoration programs must be pursued long-term.

REFERENCES

Abraham, A. 1987. The tropical botanic garden and research institute, Trivandrum, India. In: Nayar, M.P. (ed.), *Network of Botanic Gardens*. Botanical Survey of India, Culcutta, pp. 204–213.

Anonymous. 2019. Brochures. Plant Genetics Resource Division and Garden Management. Education and Information Division. Jawaharlal Nehru Tropical Botanic Garden and Research Institute, Palode, Thiruvananthapuram.

Arisdason, W. and P. Lakshminarasimhan. 2020. *Status of Plant Diversity in India: An Overview*. ENVIS Resource Partner on Biodiversity, BSI, Kolkata.

Dan Mathew, A., Rasiya Beegam, H. Biju and V. Sujatha (eds.). 2016. *Scientific Contributions of JNTBGRI, A Bibliography of Publications 1980–2015*. Jawaharlal Nehru Tropical Botanic Garden and Research Institute, Palode, Thiruvananthapuram.

Mohanan, N. and P.J. Mathew (eds.). 2016. *Live Plants of JNTBGRI*. Jawaharlal Nehru Tropical Botanic Garden and Research Institute, Palode, Thiruvananthapuram.

Namboodiri, A.N. 1994. *Professor A. Abraham, shasthrajnan, vidyabhyasavichakshanan, botanic gardente shilpi* (Malayalam). Tropical Botanic Garden and Research Institute, Palode, Thiruvananthapuram

Nayar, T.S., K.C. Koshy, C. Sathish Kumar, N. Mohanan and S. Mukuntha Kumar. 1986. *Flora of Tropical Botanic Garden, Palode, Thiruvananthapuram*. Tropical Botanic Garden and Research Institute, Palode, Thiruvananthapuram.

Rao, R.R. 2013. *Floristic Diversity in Western Ghats: Documentation, Conservation and Bio-Prospection-A Priority Agenda for Action*. Allen Institute for Artificial Intelligence, Seattle, Washington, U.S. Semantic Scholar.

9 Botanical Gardens and Their Role in Education, Research, Conservation, and Bioprospecting of Plant Diversity
Lead Botanical Garden (LBG), Shivaji University, Kolhapur – a Case Study

M.M. Lekhak and S.R. Yadav

CONTENTS

9.1 Introduction 150
9.2 Botanic Gardens 152
9.3 Lead Botanical Gardens (LBGs) 152
    9.3.1 Concept of LBGs 153
9.4 Lead Botanical Garden of Shivaji University 154
    9.4.1 Infrastructure and Facilities Available in LBG of Shivaji University 154
    9.4.2 Major Sections in LBG 157
        9.4.2.1 Fernery 157
        9.4.2.2 Rhizomatous, Cormatous, Tuberous, and Bulbous Plants 157
        9.4.2.3 The Pinetum 158
        9.4.2.4 Aquatic Plants 158
        9.4.2.5 Medicinal Plants 158
        9.4.2.6 Palmetum 158
        9.4.2.7 Orchidarium 160
        9.4.2.8 Arboretum of Medicinal Plants 160
        9.4.2.9 Arboretum of Rare, Endangered, and Threatened (RET) and Endemic Plant Species 161
        9.4.2.10 Plant Groves in LBG 161
        9.4.2.11 Climbers 162
        9.4.2.12 Indigenous, Threatened, and Endemic Plant Species Nurseries 162
        9.4.2.13 Mangrove Nursery 162
        9.4.2.14 *Barleria* and *Flemingia* Germplasm Bank 162
9.5 Role of LBG in *Ex Situ* Conservation of Plant Diversity 164
    9.5.1 Role of LBG in Bioprospecting Wild Plants as Ornamentals 164
    9.5.2 Role of LBG in Restoration of Threatened Plant Species of Western Ghats 166

DOI: 10.1201/9781003281252-9

        9.5.3   LBG Program on Seed Collection, Sapling Raising, and Distribution of
                RET/Endemic and Indigenous Tree Species... 167
9.6   Conclusions... 168
References.. 170

9.1   INTRODUCTION

We have already entered into the age of mass extinction, which is expected to happen within a short period of 100 years without giving any opportunity to the many extant species to evolve and adapt to the new climatic changes that are taking place on the planet Earth. It is predicted that by the end of the twenty-first century, we will have lost 50% of our total biodiversity (Myers, 1993; Singh, 2002). Most of the biodiversity will be lost before we discover, describe, and name the organisms. Every minute we are destroying 40.5 ha of tropical rain forest (https://www.fraserinstitute.org/sites/default/files/facts-not-fear-chapter-9.pdf) and losing two to five flowering plant species per hour from tropical forests alone (Singh, 2002). We are modifying and destroying natural habitats, which is another major reason for loss of biodiversity. With increasing population, there is heavy demand for land, water, and bio-resources, and we are utilizing our limited resources recklessly. We are burning oil and coal, which is a major reason for carbon increase in the atmosphere leading to climatic change, global warming, and biodiversity loss. The current rates of species extinction are 1,000–10,000 times higher than the background rate of 10^{-7} species/species year inferred from fossil record, and today we seem to be losing two to five species per hour from tropical forests alone, which amounts to a loss of 16 million populations per year or 1,800 populations per hour (Singh, 2002). We are losing at least 27,000 species per year from tropical forests alone, and under the current scenario, about 20% of all species are expected to be lost within 30 years and 50% or more by the end of the twenty-first century (Myers, 1993). Man has realized the effects of his activities, which are a threat to biodiversity as well as for his own survival. Therefore, conservation of biodiversity is an important issue worldwide, and it is a concern of every man, every community, and every country.

There are serious efforts taken to reduce carbon in the atmosphere, and many new programs are being initiated to bring down atmospheric carbon for which plantation of fast-growing trees is considered to be a very effective way, now known as carbon crediting. Plants will provide solutions to most important global problems as they can provide food, fodder, medicine, agriculture, and pollution control, and they can help slow down climate change. Therefore, all over the world, efforts are being made for conservation of biodiversity through various programs. *In situ* conservation is the best way to conserve biodiversity; however, every area cannot be protected where biodiversity exists, and therefore, we need to go for *ex situ* conservation measures. Among all the *ex situ* conservation measures, botanic gardens seem to be the most ideal for conservation of at least endemic, threatened, and economically important plant species of particular regions. Botanic gardens are refuges for these plants, and they have many advantages in conservation over other methods. They are a repository of living plants, a place for multiplication of threatened species, a laboratory to study and understand plant species, and a research center from where propagation, reintroduction, and restoration of endangered plant species can be undertaken and implemented. In addition to its role in conservation, a botanical garden is a laboratory for students, a seed and germplasm bank for society, and a recreational place for everyone. There is no doubt about well-placed gardens supported by government playing a crucial role in conservation of at least the most important plant species and bio-resources of the world.

We may trace the origin and foundation of botanic gardens for spices and drugs, which are universal needs of man, and the earliest mention of a botanical garden is the mention of Aristotle's Garden at Athens, which he bequeathed to Theophrastus by whom it was newly equipped and improved (Hill, 1915). The multiple and evolving roles and contributions of botanical gardens in various ways have been brought about by several researchers. Zoos, aquariums, botanical gardens, and natural history museums present an exceptional opportunity for many urban residents to see the wonders of life, and they can contribute to education and habitat preservation (Miller et al., 2004). Waylen (2006) highlighted the role of botanic gardens across the world in improving health care and nutrition, alleviating financial poverty, and improving community and social relations in many diverse ways. The role of botanical gardens in *ex situ* conservation and research has been highlighted by Stevens (2007). Public display, research, education, and conservation are the four primary roles of botanic gardens; however, in recent years, conservation has become a major focus of many botanical gardens (Ballantyne et al., 2008; Chen et al., 2009; Donaldson, 2009). Botanic gardens provide unique opportunities for the study of ecological restoration (Hardwick et al., 2011), global climate change (Donaldson, 2009; Primack and Miller-Rushing, 2009), and the relationship between humans and nature (Heyd, 2006). In addition to education and research, He and Chen (2012) and Bennett (2014) considered *ex situ* conservation as a major role of botanical gardens. Botanic gardens provide botanical education of crucial importance. They are an integral part in comprehensive training and are underutilized but invaluable resources; they have a long and well-established relationship with university education and were especially important in the development of medical education (Bennett, 2014). Education, promoting awareness, and capacity building, involving both the public and staff at botanical gardens, are vital functions of modern botanical gardens (Blackmore et al., 2011). Chang et al. (2008) emphasized the importance of landscape narrative and interpretation practice that can be applied to improve educational effectiveness. Chen et al. (2009) provide several illustrations of the effectiveness of tropical botanical gardens (TBGs) in preserving tropical biodiversity at the frontier of *in situ* ecosystem management. Crane et al. (2009) state that "at no other point in history has research in botanic gardens and arboreta, been more important." Oldfield et al. (2009) underscore the need for conservation action for globally threatened tree species to prevent the urgent loss of tree species worldwide. There is an opportunity for botanic gardens to use their living collections more effectively in global-change research and education (Donaldson, 2009; Primack and Miller-Rushing, 2009). The Global Strategy for Plant Conservation (GSPC) has been widely adopted by the botanic garden community, and the scientific contribution of botanic gardens needs to be strengthened to underpin conservation action, including the role of botanic garden horticulture, training, and international capacity building (Blackmore et al., 2011). Presently, the purpose of botanical gardens has greatly expanded to include rescuing plant biodiversity, offering serious programs of research and education to citizens of all ages and instruction for skilled botanists, creating aesthetically pleasing refuges from modern life, and maintaining storage centers both onsite and offsite for the long-term preservation of plant species against the time (Powledge, 2011). Similarly, the role of botanists in plant conservation and rescue of endangered species through biotechnological application has been emphasized by Bapat et al. (2008, 2012). The integrated conservation and restoration of threatened tree species is urgently required and presents an exciting opportunity for botanic gardens (Oldfield and Newton, 2012). The beautiful scenery and rich diversity of plant species in their living collections have been helping botanical gardens (BGs) to attract visitors and thus serve as a base for public education on biodiversity (He and Chen, 2012). The studies on environmental education on climate change in a botanical garden confirm out-of-school learning and indicate the suitability of botanical gardens as appropriate learning sites for education for sustainable development (Sellmann, 2014). Lead Botanical Gardens (LBGs) potentially offer community education about conservation and conservation attitudes, and they encourage the public to support conservation efforts (Deshpande and Yadav, 2017). With anthropogenic pressure and climate change, botanical gardens need to integrate *in situ* conservation, *ex situ* conservation, and reintroduction to effectively protect plant diversity (Ren, 2017). Mounce et al. (2017) revealed that botanic gardens manage at least 105,634 species, equating to 30% of all plant species diversity, and conserve more than 41% of known threatened species. They further revealed that botanic gardens are disproportionately temperate, with 93% of species held in the northern hemisphere, and an estimated 76% of species absent from living collections are tropical in origin. Studies on the contribution of botanic gardens to *ex situ* conservation through seed banking by O'Donnell and Sharrock (2017, 2018) suggest that institutions are increasingly conserving plant species via seed banking; around 350 botanic gardens together maintain seed collections of 57,000 taxa, and the role of botanic gardens in the conservation of crop wild relatives (CWRs) is becoming increasingly important. Chen and Weibang (2018) deliberated the future challenges and responsibilities of botanical gardens with reference to the negative effects of outbreeding and/or inbreeding depression, promoting awareness, study, and conservation of plant species diversity; accelerating global access to information about plant diversity; and increasing capacity building and training activities in this changing world. Botanic gardens are important contributors to CWR preservation, and at least 6,017 of the 6,941 socio-economically important GRIN-Global World Economic Plant (WEP) taxa (86.7%) were currently found in the living and seed collections of botanical gardens and arboreta with 1,456 taxa (21%) held in more than 40 collections (Meyer and Barton, 2019). Both living and seed collections held in botanical gardens and seed banks offer huge long-term opportunities for the conservation of wild plant diversity (Breman et al., 2021). Botanical gardens are addressing crucial issues of great significance, and there is huge literature on botanical gardens of the world. However, the purpose of this chapter is to highlight the development of Lead Botanical Gardens in India in general and at Shivaji University, Kolhapur, in particular.

9.2 BOTANIC GARDENS

According to Wyse Jackson (1999), "Botanic Gardens are institutions holding documented collections of living plants for the purpose of scientific research, conservation, display and education." The first botanic garden working in science and education was of Theophrastus and was attached to his school in Greece (Deshpande and Yadav, 2017). The credit of development of the first modern botanic garden in 1543 goes to Italian researcher Luca Ghini at the University of Pisa. The first botanic garden in India was established in 1787 in Calcutta as the Indian Botanic Garden, which is presently named the Acharya Jagadish Chandra Bose Indian Botanic Garden. Today, India includes a chain of botanic gardens distributed in all the regions of the country. After the ratification of the Convention on Biological Diversity (CBD) by India, the Ministry of Environment, Forest and Climate Change (MoEF & CC) and state forest departments launched various schemes, such as "National Action Plan on Biodiversity Conservation, Capacity Building in Taxonomy, Assistance to Botanic Gardens," etc. Presently, the Botanic Gardens Conservation International (BGCI) lists a total of 1,846 botanic gardens distributed in 148 countries. They maintain 4 million living plants, representing about 80,000 species, along with other collections in the form of herbaria and seed banks (Anonymous, 2013). The botanic gardens throughout the world have accepted the challenge of undertaking a global mission for conservation through a wide range of programs, such as collection, research, education, and public awareness. Thus, they play a crucial role as centers for rescue, recovery, and rehabilitation of rare, endangered, and extinction-prone species of vascular plants. The botanic gardens also play important roles in education as centers of training in areas such as horticulture, gardening, landscaping, *ex situ* conservation, and environmental awareness.

9.3 LEAD BOTANICAL GARDENS (LBGs)

Establishment of the Lead Botanical Garden is a brilliant concept conceived by India and is encouraged by the Ministry of Environment, Forest and Climate Change (MoEF & CC). These LBGs in India will have to shoulder crucial responsibilities of not only conserving the plant species of the region but also guiding other gardens of neighboring areas, promoting plantation of indigenous and endemic species, guiding forest departments in reintroduction and restoration of species, creating awareness among people, and providing guidelines to policy makers. Lead Botanical Gardens have the responsibility of providing saplings of appropriate species and guidance in development of small botanical gardens of the region. Thus, LBGs will be hubs of botanic activities, including research, in the near future. Similarly, ancillary botanical gardens (ABGs) are almost like a complement to

LBGs. ABGs are small gardens in urban areas that develop and are supported by a central garden, often a university garden. ABGs help with local outreach and education close to where people live. ABGs are informal versions of botanic gardens meant to offer opportunities for botanical education.

9.3.1 CONCEPT OF LBGS

The Convention on Biological Diversity (CBD), which came into action on 29 December 1993, is a comprehensive international legal framework for biodiversity conservation, sustainable use, and benefit sharing. Articles 6 (General Measures for Conservation and Sustainable Use), 7 (*Ex situ* Conservation), 12 (Education and Awareness), and 15 (Access to Genetic Resources) amply justify the need for further strengthening of the MoEF & CC's Scheme on "Assistance to Botanic Gardens" for *ex situ* conservation of plant diversity.

In view of this context, a scheme was initiated by the Ministry of Environment, Forest and Climate Change, Government of India in 1991–1992 to promote *ex situ* conservation and propagation of threatened and endemic plants through a network of botanic gardens and centers of *ex situ* conservation. The assistance for botanic sections in popular gardens on 100% grant basis is aimed at *ex situ* conservation of threatened and endemic flora, including further research promoting awareness and education.

Financial assistance was provided to various botanic gardens through 291 projects, which included 13 Lead Botanical Gardens, up to November 2013. The role of LBGs is as a referral center and a model with respect to conservation of threatened and endemic species of different phytogeographical regions of the country. The gardens, which provide the training and necessary expertise for replication at regional or local levels, could be termed as "Lead Botanical Gardens" or models that must be followed. Globally, these LBGs together form important resource centers for biodiversity conservation.

Functions of Lead Botanical Gardens

1. Help conserve natural vegetation, especially threatened and endemic species, through multiplying and rehabilitating them in natural habitats under *ex situ* conservation.
2. Undertake botanical research resulting in an excellent referral system for plants as authentically identified, classified, and labeled live collections in gardens and as dry collections (pressed, processed, and mounted specimens) in herbaria, both for monitoring and documentation of threatened and endemic plant resources of the country.
3. Study the phenology and response of the plants to climate variability/change.
4. Carry out biological studies with a view to find out ecological, biological, and genetic bottlenecks or barriers in the reproduction and survival of species.
5. Carry out rehabilitation/recovery programs for threatened and endemic species.
6. Serve as centers of training, with expertise in a focused area of subject specialization, including horticulture.
7. Compile information on *in situ* as well as *ex situ* conservation of the threatened and endemic species and their habitats.
8. Compile information on the area of occurrence, area of occupancy, number and size of populations, spatial distribution of populations, identification of important associates such as pollinators and dispersers, reproductive and breeding systems, population trends in relation to habitat changes and pattern of disturbance, etc. Prepare Red Data Sheets for the selected species as per IUCN format.
9. Promote environmental awareness of nature conservation through well-designed education programs and educational materials.
10. Develop relevant R&D expertise and capabilities in undertaking modem conservation and gene banking techniques, including in-vitro tissue banks, DNA, and cryobank.

(Anonymous, 2013)

9.4 LEAD BOTANICAL GARDEN OF SHIVAJI UNIVERSITY

Shivaji University, Kolhapur, is situated on the eastern shoulders of the Western Ghats. The University holds 344 ha of land, and the landscape itself is undulating and provides excellent opportunities to grow and conserve various kinds of species of the Western Ghats. The Department of Botany in Shivaji University was established in 1964 (Figure 9.1). Initially, a land of 2.4 ha was allotted to the department where some plantation was made, but due to lack of funds, no visible garden was established until 1985. In that year, an additional 2.4 ha of land was demanded by the department for the botanic garden, and university authorities provided the land to the department, which initiated development of the garden around the department. Over the next couple of years, the garden became visible with the introduction of various plant species and continuous enrichment of the garden. However, it took almost four to five years to give a definite shape to the garden. Developing a garden is not an overnight job but requires continuous activity for years for its growth and establishment.

Fortunately, the Ministry of Environment, Forest and Climate Change (MoEF & CC) provided funds for infrastructure in the garden in 1996, which boosted the activities of the garden, and the garden took shape. Since then, plant species, especially those that are endemic and threatened, were introduced in the garden, and the garden was enriched by plants from various parts of the country, especially from the Western Ghats of India. By 2005, more than 150 endemic plant species were introduced in the garden, which boosted research in the department. The Department of Biotechnology (DBT), Govt. of India, New Delhi, supported several projects on bioprospecting of the Western Ghats, mapping and quantification of plant diversity of the Western Ghats, and especially for conservation and restoration of 20 critically endangered endemic plant species of the Western Ghats, which resulted in establishment of good infrastructure for micro-propagation, conventional propagation, nursery development, irrigation facilities, and protection to the botanic garden.

In 2005, a team of experts deputed by the Ministry of Environment, Forest and Climate Change visited our garden (Figure 9.2). The achievements of the garden were presented, and the expert committee recommended the garden for funding. This botanic garden in India was declared as the Lead Botanical Garden for Western India with funding from MoEF & CC for additional required infrastructure. Moreover, Shivaji University provided an additional 12 ha for planting, which is also a very important achievement of the department and for the garden. Thus, the garden started improving, and many new introductions are being made every year. The botany department, with LBG as a center, aims at converting the entire 344 ha of this educational institute into biodiversity parks, which will not only act as lungs of Kolhapur but as a center of botanic activities and education; a base camp for restoration of threatened species; a source for bioprospecting of plant resources; a germplasm depository from where germplasm will flow to society, students, and researchers; and a recreation place for generations to come.

9.4.1 INFRASTRUCTURE AND FACILITIES AVAILABLE IN LBG OF SHIVAJI UNIVERSITY

Presently, LBG of Shivaji University is spread over 17 ha of land with diverse topography suited for growth of different species. Nilambari Auditorium, with a seating capacity of 180, is an important asset to the LBG where various awareness programs/workshops on biodiversity, its importance and conservation, and sustainable utilization are being conducted. The conservation strategies are not practically possible unless society or the public is actually involved and awareness programs are arranged. Nilambari Auditorium of LBG has become an important place for scientific discussions, seminars, and workshops. A recognized herbarium (SUK herbarium) is attached to the LBG, housing more than 30,000 specimens mainly from the northern Western Ghats. This is an important aspect of any botanical garden and was supported by the Department of Science and Technology (DST). The herbarium provides an advantage to the people, research scholars, and students in

Botanical Gardens and Their Role in Education, Research 155

FIGURE 9.1 Fossilized *Mesembrioxylon mahabalei*: 33.7–23.8 million-year-old petrified tree trunk in LBG.

FIGURE 9.2 Lead Botanical Garden (LBG), Department of Botany, Shivaji University, Kolhapur: (a) front view of garden, (b) aerial view of garden.

Botanical Gardens and Their Role in Education, Research 157

confirmation of the identity of species. The LBG has established good infrastructure required for education, conducting workshop/seminars, irrigation facilities, raising saplings, and maintenance of the garden. Presently, the garden has five polyhouses, three shade net houses, two wire houses, one glass house, six water tanks, two wells, three water bodies, pumping and irrigation facilities, one mini tractor, five nurseries, audio-visual aids, plasma screens, tissue culture laboratories, grass cutter machines, solar lamps, a library, computers and peripherals, rooms for gardeners and garden implements, and basic facilities like electricity, internet, drinking water, and administrative support.

9.4.2 MAJOR SECTIONS IN LBG

Although, with time, the structure of a developing garden changes, especially with regard to sections on herbaceous plants, the major sections remain more or less constant. Presently, the LBG has the following sections: Fernery; Pinetum; Conservatory of Rhizomatous, Cormatous, Tuberous, and Bulbous Plants;, Aquatic Plants; Medicinal Plants; Palmetum; Orchidarium; Arboretum of Medicinal Plants; Arboretum of Rare, Endangered, and Threatened (RET) Plants; Ficus Grove; Insectivorous Plants; Climbers; Endangered and Endemic Plant Species Nursery; Mangrove Nursery; *Barleria* Germplasm Bank; *Flemingia* Germplasm Bank; Nilambari Auditorium; Herbarium (SUK); and other facilities, such as the Library of Taxonomic Literature, and the Herbarium and Photo Gallery under the Center for Taxonomy Training.

9.4.2.1 Fernery

The fernery of the department was established in 2005 with financial support from MoEF & CC. The fernery holds germplasm collections of 61 species of which 11 are endemic and medicinal ferns. Ferns and fern allies from the northern Western Ghats, such as *Actiniopteris radiata, Adiantum capillus-veneris, A. caudatum, A. concinnum, A. incisum, A. lunulatum, A. poiretii, A. raddianum, Aleuritopteris bicolor, Angiopteris helferiana, Anogramma leptophylla, Athyrium falcatum, A. hohenackerianum, A. parasnathense, A. pectinatum, Azolla pinnata* subsp. *asiatica, Botrychium lanuginosum, Calamaria coromandelina, Ceratopteris thalictroides, Cheilanthes albomarginata, C. anceps, C. rufa, Cheilosoriate nuifolia, Christella dentate, C. parasitica, Diplazium esculentum, Equisetum ramosissimum, Huperzia hamiltonii, H. hamiltonii, Isoetes dixitii, I. panchganiensis, I. sahyadriensis, Lepisorus nudus, Leptochilus decurrens, Lindsaea ensifolia, L. heterophylla, Lygodium flexuosum, Marsilea minuta, Microsorum membranaceum, M. punctatum, Ophioglossum costatum, O. lusitanicum, O. lusitanicum* subsp. *coriaceum, O. nudicaule, O. parvifolium, O. petiolatum, O. reticulatum, Osmunda hugeliana, Pityrogramma calomelanos, Pteridium revolutum, Pteris biaurita, P. blumeana, P. camerooniana, P. linearis, P. vittata, Pyrrosia lanceolata, Regnellidium diphyllum, Salvinia adnata, Selaginella ciliaris, S. crassipes, S. delicatula, S. repanda,* and *S. tenera* were introduced in LBG.

9.4.2.2 Rhizomatous, Cormatous, Tuberous, and Bulbous Plants

A rich collection of rhizomatous, cormatous, tuberous, and bulbous wild plant species have been maintained in the garden as they are important from the point of view of food, medicine, ornamental, and alkaloid yielding plants. Some of these in the LBG are the following: *Amorphophallus paeoniifolius, Aponogeton crispus, A. natans, A. undulatus, Asparagus racemosus, Chlorophytum arundinaceum, C. belgaumense, C. bharuchae, C. borivilianum, C. breviscapum, C. glaucoides, C. gothanense, C. heynei, C. indicum, C. kolhapurense, C. laxum, C. malabaricum, C. nepalense, C. nilgheriensis, C. tuberosum, Crinum asiaticum, C. brachynema, C. latifolium, C. malabaricum, C. pratense, C. solapurense, C. viviparum, C. woodrowii, Cryptocoryne spiralis, C. ciliata, Curcuma aromatica, C. caesia, C. neilgherrensis, Dioscorea bulbifera, Dipcadi concanense, D. erythraeum, D. goaense, D. minor, D. montanum, D. saxorum, D. serotinum, D. ursulae, Drimia coromandeliana, D. govindappae, D. indica, D. nagarjunae, D. polyantha, D. raogibikei, D. razii,*

D. wightii, Eulophia nuda, Gloriosa superba, Hellenia speciosa, Iphigenia indica, Iphigenia magnifica, I. mysorensis, I. pallida, I. ratnagirica, I. stellata, Ledebouria revoluta, Pancratium donaldii, P. longiflorum, P. nairii, P. parvum, P. st-mariae, P. triflorum, P. verecundum, P. zeylanicum, Sauromatum venosum, Stemona tuberosa, and *Tacca leontopetaloides.* More than 75 species of rhizomatous, cormatous, tuberous, and bulbous wild plant species are maintained in LBG; many of them are endemic, and some of them are threatened plant species of peninsular India.

9.4.2.3 The Pinetum

The Pinetum is situated in front of the department. *Mesembrioxylon mahabalei*, a petrified fossil of a member of Podocarpaceae, 33.7–23.8 million years old, is in the center of the Pinetum and forms a major attraction for students and visitors (Figure 9.1). The Pinetum has about 30 species of gymnosperms. Some notable species are *Agathis alba, Araucaria bidwillii* Hook., *Araucaria columnaris, Cupressus* sp., *Cycas pschannae, C. beddomei, C. rumphii, C. revoluta, Gnetum ula, Pinus roxburghii, Dioon spinulosum, Podocarpus macrophyllus, P. neriifolius,* and four species of *Zamia*.

9.4.2.4 Aquatic Plants

Remarkable aquatic plant species in this section are *Acorus calamus* L., *Aponogeton bruggennii, A. crispus, A. natans, A. undulatus, Cryptocoryne ciliata, C. spiralis, Limnocharis flava, Limnophyton obtusifolium, Ludwigia sedoides, Nelumbo nucifera, Nymphaea nouchali, N. pubescens, N. rubra, Nymphoides indica, Nypa fruticans, Ottelia alismoides, Pontederia hastata, Rotala densiflora, Rotala macrandra, Victoria amazonica* (The giant water lily), *Wiesneria triandra*, etc.

9.4.2.5 Medicinal Plants

This section was developed for live demonstration of especially herbaceous or shrubby medicinal plants to visitors. More than 75 species of herbaceous medicinal plants have been maintained in this section. Some of them are *Aloe vera, Andrographis paniculata, Ardisia solanacea, Asparagus racemosus, Baliospermum solanifolium, Barleria prionitis, Catharanthus roseus, Celastrus paniculatus, Cissus quadrangularis, Costus pictus, Curcuma aromatica, Curcuma caesia, Decalepis arayalpathra, Decalepis salicifolia, Embelia ribes, Gloriosa superba, Gymnema sylvestre, Hellenia speciosa, Hemidesmus indicus, Kalanchoe pinnata, Lobelia nicotianifolia, Ocimum tenuiflorum, Pandanus amaryllifolius, Piper diurnum, Piper betle, Piper longum, Piper sarmentosum, Plumbago zeylanica, Rauvolfia serpentina, Rotheca serrata, Rubia cordifolia, Ruta chalepensis, Scaevola taccada, Spermadictyon suaveolens, Tinospora cordifolia, Trichopus zeylanicus, Vitex negundo, Withania somnifera,* and *Zingiber zerumbet*.

9.4.2.6 Palmetum

Palms stand second to Poaceae in their economic importance and have great ornamental beauty. Griffith et al. (2021) consider them as living treasures for research and education. Therefore, special efforts were made to introduce them in the garden. The diversity of palms in the Palmetum provides a tropical look to LBG, and it is an important attraction for visitors (Figure 9.3). The Palmetum has a collection of more than 110 species of both indigenous and exotic palms. Of the 37 indigenous palms introduced in LBG, 8 are endemic to India. The palms of the Palmetum include the following: *Acoelorrhaphe wrightii, Adonidia merrillii, Aiphenes horrida, Archontophoenix alexandrae, Areca catechu, A. triandra, Arenga engleri, A. micrantha, A. pinnata, A. wightii, Attalea cohune, Bentinckia condapanna, B. nicobarica, Bismarckia nobilis, Borassus flabellifer, Brahea armata, Butea capitata, Calamus kingianus, C. nagbettai, Carpentaria acuminata, Caryota mitis, C. obtusa, C. urens, Chamaedorea elegans, C. metallica, C. microspadix, Chamaerops humilis, Coccothrinax baebadensis, Coccothrinax crinita, Coccothrinax macroglossa, Cocos nucifera, Copernicia prunifera, Corypha macropoda, C. umbraculifera, C. utan, Cyrtostachys album* var. *aureum, Cyrtostachys renda, Dictyosperma album, Dypsis decaryi, D. madagascariensis,*

FIGURE 9.3 Some important palms in LBG: (a) *Hyphaene dichotoma*, (b) *Nypa fruticans*, (c) *Corypha umbraculifera*, and (d) *Pinanga manii*.

D. lutescens, Elaeis guineensis, Euterpe oleracea, Howea foesteriana, Hyophorbe lagenicaulis, H. verschaffeltii, Hyphaene dichotoma, H. thebaica, Kerriodoxa elegans, Korthalsia rogersii, Latania loddigesii, L. lontaroides, L. verschaffeltii, Leopoldia maritime, Licuala rumphii, L. grandis, L. lauterbachii, L. peltata, L. spinosa, Livistona australis, Livistona decora, L. rotundifolia, L. jenkinsiana, Nannorrhops ritchieana, Normanbya normanbyi, Nypa fruticans, Oenocarpus bacaba, Phoenix acaulis, P. canariensis, P. dactylifera, P. hanceana, P. loureiroi, P. loureiroi var. *pedunculata, P. paludosa, P. pusilla, P. reclinate, P. roebelenii, P. rupicola, P. sylvestris, Pinanga coronata, P. dicksonii, P. gracilis, P. manii, P. sylvestris, Pritchardia pacifica, Ptychosperma macarthurii, Rhapaloblaste augusta, Rhapis excelsa, R. humilis, R. laosensis, R. multifida, Roystonea regia, Sabal mauritiiformis, S. minor, S. palmetto, Saribus rotundifolius, Serenoa repens, Syagrus coronata, Syagrus romanzoffiana, Syagrus schizophylla, Thrinax excelsa, T. parviflora, T. radiata, Trachycarpus fortunei, T. martianus, T. takil, Wallichia disticha, W. oblongifolia, Washingtonia filifera, W. robusta,* and *Wodyetia bifurcata*. Important endemic species in the garden are *Arenga wightii, Bentinckia condapanna, B. nicobarica, Calamus kingianus, Pinnanga dicksonii, P. manii, Rhopaloblaste angusta,* and *Trachycarpus takil. Arenga wightii, Corypha umbraculifera,* and *C. utan* are some of the most elegant indigenous palms of the garden (Figure 9.3).

9.4.2.7 Orchidarium

Attempts have been made to introduce about 88 wild species of orchids from the Western Ghats in LBG. They include the following: *Acampe praemorsa, Aerides crispa, A. maculosa, A. ringens, Bulbophyllum fimbriatum, B. sterile (Phyllorkis nilgherensis), B. stocksii (Trias stocksii), Cottonia peduncularis, Crepidium versicolor (Malaxis rheedii, Malaxis versicolor), Dendrobium aqueum, D.barbatulum, D. crepidatum, D. herbaceum, D. macraei, D. macrostachyum, D. microbulbon, D. nanum, D. ovatum, D. peguanum, Eulophia andamanensis, E. dabia, E. graminea, E. nuda, E. ochreata, Gastrochilus flabelliformis, Geodorum densiflorum, Habenaria brachyphylla, Habenaria commelinifolia, H. crinifera, H. dentata, H. digitata, H. diphylla, H. foliosa, Habenaria furcifera, H grandifloriformis, H. heyneana, H. longicorniculata, H. marginata, H. multicaudata, H. plantaginea, Habenaria roxburghii, H. stenopetala, H. suaveolens, H. suaveolens, Liparis nervosa, Luisia macrantha, L. tenuifolia* Blume, *L. tristis, Nervilia concolor, N. infundibulifolia, N. plicata, N. simplex (Nervilia crociformis), Oberonia brunoniana, O. mucronata, Oberonia recurva, Pecteilis gigantea, Peristylus densus, P. goodyeroides, P. lawii, P. plantagineus, Phaius tankervilleae, Pholidota pallida, Platanthera unalascensis, Polystachya concreta, Porpax filiformis, P. jerdoniana, P. reticulata, Rhynchostylis retusa, Smithsonia maculata, S. viridiflora, Spathoglottis plicata, Taprobanea spathulata (Vanda spathulata), Thunia alba* var. *bracteata, Vanda tessellata, Vanda testacea, Vanilla walkerae, Zeuxine gracilis, Z. longilabris,* and *Z. strateumatica.*

9.4.2.8 Arboretum of Medicinal Plants

Tree species play important roles in ecosystems. There are about 60,000 tree species belonging to diverse families worldwide. Extinction risk information available on 58,497 tree species worldwide revealed that 30%, i.e., 17,510, are threatened with extinction and at least 142 are recorded as extinct in the wild (BGCI, 2021). In India, there are about 2,603 tree species of which 650 are endemic and 469 (18%) are threatened. Conservation of tree species is of crucial importance as 200–300 other species depend on single tree species for their survival. Major emphasis is given on conservation of medicinal, indigenous, and endemic trees of the Western Ghats in the Lead Botanical Garden.

An arboretum of tree species with established medicinal value in LBG harbors more than 125 species, including the following: *Acrocarpus fraxinifolius, Adansonia digitata, Adina cordifolia, Aegle marmelos, Ailanthus excelsa, Alangium salviifolium, Alstonia scholaris, Ancistrocladus heyneanus, Antiaris toxicaria, Aphanamixis polystachya, Artocarpus heterophyllus, Atalantia racemosa, Averrhoa bilimbi, Averrhoa carambola, Azadirachta indica, Balanites aegyptiaca, Bauhinia racemosa, Bixa orellana, Bombax ceiba, Boswellia serrata, Buchanania cochinchinensis, Butea monosperma, Calophyllum inophyllum, Canarium strictum, Caryota urens, Casearia tomentosa, Cassia fistula, Castanospermum australe, Chloroxylon swietenia, Chukrasia tabularis, Cinnamomum camphora, Cochlospermum religiosum, Cocos nucifera, Commiphora wightii, Cordia dichotoma, Crateva adansonii, Dichrostachys cinerea, Dillenia indica, Dolichandrone falcata, Dysoxylum gotadhora, Dysoxylum malabaricum, Elaeocarpus tuberculatus, Embelia ribes, Erythrina variegata, Erythroxylum monogynum, Eucalyptus globulus, Ficus benghalensis, F. carica, F. hispida, F. racemosa, F. religiosa, Flacourtia montana, Garcinia cambogioides, G. gummi-gutta, G. indica, G. xanthochymus, Gardenia resinifera, Garuga pinnata, Gmelina arborea, Grewia asiatica, Helicteres isora, Heterophragma quadriloculare, Holarrhena pubescens, Hopea ponga, Hydnocarpus pentandrus, Limonia acidissima, Litsea ligustrina, Madhuca longifolia, Magnolia champaca, Mallotus philippensis, Mammea suriga, Mangifera indica, Mappia nimmoniana, Mesua ferrea, Mimusops elengi, Mitragyna parvifolia, Morinda citrifolia, Moringa concanensis, Murraya koenigii, Myristica fragrans, M. malabarica, Nothopegia castaneifolia, Nyctanthes arbor-tristis, Oroxylum indicum, Ougeinia oojeinensis, Pandanus furcatus, Parkinsonia aculeata, Phyllanthus emblica, Pittosporum dasycaulon, Plumeria alba, Premna tomentosa, Pterocarpus marsupium, P. santalinus, Pterospermum acerifolium, Putranjiva roxburghii, Radermachera xylocarpa, Salacia chinensis, Santalum album, Saraca asoca, Schrebera swietenioides, Semecarpus anacardium, Senegalia catechu, Soymida febrifuga, Spondias pinnata, Sterculia urens, Stereospermum chelonoides, Streblus asper, Strychnos nux-vomica, S. potatorum, Syzygium cumini, Tabernaemontana alternifolia, Tamarindus indica, Tectona grandis, Terminalia arjuna, Terminalia bellirica, Terminalia chebula, Thespesia populnea, Vachellia nilotica, Wrightia tinctoria, Xantolis tomentosa, Xylia xylocarpa, Zanthoxylum rhetsa,* and *Ziziphus mauritiana.*

9.4.2.9 Arboretum of Rare, Endangered, and Threatened (RET) and Endemic Plant Species

Presently, more than 125 RET and endemic trees species from the Western Ghats, the Eastern Ghats, Northeast India, and the Andaman Nicobar Islands have been introduced in the arboretum of LBG. Threatened species introduced in LBG include the following: *Actinodaphne lanceolata, Ailanthus triphysa, Arenga wightii, Artocarpus gomezianus, A. hirsutus, Atalantia wightii, Baccaurea courtallensis, Beilschmiedia dalzellii, Bentinckia condapanna, B. nicobarica, Blepharistemma membranifolium, Bombax insigne, Calophyllum apetalum, Calophyllum inophyllum, Canarium strictum, Carallia brachiata, Caryota urens, Chionanthus mala-elengi, Cinnamomum goaense, Cinnamomum malabatrum, Cinnamomum sulphuratum, Clausena anisata, Cordia domestica, Corypha umbraculifera, Cullenia exarillata, Cynometra bourdillonii, C. travancorica, Diospyros buxifolia, D. crumenata, D. paniculata, Drypetes venusta, Duabanga grandiflora, Dysoxylum gotadhora, D. malabaricum, Elaeocarpus munroi, E. serratus, Erinocarpus nimmonii, Flacourtia montana, Garcinia cambogioides, G. gummi-gutta, G. indica, G. rubroechinata, G. talbotii, G. wightii, G. xanthochymus, Garuga pinnata, Glochidion ellipticum, G. zeylanicum* var. *tomentosum, Gluta travancorica, Hardwickia binata, Harpullia arborea, Heptapleurum bourdillonii, Hildegardia populifolia, Holigarna arnottiana, H. grahamii, Hopea canarensis, H. parviflora, H. ponga, H. racophloea, Humboldtia bourdillonii, H. brunonis, H. decurrens, H. sanjappae, Hydnocarpus pentandrus, Hyphaene dichotoma, Knema attenuata, Lagerstroemia microcarpa, Lagerstroemia parviflora, Lophopetalum wightianum, Macaranga peltata, Machilus glaucescens (Persea macrantha), Mallotus resinosus, Mangifera andamanica, Mastixia arborea, Meiogyne pannosa, Memecylon randerianum (Memecylon malabaricum), M. talbotianum, Mesua ferrea, M. ferrea* var. *coromandeliana, Mitragyna parvifolia, Monoon fragrans (Polyalthia fragrans), Myristica dactyloides, M. magnifica, M. malabarica, Neolamarckia cadamba, Nothopegia castaneifolia, Otonephelium stipulaceum, Palaquium ellipticum, Phoenix loureiroi* var. *pedunculata (Phoenix robusta), Piliostigma foveolatum, Pinanga dicksonii, P. manii, Pittosporum dasycaulon, Poeciloneuron indicum, Polyspora obtusa (Gordonia obtusa), Prioria pinnata, Psydrax dicoccos, Pterocarpus marsupium, P. santalinus, Pterospermum reticulatum, Rhopaloblaste augusta, Sageraea laurina (Sageraea laurifolia), Semecarpus kathalekanensis, Semecarpus kurzii, Shorea robusta, Solenocarpus indicus (Spondias indica), Stereospermum chelonoides (Stereospermum suaveolens), Strychnos nux-vomica, Symplocos racemosa, Syzygium laetum, Syzygium stocksii, Tabernaemontana alternifolia, Terminalia bialata, Terminalia paniculata, Trachycarpus takil, Turpinia malabarica, Vateria indica, Vatica chinensis, Vepris bilocularis,* and *Vitex altissima.*

9.4.2.10 Plant Groves in LBG

Attempts have been made to introduce 35 local varieties of mangoes in LBG in *Mangifera indica* germplasm grove. Clusiaceae and *Garcinia* grove includes *Calophyllum apetalum, C. inophyllum, Garcinia gummi-gutta, G. imberti, G. indica, G. livingstonei, G. mangostana, G. morella, G. spicata, G. talbotii,* and *G. xanthochymus. Ficus* species are keystone species, and about 12 species, viz. *Ficus benghalensis, F. auriculata, F. benghalensis* var. *krishnae, F. callosa, F. drupacea, F. elastica, F. lyrata, F. microcarpa, F. pumila, F. racemosa, F. religiosa, F. benjamina,* have been planted and maintained in the Ficus Grove of LBG. *Victoria amazonica* and *Nepenthes khasiana* are also introduced in the garden.

9.4.2.11 Climbers

Important climbers of the Western Ghats and some exotic climbers have been planted all along the compound wall of the Lead Botanical Garden. These climbers are *Aganosma cymosa*, *A. cathartica*, *Anredera cordifolia*, *Capparis zeylanica*, *Aristolochia littoralis*, *A. ringens*, *A. tagala*, *Asparagus racemosus*, *Beaumontia jerdoniana*, *Celastrus paniculatus*, *Centrosema pubescens*, *Chonemorpha fragrans*, *Clitoria ternatea*, *Combretum razianum*, *Coscinium fenestratum*, *Cryptolepis buchananii* (*Cryptolepis dubia*), *Elaeagnus latifolia* L., *Gymnema latifolium*, *G. sylvestre*, *Hemidesmus indicus*, *Hiptage benghalensis*, *Kunstleria keralensis*, *Mucuna atropurpurea*, *M. monosperma*, *M. pruriens*, *M. pruriens* var. *utilis*, *M. sanjappae*, *Rubia cordifolia*, *Tarlmounia elliptica*, *Thunbergia alata*, *T. coccinea*, *T. erecta*, *T. fragrans*, *T. grandiflora*, *T. laurifolia*, *T. mysorensis*, *Tinospora cordifolia*, *T. sinensis*, *Vallaris solanacea*, and *Vincetoxicum indicum* (*Tylophora indica*). Rare endemic climbers include *Coscinium fenestratum*, *Kunstleria keralensis*, *Mucuna sanjappae*, and *Thunbergia mysorensis*.

9.4.2.12 Indigenous, Threatened, and Endemic Plant Species Nurseries

Lead Botanical Garden has established four nurseries where saplings of various indigenous, endemic, and economically important plant species are raised in the thousands. The activities are mainly targeted at raising of indigenous, threatened, and endemic species. Some important species are *Canarium strictum*, *Dysoxylum binectariferum*, *Hopea ponga*, *Knema attenuata*, *Manilkara littoralis*, *Myristica dactyloides*, *Polyalthia fragrans*, *Sageraea laurifolia*, *Syzygium laetum*, and *Vateria indica*.

9.4.2.13 Mangrove Nursery

The department is supported by UGC–SAP (University Grants Commission-Special Assistance Programme), under which a nursery of mangroves is constructed. The seedlings are raised, maintained, and reintroduced in their natural habitats in collaboration with the State Forest Department. The mangroves and associated species collections include *Rhizophora mucronata*, *Bruguiera gymnorhiza*, *B. cylindrica*, *Ceriops tagal*, *Kandelia candel*, *Cynometra iripa*, *Xylocarpus granatum*, *Heritiera littoralis*, *H. fomes*, *Dolicandrone spathacea*, *Excoecaria agallocha*, *Acanthus ilicifolius*, *Cerbera odollam*, *Avicennia marina*, *Sesuvium portulacastrum*, *Ixora coccinea*, *Ipomoea pes-caprae*, *Derris trifoliata*, *Morinda citrifolia*, *Calophyllum inophyllum*, *Caesalpinia crista*, *Volkameria inermis*, *Barringtonia racemosa*, *Hibiscus tiliaceus*, *Spinifex littoreus*, *Carallia brachiata*, and *Manilkara littoralis*.

9.4.2.14 *Barleria* and *Flemingia* Germplasm Banks

In India, there are about 30 species of *Barleria*, of which 17 are endemic to the country. Presently, LBG maintains *Barleria acanthoides*, *B. acuminata*, *B. courtallica*, *B. cristata* (pink, Purple, White flowered), *B. cuspidata*, *B. gibsonii*, *B. grandiflora*, *B. hochstetteri*, *B. involucrata* var. *elata*, *B. lawii*, *B. longiflora*, *B. lupulina*, *B. montana*, *B. mysorensis*, *B. nitida*, *B. noctiflora*, *B. prattiana*, *B. prionitis*, *B. prionitis* subsp. *pubiflora*, *B. repens*, *B. sepalosa*, *B. strigosa*, *B. stocksii*, *B. terminalis*, *B. tomentosa*, *B. buxifolia*, *B. durairajii*, and *B. pilosa* (Figure 9.4). Similarly, there are about 28 species of *Flemingia*, of which 6 taxa are endemic to India. This germplasm has importance as a peripheral germplasm for the economically important genus *Cajanus*. Therefore, attempts have been made to introduce them in LBG, and presently, *Flemingia chappar*, *F. lineata*, *F. nana*, *F. paniculata*, *F. praecox* var. *robusta*, *F. procumbens*, *F. semialata*, *F. sootepensis*, *F. stricta*, *F. strobilifera*, *F. wallichii*, and *F. wightiana* are maintained in LBG. They have great potential as

FIGURE 9.4 *Barleria* species collection and their utilization as ornamentals in LBG: (a) *B. cristata*, (b) *B. gibsonii*, (c) *B. grandiflora*, (d) *B. involucrata* var. *elata*, (e) *B. morrisiana*, (f) *B. nitida*, (g) *B. prattiana*, (h) *B. prionitis*, (i) *B. repens*, (j) *B. sepalosa*, (k) *B. strigosa*, (l) *B. tomentosa*.

medicinal and ornamental plants. *Crotalaria shevaroyensis* and *Moullava spicata* are some of the important species in the garden.

9.5 ROLE OF LBG IN *EX SITU* CONSERVATION OF PLANT DIVERSITY

Botanical gardens are of significant importance in *ex situ* conservation of especially endemic and threatened plant species worldwide. During the last two decades, LBG of Shivaji University has played a significant role not only in *ex situ* conservation but also in restoration of some of the threatened plant species of the Western Ghats. Lead Botanical Garden served as the most important asset in undertaking restoration species recovery programs.

A program supported by Rajiv Gandhi Science and Technology Commission (RGSTC), Government of Maharashtra, on *ex situ* conservation and development of an arboretum of a hundred rare, endangered, and threatened (RET) and endemic tree species of angiosperms of the Western Ghats was initiated in 2014, and more than 90 species (usually with each species represented by nine individuals) were introduced in an arboretum. This is probably the only arboretum in the country harboring about 100 RET and endemic tree species from the Western Ghats. Important tree species include *Arenga wightii, Artocarpus hirsutus, Atalantia wightii, Piliostigma foveolatum, Bentinckia condapanna, Bentinckia nicobarica, Blepharistemma membranifolium, Calophyllum apetalum, Chionanthus mala-elengi* subsp. *mala-elengi, Cinnamomum goaense, Dysoxylum malabaricum, Elaeocarpus serratus, Erinocarpus nimmonii, Flacourtia montana, Garcinia gummi-gutta, G. indica, G. cambogioides, G. talbotii, Gluta travancorica, Hardwickia binata, Hildegardia populifolia, Holigarna arnottiana, H. grahamii, Hopea parviflora, H. ponga, Humboldtia brunonis, H. decurrens, H. sanjappae, Hydnocarpus pentandrus, Hyphaene dichotoma, Pinanga dicksonii, P. manii, Prioria pinnata, Knema attenuata, Lagerstroemia microcarpa, Myristica malabarica, Nothopegia castaneifolia, Otonephelium stipulaceum, Pittosporum dasycaulon, Poeciloneuron indicum, Monoon fragrans, Pterocarpus santalinus, Pterospermum reticulatum, Sageraea laurina, Semecarpus travancoricus, Solenocarpus indicus, Syzygium laetum, S. stocksii, Tabernaemontana alternifolia, Vateria indica, Vatica chinensis*, etc. Of the six monotypic tree genera of Peninsular India, five, viz. *Blepharistemma membranifolium, Erinocarpus nimmonii, Hardwickia binata, Otonephelium stipulaceum,* and *Poeciloneuron indicum,* have been introduced in the garden.

Alangium salviifolium, Buchanania cochinchinensis, Carallia brachiata, Harpullia arborea, Shorea robusta, and *Terminalia bialata* are some of the other tree species of the garden. Palms are the main attraction in the garden, with more than 110 species of indigenous and exotic palms currently present. Important indigenous palms of the garden are *Areca triandra, Arenga wightii, Bentinckia condapanna, B. nicobarica, Borassus flabellifer, Caryota mitis, C. urens, Corypha umbraculifera, C. utan, Hyphaene dichotoma, Livistona jenkinsiana, Nypa fruticans,* five species of *Phoenix,* three species of *Licuala, Trachycarpus fortunei, T. takil,* and *Wallichia disticha* (Figure 9.5).

9.5.1 ROLE OF LBG IN BIOPROSPECTING WILD PLANTS AS ORNAMENTALS

Lead Botanical Garden plays a significant role in primary domestication of wild plants of ornamental value. Some six plants were selected for primary domestication under the DBT project, of which *Delphinium malabaricum*, with metallic, blue long-lasting flowers in racemes, was released as a floricultural crop for farmers after primary domestication (Figure 9.6). Similarly, methods for propagation of Barlerias were standardized, and *Barleria grandiflora* and *Barleria gibsonii* were introduced as ornamentals in the garden. More than 20 *Barleria* species have been introduced in the garden for their ornamental value, such as growth forms, foliage, and flowers.

Botanical Gardens and Their Role in Education, Research 165

FIGURE 9.5 *Ex situ* conservation of tree species in LBG: (a) *Ailanthus triphysa*, (b) *Artocarpus hirsutus*, (c) *Vateria indica*, (d) *Humboldtia brunonis*, (e) *Pterospermum reticulatum*, (f) *Carallia brachiata*, (g) *Calophyllum apetalum*, (h) *Harpullia arborea*, (i) *Cycas circinalis*.

FIGURE 9.6 *Delphinium malabaricum* (Nilambari) domesticated in LBG and released as a new floricultural crop for farmers (inset shows the inflorescence).

9.5.2 ROLE OF LBG IN RESTORATION OF THREATENED PLANT SPECIES OF THE WESTERN GHATS

LBG has paramount value in restoration and species recovery programs as it provides a base camp for establishment of standardized techniques for seed collection, seed germination, and seedling raising, and to overcome associated problems. During the last two decades, 20 species were selected for restoration in their natural habitats under the restoration program of threatened plants funded by the Department of Biotechnology (DBT) where facilities in LBG made it possible to undertake the work. Attempts have made to restore *Abutilon ranadei, Ceropegia anantii, C. anjanerica, C. attenuata, C. bhatii, C. concanensis, C. evansii, C. fantastica, C. huberi, C. jainii, C. juncea, C. lawii, C. maccannii, C. mahabalei, C. media, C. mohanramii, C. noorjahaniae, C. oculata, C. odorata, C. panchganiensis, C. rollae, C. sahyadrica, C. santapaui, C. spiralis, C. vincifolia, Erinocarpus nimmonii, Hubbardia heptaneuron*, and *H. diandra*. Usually fruit/seed setting and seedling establishment of the endangered *Ceropegia* species of the Western Ghats is rare. Therefore, both conventional and tissue culture methods were employed for propagation (Chandore et al., 2010; Chavan et al., 2011a, 2011b, 2013a, 2013b, 2014, 2017; Desai et al., 2014; Nalawade et al., 2016); LBG played an important role as a base camp with micropropagation and hardening facilities and reintroduction of the species.

Each one of the threatened species was propagated by conventional methods as well as through tissue culture methods, and saplings were reintroduced in their natural habitats with the help of students and in collaboration with the forest department. One of the successful stories of restoration of critically endangered species is that of *Hubbardia heptaneuron* and *H. diandra*. These species have been reintroduced in 25 Ghat areas at 100 suitable locations, and more than 10,000 individuals have been established (Yadav et al., 2009, 2010). The populations are self-perpetuating both

Botanical Gardens and Their Role in Education, Research 167

FIGURE 9.7 Restoration of threatened plant species involving students using LBG as base camp: (a–e) plantation of *Ceropegia* species, (f–g) reintroduction of *Hubbardia heptaneuron*.

vegetatively and sexually since their reintroduction. It is a successful story from India in restoration of the *Hubbardia* species, which is ecologically uniquely adapted to grow on vertical dripping rocks in the Western Ghats (Figure 9.7).

Rehabilitation of *Dipcadi concanense* and *Iphigenia ratnagirica* by research scholars and students is an activity of LBG worth mentioning. More than 20,000 individuals of both species from an area under building construction were transplanted at ten other localities in 2008, establishing their populations. Transplanted plants are growing in the areas of introduction and are sexually reproducing and self-perpetuating.

9.5.3 LBG Program on Seed Collection, Sapling Raising, and Distribution of RET/Endemic and Indigenous Tree Species

LBG has a regular yearly program on seed collection, especially from January to June from the Western Ghats and other botanical gardens in the country. Every year seeds of 100–200 indigenous plant species are collected from especially the Western Ghats, and saplings are raised in thousands for each one of the species. The seedlings are distributed to schools, colleges, universities, other botanical gardens, research institutes, forest departments, gram panchayats, etc. The saplings are also sold to different organizations throughout the region. Samples of carpological collections of all the species are maintained in LBG. This is a very fruitful program as local people are presently more interested in growing indigenous and rare plants. During the past two decades, LBG has raised and distributed hundreds of thousands of seedings of several hundred species to various organizations, thereby promoting and achieving goals of conservation, education, awareness, bioprospecting, and sustainable use of plant diversity (Figure 9.8).

LBG provides saplings of indigenous, threatened, and endemic plant species for botanical gardens of municipal corporations, schools, and colleges affiliated with Shivaji University, thereby achieving their larger goal of conservation of local plant species.

Indeed, the Lead Botanical Garden has become an important center for propagation and multiplication of native plant species, establishment of nursery techniques for threatened species, conservation and restoration of threatened plant species, bioprospecting of plant resources of the Western Ghats, education and learning, creating environmental awareness, and thus contributing significantly in *ex situ* as well as *in situ* conservation of plants of the Western Ghats. LBG serves as an important field laboratory to post-graduate students and research scholars. More than 10,000 visitors, including college students, research scholars, and scientists, pay visits to LBG each year, which is indicative of its role in learning, education, research, environmental awareness, conservation, and bioprospecting of plant resources. The Lead Botanical Garden of Shivaji University, supported by the Ministry of Environment, Forest and Climate Change and other funding agencies, is one of the significant botanical gardens of the country, where more than 1,000 species have been introduced, many of which are endemic and RET species of our country.

9.6 CONCLUSIONS

The Lead Botanical Garden concept by the Government of India is a well-conceived program under which LBGs are being supported. These LBGs are usually associated with academic, educational, and research institutes that have required expertise with them. These LBGs belong to government, semi-government, or public organizations. In other words, LBGs are public properties serving communities, and every citizen has access to the germplasm. The LBGs are playing important roles in training, education, research, and creation of general awareness in society about importance of biodiversity in the welfare of mother earth and peoples. These gardens are of crucial importance in *ex situ* conservation of critically endangered and economically important plant species. LBGs are biological laboratories for students and research scholars of plant sciences and recreation places for general public. LBGs and BGs have a significant role in bioprospecting wild plant resources into useful products. LBGs are primary grounds for domestication of plant species as ornamentals, crop-fruit yielding plants, medicines, and many other purposes. LBGs also have a role to identify threatened, economically important plant species and contribute to their multiplication and conservation by providing saplings of the right species in precise areas. They are important in developing nursery techniques for propagation of indigenous species and making them available to the public, the forest department, and other organizations. These activities help greatly in overall conservation of plant diversity and helping to mitigate climate change. To overcome all the difficulties, such as selection of species for conservation, multiplication, and seed and propagule collection; overcoming seed dormancy and reproductive barriers; raising and transplantation of saplings; and restoration of threatened species, organizations need expertise in various disciplines of biology, which are usually available in educational, academic, and research institutes. Therefore, LBGs need to be associated with these institutes. These institutes have great strengths in the students who are the owners of biodiversity. Therefore, educational and research institutes have profound significance in documentation, conservation, bioprospecting, and sustainable utilization of our plant resources. BGs and LBGs play multidimensional roles and need to be generously supported.

Botanical Gardens and Their Role in Education, Research

FIGURE 9.8 Seed collection, seedling raising, and developing nursery techniques for wild plant species in LBG: (a–b) seed collection from different localities, (c–d) seedling raising and developing nursery.

REFERENCES

Anonymous. 2013. *Guidelines for Assistance to Botanic Gardens*. Ministry of Environment and Forest, Government of India, pp. 1–37.

Ballantyne, R., J. Packer and K. Hughes. 2008. Environmental awareness, interests and motives of botanic gardens visitors: Implications for interpretive practice. *Tourism Management* 29: 439e444.

Bapat, V.A., G.B. Dixit and S.R. Yadav. 2012. Plant biodiversity conservation and role of Botanists. *Current Science* 102: 1366–1369.

Bapat, V.A., S.R. Yadav and G.B. Dixit. 2008. Rescue of endangered plants through biotechnological applications. *National Academy of Science Letters* 31(7–8): 201–210.

Bennett, B. 2014. Learning in paradise: The role of botanic gardens in university education. In: Quave, C. (eds.), *Innovative Strategies for Teaching in the Plant Sciences*. Springer, New York, NY. https://doi.org/10.1007/978-1-4939-0422-8_13

BGCI. 2021. *State of the World's Trees*. BGCI, Richmond, UK.

Blackmore, S., M. Gibby and D. Rae. 2011. Strengthening the scientific contribution of botanic gardens to the second phase of the global strategy for plant conservation. *Botanical Journal of Linnean Society* 166: 267–281.

Breman, E., D. Ballesteros, E. Castillo-Lorenzo, C. Cockel, J. Dickie, A. Faruk, K. O'Donnell, C.A. Offord, S. Pironon, S. Sharrock and T. Ulian. 2021. Plant diversity conservation challenges and prospects–The perspective of botanic gardens and the millennium seed bank. *Plants* (Basel). 10(11): 2371.

Chandore, A.N., M.S. Nimbalkar, R.V. Gurav, V.A. Bapat and S.R. Yadav. 2010. A protocol for multiplication and restoration of *Ceropegia fantastica* Sedgw.: A critically endangered plant species. *Current Science* 99(11): 1593–1596.

Chang, L., R.J. Bisgrove and M. Liao. 2008. Improving educational functions in botanic gardens by employing landscape narratives. *Landscape and Urban Planning* 86: 233–247.

Chavan, J.J., N.B. Gaikwad, G.B. Dixit, S.R. Yadav and V.A. Bapat. 2017. Biotechnological interventions for propagation, conservation and improvement of 'Lantern Flowers' (*Ceropegia* spp.). *South African Journal of Botany* 114: 192–216.

Chavan, J.J., N.B. Gaikwad, P.R. Kshirsagar, S.D. Umdale, K.V. Bhat, G.B. Dixit, S.R. Yadav. 2013b. Application of molecular markers to appraise the genetic fidelity of *Ceropegia spiralis*, a threatened medicinal plant of South India. *Current Science* 105(10): 1348–1350.

Chavan, J.J., N.B. Gaikwad and S.R. Yadav. 2013a. High multiplication frequency and genetic stability analysis of *Ceropegia panchganiensis*, a threatened ornamental plant of Western Ghats: Conservation implications. *Scientia Horticulturae* 161(4): 134–142.

Chavan, J.J., A.S. Nalawade, N.B. Gaikwad, R.V. Gurav, G.B. Dixit and S.R. Yadav. 2014. An efficient in vitro regeneration of *Ceropegia noorjahaniae*: An endemic and critically endangered medicinal herb of the Western Ghats. *Physiology and Molecular Biology of Plants*. DOI: 10.1007/s12298-014-0236-4.

Chavan, J.J., M.S. Nimbalkar, A.A. Adsul, S.S. Kamble, N.B. Gaikwad, G.B. Dixit, R.V. Gurav, V.A. Bapat and S.R. Yadav. 2011a. Micropropagation and in vitro flowering of endemic and endangered plant *Ceropegia attenuata* Hook. *Journal of Plant Biochemistry and Biotechnology* 20(2): 276–282.

Chavan, J.J., M.S. Nimbalkar, N.B. Gaikwad, G.B. Dixit and S.R. Yadav. 2011b. In vitro propagation of *Ceropegia spiralis* Wight: An endemic and rare potential ornamental plant of peninsular India. *Proc. Natl. Acad. Sci., India* 81(1): 120–126.

Chen, G. and S. Weibang. 2018. The role of botanical gardens in scientific research, conservation, and citizen science. *Plant Diversity* 40(4): 181–188.

Chen, J., C.H. Cannon and H.B. Hu. 2009. Tropical botanical gardens: At the *in-situ* ecosystem management frontier. *Trends Plant Sci.* 14: 584–589.

Crane, P.R., S.D. Hopper, P.H. Raven and D.W. Stevenson. 2009. Plant science research in botanic gardens. *Trends in Plant Science* 14: 575–577.

Desai, M.T., S.S. Kambale, A.S. Nalawade, N.B. Gaikwad, R.V. Gurav, G.B. Dixit and S.R. Yadav. 2014. In vitro propagation of *Ceropegia fimbriifera* Bedd., an endangered, endemic plant of South India. *J. Intern. Academic Res. for Multidisciplinary* 2(1): 124–132.

Deshpande, S.M. and S.R. Yadav. 2017. Role of botanical gardens in conservation of rare plants: A case study of lead botanic garden of Shivaji University. *Indian Forester* 143(5): 471–482.

Donaldson, J.S. 2009. Botanic gardens science for conservation and global change. *Trends in Plant Science* 14: 608–613.

Griffith, M.P., A. Meyer and A. Grinage. 2021 Global *ex situ* conservation of palms: Living treasures for research and education. *Front. For. Glob. Change* 28 September 2021. https://doi.org/10.3389/ffgc.2021.711414

Hardwick, K.A., P. Fiedler, L. Lee, B. Pavlik, R. Hobbs, J. Aronson, M. Bidartondo, Eric Black, D. Coates, M. Daws, K. Dixon, S. Elliott, K. Ewing, George D. Gann, D. Gibbons, Joachim Gratzfeld, M. Hamilton, D. Hardman, Jim A. Harris, P. Holmes, Meirion Jones, D. Mabberley, A. MacKenzie, C. Magdalena, R. Marrs, W. Milliken, A. Mills, E.N. Lughadha, M. Ramsay, Paul L. Smith, N. Taylor, C. Trivedi, M. Way, Oliver Q. Whaley and S. Hopper 2011. The role of botanic gardens in the science and practice of ecological restoration. *Conservation Biology* 25: 265–275.

He, H. and J. Chen. 2012. Educational and enjoyment benefits of visitor education centers at botanical gardens. *Biological Conservation* 149: 103–112.

Heyd, T. 2006. Thinking through botanic gardens. *Environmental Values* 15: 197–212.

Hill, A.W. 1915. The history and functions of botanic gardens. *Annals of Missouri Botanical Garden* 2: 185–240. https://lisbdnet.com› how-much-tropical-rainforest-is-cleared-every-minute

Meyer, A. and N. Barton. 2019. Botanic gardens are important contributors to crop wild relative preservation. *Crop Science* 59: 2404–2412.

Miller, B., W. Conway and R.P. Reading. 2004. Evaluating the conservation mission of zoos, aquariums, botanical gardens, and natural history museums. *Conservation Biology* 18: 86–93.

Mounce, R., P. Smith and S. Brockington. 2017. *Ex situ* conservation of plant diversity in the world's botanic gardens. *Nature Plants*. 3: 795–802.

Myers, N. 1993. Biodiversity and the precautionary principle. *Ambio* 22: 74–79.

Nalawade, A.S., M.T. Desai, N.B. Gaikwad, R.V. Gurav, G.B. Dixit and S.R. Yadav. 2016. In vitro propagation of *Ceropegia anjanerica* Malpure et al.: A rare endemic plant from Maharashtra. *Proceedings of National Academy of Sciences, India* 86: 275–281.

O'Donnell, K. and S. Sharrock. 2017. The contribution of botanic gardens to *ex situ* conservation through seed banking. *Plant Divers*. 39: 373–378.

O'Donnell, K. and S. Sharrock. 2018. Botanic gardens complement agricultural gene bank in collecting and conserving plant genetic diversity. *Biopreservation and Biobanking* 16(5): 384–390.

Oldfield, S.F., P.R. Crane and S.D. Hopper. 2009. Botanic gardens and the conservation of tree species. *Trends in Plant Science* 14: 581–583.

Oldfield, S. and A.C. Newton. 2012. *Integrated Conservation of Tree Species by Botanic Gardens: A Reference Manual*. Botanic Gardens Conservation International. Richmond, Surrey, UK.

Powledge, F. 2011. The evolving role of botanical gardens: Hedges against extinction, showcases for botany? *BioScience* 61(10): 743–749.

Primack, R.B. and A.J. Miller-Rushing. 2009. The role of botanical gardens in climate change research. *New Phytologist* 182: 303–313.

Ren, H. 2017. The role of botanical gardens in reintroduction of plants. *Biodiversity Science* 25(9): 945–950.

Sellmann, D. 2014. Environmental education on climate change in a botanical garden: Adolescents' knowledge, attitudes and conceptions. *Environmental Education Research* 20: 286–287.

Singh, J.S. 2002. The biodiversity crisis: A multifaceted review. *Current Science* 82(6): 638–647.

Stevens, A.D. 2007. Botanical gardens and their role in *ex situ* conservation and research. *Phyton*. 46: 211–214.

Waylen, K. 2006. Botanic gardens: Using biodiversity to improve human wellbeing. *Medicinal Plant Conservation* 12: 4–8.

Wyse Jackson, P.S. 1999. Experimentation on a large scale: An analysis of the holdings and resources of botanic gardens. *Botanic Garden Conservation International* 3: 53–57.

Yadav, S.R., A.N. Chandore, S.M. Gund, M. Nandikar and M. Lekhak. 2010. Relocation of *Hubbardia heptaneuron* Bor, from its type locality. *Current Science* 98(7): 884.

Yadav, S.R., A.N. Chandore, M.S. Nimbalkar and R.V. Gurav. 2009. Reintroduction of *Hubbardia heptaneuron* Bor, a critically endangered endemic grass in Western Ghats. *Current Science* 96(7): 880.

10 Lead Botanical Garden of Yogi Vemana University, Kadapa, India, and Its Role in Plant Conservation

A. Madhusudhana Reddy and C. Nagendra

CONTENTS

10.1 Introduction 173
10.2 Role of Botanical Gardens in Plant Conservation 174
10.3 Lead Botanical Garden of Yogi Vemana University (YVUBG) 176
10.4 Conclusions 186
Acknowledgments 186
References 186

10.1 INTRODUCTION

Worldwide destruction of the natural environment has alarming consequences, especially in developing countries. Populations of endemic, threatened, and native species are decreasing day to day in the wild due to various factors, such as anthropogenic activities and exogenous and endogenous factors. This may cause extinction of many species in the future. The Indian subcontinent, with its rich biodiversity, is one of the 12 mega-diversity (centers) regions of the world. The Western Ghats, the Eastern Ghats, the Himalayas, the northeastern hills, and the Andaman and Nicobar Islands are the main biodiversity hotspots of India. The Eastern Ghats are generally considered as the hill ranges lying on the eastern side of the Deccan Plateau of Peninsular India. The Eastern Ghats are home for several flowering plants, and the area contains 3,417 species belonging to 1,172 genera and 180 families (Pullaiah and Karuppusamy, 2020); of these, 188 species belonging to 120 genera are endemic. The core forest area of the Eastern Ghats, which spreads through Andhra Pradesh, Odisha, Telangana, Tamil Nadu, and parts of Chattisgarh and Karnataka, has decreased by almost 32,200 km^2 between 1920 and 2015. The overall forest cover, spread over 43.4% of the Eastern Ghats in 1920, reduced to 27.5% by 2015, and it's not just loss of forest cover but also loss of native biodiversity. The Eastern Ghats have lost 15.83% of forest area over a span of ~100 years (Ramachandran et al., 2018). Changes in land use and land cover have significantly affected the extent and condition of forests in the Eastern Ghats of India, causing a decline in the forest cover as well as disturbing the natural habitats. The endemic, threatened, rare, medicinal, economically useful, and native species to the Eastern Ghats are now facing the risk of extinction. In the Eastern Ghats, growing human population, overexploitation, and endangerment of biological resources are leading to habitat loss of flora and fauna and, in turn, loss of biodiversity. In the Eastern Ghats, immediately appropriate conservation strategies should be initiated on these threatened areas to prevent further decline in the extent and habitat quality of the flora and fauna in the Eastern Ghats. For their long-term survival, they need to be conserved through *in situ* and *ex situ* conservation

DOI: 10.1201/9781003281252-10

strategies. Apart from *in situ* conservation, *ex situ* conservation and mass propagation of plants are important not only to conserve but also to sustainably utilize these resources.

10.2 ROLE OF BOTANICAL GARDENS IN PLANT CONSERVATION

Botanical gardens offer the opportunity to conserve and mange a wide range of plant diversity, including endemic, threatened, medicinal, ornamental, and edible species in *ex situ*. In India, several large- and small-scale botanical gardens have been established for *ex situ* conservation. Several of these botanical gardens have also become popular travel destinations and entertainment sites for local people. In recent years, the botanical gardens also play major roles in public education on environment, *ex situ* conservation, and scientific research in order to facilitate plant conservation and sustainability. In view of this background and context, a scheme was initiated by the Ministry of Environment, Forest and Climate Change (MoEF & CC) Government of India in 1991 to promote *ex situ* conservation and propagation of endemic and threatened plants through a network of botanic gardens (BGs) and centers of *ex situ* conservation. Financial assistance has been provided through the Lead Botanical Gardens scheme by MoEF.

Botanic gardens are institutions holding documented collections of living plants for the purpose of scientific research, conservation, display, and education. Botanic gardens have had a changing role throughout history, and they continue to adapt and serve the needs of society as new challenges arise. They host unique resources, including diverse collections of plant species growing in natural conditions, historical records, and expert staff, and they attract large numbers of visitors and volunteers. Being major tourist destinations, attracting an estimated 500 million visitors each year, they are important contributors to local and national economies. They also provide many benefits to society, such as having a positive impact on mental and physical health, particularly in urban settings where the majority of botanic gardens are situated. Botanic gardens have in more recent years become key players in both the conservation of plants and in the education of the people who come to see them. Like zoos, botanical gardens often work in tandem with each other, exchanging seeds, pollen, and other genetic information to preserve RET and endemic species. Several botanic garden activities are important for the conservation of plants around the world, for the following reasons:

- Living collections and seed banks safeguard species and enable the restoration and rehabilitation of degraded habitats.
- They conduct research and development into plant taxonomy and genetics, phytochemistry, useful plant properties, and informing selection of plants that can withstand degraded and changing environments (especially important with the threat posed by climate change).
- Education is a strength of botanic gardens that allows them to communicate the importance of conserving plants, reaching out to diverse audiences, and also to communicate how this may be achieved.
- Botanic gardens link plants with the well-being of people and also help to conserve indigenous and local knowledge and to encourage the sustainable use of plant resources for the benefit of all, as part of sustainable development.

The cultivation of plants has been around for thousands of years with the first examples dating to around 3,000 years ago in ancient Egypt and Mesopotamia. However, the first "true" botanic gardens with an underlying scientific foundation were the physic gardens of Italy created in the sixteenth and seventeenth centuries. The first of these physic gardens was the garden of the University of Pisa, which was created by Luca Ghini in 1543. Following this, other Italian universities followed suit, and gardens were created in Padova (1545), Firenze (1545), and Bologna (1547). Aromatic and medicinal herbs still exist in the Botanical Garden of Padova. These gardens were purely for the academic study of medicinal plants. The first US botanical garden was established by John Bartram in Philadelphia in 1728. Botanic gardens then experienced a change in usage during the sixteenth and seventeenth centuries. This was the age of exploration and the beginnings of international trade. Gardens such as the Royal Botanic Gardens, Kew and the Real Jardín Botánico de Madrid were set up to try and cultivate new species that were being brought back from expeditions to the tropics. Not only did these gardens promote and encourage botanical exploration in the tropics, they also helped found new gardens in the tropical regions to help cultivate these newly discovered plant species. During the nineteenth and twentieth centuries, municipal and civic gardens were created throughout Europe and the British Commonwealth. Nearly all of these gardens were pleasure gardens with very few of them having any scientific programs. Missouri Botanical Garden is an exception to this and was among the first botanic gardens to be established in the USA in 1859. In the last 50 years, botanic gardens have seen a revival as scientific institutions due to the emergence of the conservation movement. They are recognized as being extremely important to conservation due to their existing collections and the scientific knowledge they possess in the propagation of plant species. There are now currently 1,775 botanic gardens and arboreta in 148 countries around the world with many more under construction or being planned (www.bgci.org). They maintain 3.2 million living plants (Hawksworth, 1995), representing about 100,000 species (Heywood, 1991), along with other collections in the form of herbaria and seed banks. Some of the more famous botanical gardens in the world are Royal Botanic Garden, Kew (England); Montreal Botanical Garden (Canada); Singapore Botanical Garden (Singapore); Adelaide Botanical Garden (Australia); Desert Botanical Garden, central Arizona (USA); New York Botanical Garden (NYBG) (USA); Hawaii Tropical Bioreserve and Garden (USA); Royal Botanic Garden, Sydney (Australia), etc. (www.bgci.org).

The existence of ancient Indian botanical gardens can be dated back to 5000 BC. It is believed that the Sumerians of Mesopotamia and Dravidians settled in the Indus valley around 4000 BC and were conversant with agriculture and horticultural practices. Thus, the ancient Sumerians were the first to develop gardens in the Indus valley (Randhawa, 1976). The epics of the Aryans, the Ramayan, and the Mahabharata mentioned many forests or vatika, such as the Panchvati, Ashoka-Kanan, and Kamyak, and in the Charaka Samhita (Maharshi Charak), Charaka Udyan was mentioned. Botanical garden information is also mentioned in the Gupta period (AD 300–600) and in well-known Indian epics, such as Abhijnana Sakuntalam, Meghadutam, Raghuvamsa Kumarasambhava Sudrak, Ritusamhara (Kalidasa), and Kadambari (Banabhatt) (Majumder, 1935). Garden architecture is mentioned by historians of the Delhi Sultante (AD 602–932 and 1206–1526), and the gardens were developed across northern India in Persian style. Many other gardens were also developed under patronage of local rulers in North and South India.

The British in India were instrumentational in setting up a number of botanic gardens for economic and germplasm exchange purposes. Various agri-horticultural societies were also formed in different parts of India through the initiatives of Rev. William Carey. These societies helped immensely in popularizing gardens and gardens plants. Some important gardens established during the British rule include the Royal Botanic Garden, Calcutta (1787); Botanic Garden, Saharanpur (1817); Lloyd Botanic Garden, Darjeeling (1878); Government Garden, Ooty (1947); Sim's Park, Coonoor (1974); Botanic Garden, Coimbatore (1908); Bryant Park, Kodaikanal (1900); and Forest Research Institute and Colleges, Dehradun (1934), and a large number of parks and gardens were comparable to organized botanical gardens (Chakraverty and Mukhopadhyay, 1990). The first Western concept of a botanical garden in India started in 1787 with the establishment of Royal Botanic Garden by the East India Company in Howrah. Initially, this garden was established with the intention of supplying live plant material to European gardens. The Royal Botanical Garden, Calcutta (1787) has been an important place for plant introductions from different areas. After independence from Britain, it was renamed the Indian Botanical Garden; very recently, it was again renamed as Acharya Jagdish Chandra Bose Indian Botanic Garden (AJCBIBG). Until the end of the eighteenth century, many imperial botanical gardens of India were developed, namely Lalbagh Botanical Garden, Gardens of Agri-Horticultural Society, Calcutta and Madras, Eden Garden, King Narendra Narayana Park, Cooch Behar, Sims Park, Coonoor, Lloyd Botanic Garden, Darjeeling, Sikandar Bagh, Lucknow, Byculla Botanical Garden, Bombay, etc.

The focus shifted to conservation during the post-independence period, and establishment of new botanical gardens in the country was taken up. The Botanical Survey of India is a scientific department under MoEF & CC that has about 11 botanical gardens in different pytogeographical regions of India, namely AJC Bose Indian Botanical Garden, Howrah, Experimental Gardens at Barapani, Dhanikheri, Gangtok, Jodhpur, Mundhwa, Allahabad, National Orchidarium and Experimental Garden, Yercaud, National Gymnosperm collection cum Botanic Garden, Pauri and Botanic Garden of Indian Republic, Noida. The most famous or Lead Botanical Gardens in India are AJC Bose Indian Botanical Garden (Howrah), National Botanical Research Institute (Lucknow), Botanic Garden at Guru Nanak Dev University (Amritsar), Jawaharlal Nehru Tropical Botanical Garden and Research Institute (JNTBGRI) (Thiruvanthapuram), Forest Research Institute (Jabalpur), Kodaikanal Botanical Garden (Kodaikanal), Lead Botanical Garden of Shivaji University (Kolhapur), GKVK Botanical Garden (Bangalore), Calicut University Botanical Garden (CUBG) (Calicut), Van Vigyan Kendra (Chessa), Centre of Biodiversity Studies at Baba Ghulam Shah Badshah University (Rajouri), G.B. Pant Institute of Himalayan Environment and Development (Almora), Institute of Forest Productivity (Ranchi), and Malabar Botanic Garden (Kozikode). In the Eastern Ghats, only a few botanical gardens have been established, and the famous gardens are Regional Plant Resource Centre (RPRC) (Bhubaneswar), CSIR-IMMT (Bhubaneswar), Botanical gardens at Andhra University (Visakhapatnam), Sri Krishnadevaraya University (Ananthapuramu), and Yogi Vemana University (Kadapa).

10.3 LEAD BOTANICAL GARDEN OF YOGI VEMANA UNIVERSITY (YVUBG)

The Lead Botanical Garden (Plate 10.1) of Yogi Vemana University, Kadapa (AP) is located within the YVU campus, was established in 2008, and is being maintained by the Department of Botany. The botanic garden covers an area of about 25 acres with a good collection of living plant species and is located in the southern Eastern Ghats of Andhra Pradesh (14° 28′ 09″ N, 78°42′ 43″ E, elevation 146 m ASL). Initially, it was set up with the objective of providing an additional field facility for studies on plant biology and plant conservation. The university campus has alkaline soil, and water scarcity is found in the region. The climate is warm tropical, and the maximum average temperature is 42°C. Average annual rainfall is 750 mm, and the number of rainy days averages 25 to 40 per year. Despite such climatic conditions, the efforts of the university for the plantation and development of green spaces within the campus is commendable. The main green spaces in the campus consist of the botanic garden, social forestry, and other lawns that are properly grown to conserve biodiversity. The garden is home for a number of animals, birds, reptiles, butterflies, and insects.

This garden is a living repository of indigenous and exotic, cultivated and wild plants. At present, species collected from different parts of the Eastern Ghats of Andhra Pradesh and elsewhere are being maintained in the botanical garden and in greenhouses. The garden is currently harboring about 900 indigenous and exotic taxa (445 herbs, 89 climbers and lianas, 98 shrubs, and 268 trees) belonging to 118 families (more than 10,000 individual plants), of which includes about 150 endemic and threatened species. All the plants in the garden are properly labeled with botanical name, local name, family name, and known uses. Giant water lily, lotus, and lily ponds are developed for water plants. The YVU Botanic Garden is one of the largest botanic gardens in the state of Andhra Pradesh for *ex situ* conservation of rare, endangered, and threatened (RET) and endemic plants of the Eastern Ghats of Andhra Pradesh.

The existing infrastructural facilities in the YVUBG are net houses (465 m^2), a glass house (279 m^2) store room, water sump (45,000 liters capacity), power supply facility, and 3.5 km of internal roads. The whole garden area has a barbed wire fence. Drip irrigation facility is spread over an area of about 20 acres in the garden. The irrigation facility is provided with an underground pipeline system accessible throughout garden. A vermi-compost unit has been established within the garden for recycling leaf litter and other waste material toward preparation of organic manure for garden plants. The garden entrance area has been beautified with paintings of RET plants. Rain

Lead Botanic Garden of Yogi Vemana University, Kadapa, India

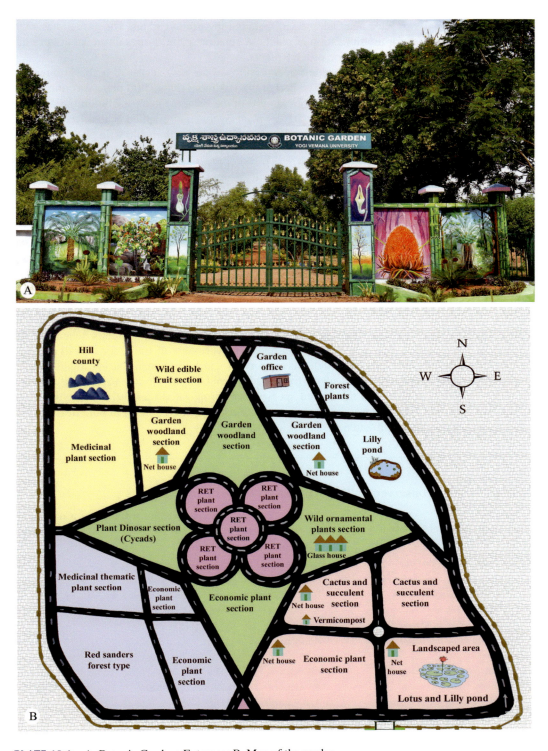

PLATE 10.1   A. Botanic Gardens Entrance; B. Map of the garden

water harvesting methods have been adopted to recharge the ground water. The garden has well-maintained greenhouses for multiplication of plants. A separate plant tissue culture laboratory has been established for *in situ* propagation of RET plants.

An interactive user-friendly database has been developed for easy access of herbarium records and live plant collections of the garden. The herbarium has more than 5,000 plant specimens, and the plant specimens were collected from different parts of the Eastern Ghats and South India. A digital library has also been established; so far, 50,000 plants were photographed from different forests of the Eastern Ghats. The seed bank was established and holds more than 400 wild plant seeds. The lichen herbarium was established for the first time in the state of Andhra Pradesh in the Department of Botany through a collaborative project with CSIR – National Botanical Research Institute (NBRI Lucknow) funded by the CSIR, New Delhi. This herbarium has around 5,200 specimens. The visitors to the garden mostly comprise students from schools, colleges, and universities from near and far; other visitors include researchers, botanists, tree farmers, medicinal pant growers, nature lovers and environmentalists, and the general public. The university conducts educational programs on environmental awareness to students especially. It also undertakers training programs for garden professionals and foresters. It supplies seedlings to other botanical gardens and provides basic facilities for research to students, researchers, and faculty of other institutes or universities. The Lead Botanical Garden received the "Andhra Pradesh State Biodiversity Conserver Award," which is a prestigious award that was granted on the occasion of International Day of Biological Diversity 2021, and the university also received "AP Greenery Award, 2018."

In 2017, this garden obtained support under the Assistance to Botanic Gardens Scheme of the Ministry of Environment, Forest and Climate Change, Government of India, for overall development as a Lead Botanical Garden. Through this support, the garden has been able to establish for the propagation and conservation of RET plants from the Eastern Ghats of Andhra Pradesh (Plates 10.2 and 10.3). Selected endemic and threatened species were propagated massively (Plate 10.4), and saplings were introduced to natural forests, educational institutes, and other lead botanical gardens in the country. The selected threatened species are listed in Table 10.1.

TABLE 10.1

List of RET Plants Conserved at YVUBG: Plates 10.2 & 10.3

S. N.	Name of the Taxon	Family	Distribution Status at Different Levels	Conservation Status
			Distribution in Andhra Pradesh	
1	*Andrographis glandulosa*	Acanthaceae	KDP, ATP, & NLR	---
2	*Andrographis nallamalayana*	Acanthaceae	KDP, KNL, & NLR	---
3	*Boswellia ovalifoliolata*	Burseraceae	KDP, CTR, KNL, NLR, & PKSM	VU
4	*Brachystelma annamacharyae*	Apocynaceae	KDP	CR
5	*Brachystelma kadapense*	Apocynaceae	KDP	CR
6	*Brachystelma maculatum* var. *breviflorum*	Apocynaceae	ATP	DD
7	*Brachystelma nigidianum*	Apocynaceae	ATP	CR
8	*Brachystelma pullaiahii*	Apocynaceae	KDP, KNL, & PKSM	CR
9	*Brachystelma seshachalamense*	Apocynaceae	KDP	CR
10	*Brachystelma vemanae*	Apocynaceae	NLR	CR
11	*Ceropegia pullaiahii*	Apocynaceae	ATP	CR
12	*Croton scabiosus*	Euphorbiaceae	KDP, KNL, & ATP	VU
13	*Cycas beddomei*	Cycadaceae	KDP, CTR, & NLR	EN
14	*Cycas seshachalamensis*	Cycadaceae	CTR	CR
15	*Decaschistia cuddapahensis*	Malvaceae	KDP, CTR, & ATP	---

S. N.	Name of the Taxon	Family	Distribution Status at Different Levels	Conservation Status
16	*Dipcadi krishnadevarayae*	Asparagaceae	KDP & ATP	---
17	*Drimia nagarjunae*	Asparagaceae	KDP, NLR, & PKSM	EN
18	*Euphorbia seshachalamensis*	Euphorbiaceae	KDP	---
19	*Glochidion talakonense*	Phyllanthaceae	CTR	---
20	*Lepidagathis rajasekharae*	Acanthaceae	CTR	CR
21	*Ophiorrhiza chandrasekharanii*	Rubiaceae	VSKP	---
22	*Phyllanthus narayanswamii*	Phyllanthaceae	EG, CTR, & VSKP	---
23	*Pimpinella tirupatiensis*	Apiaceae	CTR & NLR	EN
24	*Pterocarpus santalinus*	Fabaceae	KDP, CTR, KNL, NLR, & PKSM	EN
25	*Rhynchosia ravii*	Fabaceae	KDP & ATP	CR
26	*Tripogon tirumalae*	Poaceae	CTR	---
	Distribution in South India			
27	*Abutilon neelgherrense*	Malvaceae	KA, TN, & AP	---
28	*Alphonsea madraspatana*	Annonaceae	OD, TN, & AP	---
29	*Andrographis beddomei*	Acanthaceae	KA & AP	---
30	*Andrographis serpyllifolia*	Acanthaceae	KA, TN, & AP	---
31	*Barleria stocksii*	Acanthaceae	KA & AP	---
32	*Boucerosia diffusa*	Apocynaceae	KL, TN, & AP	---
33	*Boucerosia indica*	Apocynaceae	KA, TN, & AP	---
34	*Brachystelma ciliatum*	Apocynaceae	KA & AP	CR
35	*Brachystelma kolarense*	Apocynaceae	KL, KA, & AP	CR
36	*Brachystelma maculatum*	Apocynaceae	KA, TN, & AP	CR
37	*Ceropegia candelabrum* var. *biflora*	Apocynaceae	KA, TN, OD, & AP	---
38	*Crotalaria sandoorensis*	Fabaceae	KA & AP	---
39	*Cycas sphaerica*	Cycadaceae	OD & AP	DD
40	*Decalepis hamiltonii*	Apocynaceae	KA, TN, KL, & AP	EN
41	*Eriolaena lushingtonii*	Malvaceae	KA, KL, TN, TS, & AP	VU
42	*Habenaria longicornu*	Orchidaceae	KA, KL, TN, & AP	---
43	*Hildegardia populifolia*	Malvaceae	TN, KA, & AP	CR
44	*Memecylon lushingtonii*	Melastomataceae	KA, KL, TN, & AP	---
45	*Meyenia hawtayneana*	Acanthaceae	KA, TN, & AP	---
46	*Phyllanthus indofischeri*	Phyllanthaceae	KA, TN, & AP	VU
47	*Rhynchosia beddomei*	Fabaceae	KA, TN, & AP	---
48	*Shorea roxburghii*	Dipterocarpaceae	TN & AP	VU
49	*Shorea tumbuggaia*	Dipterocarpaceae	TN & AP	EN
50	*Syzygium alternifolium*	Myrtaceae	KA, TN, & AP	EN
51	*Tephrosia calophylla*	Fabaceae	TN, KA, & AP	---
52	*Terminalia pallida*	Combretaceae	TN & AP	VU
	Distribution in Peninsular India			
53	*Actinodaphne madraspatana*	Lauraceae	OD, TN, KL, & AP	---
54	*Amorphophallus konkanensis*	Araceae	Goa, KA, MH, & AP	---
55	*Ceropegia spiralis*	Apocynaceae	MH, KL, KA, TN, & AP	---
56	*Ficus dalhousiae*	Moraceae	MH, KA, TN, KL, & AP	---
57	*Habenaria panigrahiana*	Orchidaceae	OD, TN, & AP	---
58	*Habenaria rariflora*	Orchidaceae	KL, MH, TN, KA, & AP	---
59	*Habenaria roxburghii*	Orchidaceae	KL, MH, TN, KA, & AP	---
60	*Santalum album*	Santalaceae	KA, TN, MP, OD, & AP	VU
61	*Sophora interrupta*	Fabaceae	KA, MH, TN, & AP	---

(Continued)

TABLE 10.1
(Continued)

S. N.	Name of the Taxon	Family	Distribution Status at Different Levels	Conservation Status
		Distribution in India and Outside India		
62	*Aphyllorchis montana*	Orchidaceae	Taiwan, Thailand, Japan, & India	VU
63	*Ceropegia hirsuta*	Apocynaceae	Thailand & India	---
64	*Dalbergia latifolia*	Fabaceae	Nepal, Malaysia, & India	VU
65	*Oroxylum indicum*	Bignoniaceae	Bangladesh, China, Myanmar, & India	VU
66	*Saraca asoca*	Fabaceae	Sri Lanka, Bangladesh, Nepal, & India	VU

Distribution in Andhra Pradesh: ATP – Ananthapuramu; CTR – Chittoor; EG – East Godavari; KDP – Kadapa; KNL – Kurnool; NLR – Nellore; PKSM – Prakasam; VSKP – Visakhapatnam. Distribution in India: AP – Andhra Pradesh; KA – Karnataka; KL – Kerala; TN – Tamil Nadu; OD – Odissa; MH – Maharashtra; GO – Goa. Conservation status: CR– Critically Endangered; EN – Endangered; VU – Vulnerable; DD – Data deficient.

Depending upon the mandate/priorities, the garden area is divided into 25 blocks, and all the blocks are provided with pathways for easy access to all parts of the garden. Additionally, each block has a different name, and each block contains different types of plants.

1. **Lotus and lily pond section**: Two sections were established in different locations. These sections have more than 50 water and other plants. The important plants are *Nelumbo nucifera, Nymphaea nouchali, N. pubescens, N. rubra, Bacopa monnieri, Canna indica, Oxystelma esculentum*, etc.
2. **Economic plant section**: In this category, two sections were established, and more than 150 species are growing in these sections. These sections have medicinal, timber, ornamental, and other useful plants. Some economically useful plants are *Aristolochia indica, Artabotrys hexapetalus, Butea monosperma* var. *lutea, Caesalpinia sappan, Cissus quadrangularis, Commiphora caudata, Cassine glauca, Cymbopogon citratus, Dalbergia paniculata, Ficus racemosa, Givotia moluccana, Hardwickia binata, Lepisanthes tetraphylla, Limonia acidissima, Rauvolfia serpentina, R. tetraphylla, Salacia chinensis, Terminalia arjuna, Tylophora indica, Zamia furfuracea*, etc.
3. **Cactus and succulent plant section:** This section has a number of herbaceous succulent plants. A few of the more important plants are *Barleria cuspidata, Boucerosia diffusa, Brachystelma ciliatum, Caralluma stalagmifera, Ceropegia juncea, Chlorophytum tuberosum, Cynanchum acidum, Kalanchoe daigremontiana*, etc.
4. **Giant water lily pond**: (Plate 10.5) This pond has the world's largest water lily plant, *Victoria amazonica. Victoria amazonica* is well known for its huge circular leaves, which are often pictured with a small child sitting supported in the center as a demonstration of their size and strength. This section has other ornamental and native plants, which are growing along the margins of the pond.
5. **Medicinal plant sections**: In the botanical garden, three sections were established for conservation of medicinal plants. These sections have more than 60 medicinal plants, and the main plants are *Abrus precatorius, Azadirachta indica, Canavalia gladiata, Capparis grandis, C. zeylanica, Centella asiatica, Costus speciosus, Dioscorea pentaphylla, Ficus hispida, Gymnema sylvestre, Mentha arvensis, Oroxylum indicum, Phyllanthus emblica, Piper nigrum, Stemonia tuberosa, Terminalia chebula, Wrightia tinctoria*, etc.

Lead Botanic Garden of Yogi Vemana University, Kadapa, India 181

PLATE 10.2 Some RET plants of Eastern Ghats conserved in the Botanical garden. A. *Andrographis beddomei* B. *Barleria stocksii* C. *Boucerosia diffusa* D. *Brachystelma annamacharyae* E. *Brachystelma kolarense* F. *Brachystelma maculatum* G. *Brachystelma vemanae* H. *Ceropegia pullaiahii* I. *Ceropegia spiralis* J. *Decaschistia cuddapahensis* K. *Ficus heterophylla* L. *Lepidagathis rajasekharae* M. *Litsea glutinosa* N. *Phyllanthus narayanaswamii* O. *Pimpinella tirupatiensis*

PLATE 10.3 Some RET plants of Eastern Ghats conserved in the Botanical Garden. A. *Boswellia ovalifoliolata* B. *Cycas seshachalamensis* C. *Cycas zeylanica* D. *Dalbergia latifolia* E. *Eriolaena lushingtonii* F. *Ficus dalhousiae* G. *Hildegardia populifolia* H. *Mesua ferrea* I. *Phyllanthus indofischeri* J. *Pterocarpus santalinus* K. *Radermachera xylocarpa* L. *Rhynchosia beddomei* M. *Spondias pinnata* N. *Suregada multiflora* O. *Trewia nudiflora*

PLATE 10.4 Propagation of some rare plants. Seedlings: A. *Croton scabiosus* B. *Cycas beddomei* C. *Cycas sphaerica* D. *Hildegardia populifolia* E. *Oroxylum indicum* F. *Pterocarpus santalinus* G. *Saraca asoca* H. *Shorea roxburghii* I. *Shorea tumbuggaia* l.*Syzygium alternifolium*

6. **Red sander forest section**: The Red sander plants, *Pterocarpus santalinus*, have high commercial value as a timber, medicinal, and dye. Along with Red sander plants, more than 20 other trees are growing in this section.
7. **Plant dinosaur section (*Cycad* section):** Cycads are considered to be "living fossils" and have great significance from an evolutionary and research point of view. *Cycad* species present are *Cycas beddomei, C. circinalis, C. revoluta, C. rumphii, C. seshachalamensis, C. sphaerica,* and *C. zeylanica*.
8. **Garden woodland section (cactus and succulent plot)**: This unit has a number of cactus and other species, namely *Adenanthera pavonina, Agave angustifolia, A. sisalana, Cereus pterogonus, C. repandus, Euphorbia antiquorum, E. caducifolia, E. tirucalli, E. tortilis, Opuntia dillenii*, etc.
9. **Wild ornamental plants**: Important plants in this section are *Butea monosperma, Cassia roxburghii, Delonix regia, Ficus benghalensis, Holoptelea integrifolia, Hymenodictyon orixense, Madhuca indica, Spathodea campanulata, Tamarindus indica, Terminalia bellirica*, etc.
10. **RET Plant Sections**: *Ex situ* conservation of RET and endemic species is one of the most significant steps toward conserving the genetic diversity and germplasm. *Ex situ* conservation provides an effective conservation measure for research and restoration. The YVUBG has focused on RET and endemic species of the Eastern Ghats and established five sections for these plants. A total of 152 significant RET and endemic species are conserved in the botanical garden, and some important species are as follows: *Amorphophallus konkanensis, Andrographis beddomei, Boswellia ovalifoliolata, Croton scabiosus, Curcuma neilgherrensis, Dalbergia latifolia, Decalepis hamiltonii, Decaschistia cuddapahensis, Eriolaena lushingtonii, Hildegardia populifolia, Phyllanthus narayanswamii, Pimpinella tirupatiensis, Rhynchosia beddomei, Syzygium alternifolium, Shorea tumbuggaia*, etc.
11. **Garden woodland section:** Two sections were established, and these sections have significant plant resources, namely, *Abutilon neelgherrense, Anogeissus latifolia, Asparagus gonoclados, Boswellia serrata, Byttneria herbacea, Cissus rotundifolia, Crotalaria sandoorensis, Dicliptera cuneata, Ficus mollis, Indigofera mysorensis, Ochna obtusata, Strobilanthes cordifolia*, etc. In this section, one more living fossil in the order Ginkgoales (Gymnosperms) is growing, which first appeared over 290 million years ago; the plant name is *Ginkgo biloba*.
12. **Hill county section:** A unique aspect of this botanic garden is the development of one acre of land with artificial hilly mounds established to maintain a natural "hill forest ecosystem." This section has more than 25 native species, and some notable species are *Breynia vitis-idaea, Caesalpinia bonduc, Cleistanthus collinus, Cordia dichotoma, Ehretia aspera, Sterculia urens, Stereospermum tetragonum, Tectona grandis, Vachellia eburnea, V. nilotica*, etc.
13. **Wild edible fruit section**: The important collections here include more than 25 species of wild edible fruit plants, including *Annona reticulata, Bambusa bambos, Borassus flabellifer, Diospyros melanoxylon, Ziziphus mauritiana, Ziziphus oenoplia*, etc.
14. **Garden office section:** Important species in this section are *Abrus precatorius, Jasminum sambac, Pentalinon luteum, Pongamia pinnata, Senna siamea, Yucca aloifolia*, etc.
15. **Forest plants section:** The noteworthy species are *Butea monosperma, Delonix regia, Swietenia mahagoni, Tamarindus indica, Terminalia bellirica, Terminalia catappa*, etc.

PLATE 10.5 *Victoria amazonica* A-C. Leaf different stages D-F. Flower different stages G. Total habit H. Leaf back side view I & J. Leaf carrying weights

10.4 CONCLUSIONS

At present, many plant species are gradually becoming depleted due to different anthropogenic activities, such as habitat destruction, afforestation programs, and lack of suitable management practices. In this context, the value of botanic gardens in conservation of native flora is highly appreciated. Moreover, such botanic gardens are also involved in active exploration of collecting plants and identifying and documenting wild and native flora. In this context, the Lead Botanical Garden of Yogi Vemana University has played s vital role in conservation of plant resources from the Eastern Ghats of India. It has a great significance in educational and scientific fields and also in creating awareness to the public. The garden is currently harboring about 900 indigenous and exotic taxa (445 herbs, 89 climbers and lianas, 98 shrubs, and 268 trees) belonging to 118 families (more than 10,000 individual plants), of which includes about 150 endemic and threatened species. It is our responsibility to conserve the plant wealth of the Eastern Ghats of India for sustainable utilization. The nursery-raised plants in the botanic garden are distributed to the general public and for rehabilitation in forest areas. So far, about 400,000 plants were raised and 300,000 plants distributed to all over India.

ACKNOWLEDGMENTS

The author gratefully acknowledges the Ministry of Environment, Forest and Climate Change (MoEF & CC) (no. 10/16/2016-CS/BG. Dated: 31–03–2017) for financial support under ABG scheme and Andhra Pradesh Forest department.

REFERENCES

Chakravarty, R.K. and D.P. Mukhopadhyay. 1990. *A Directory of Botanic Garden and Parks in India*. Calcutta.

Hawksworth, D. 1995. The resource base for biodiversity assessments. *Global Biodiversity Assessment*: 548–605 Botanical Survey of India.

Heywood, V.H. 1991. Developing a strategy for germplasm conservation in botanic gardens. In: Heywood, V.H. and P.S. Wyse Jackson (eds.), *Tropical Botanic Gardens – Their Role in Conservation and Development*. Academic Press, London, pp. 11–23.

Pullaiah, T. and S. Karuppusamy. 2020. *Flora of Eastern Ghats*, Vol. 7. Daya Publishing House, New Delhi.

Ramachandran, R.M., R. Parth Sarathi, V. Chakravarthia, J. Sanjay and P.K. Joshi. 2018. Long-term land use and land cover changes (1920–2015) in Eastern Ghats, India: Pattern of dynamics and challenges in plant species conservation. *Ecological Indicators* 85: 21–36.

Randhawa, M.S. 1976. *Gardens Through Ages*. Macmillan, New Delhi.

Śārṅgadhara, D. and Majumder, G.P. 1935. *Upavana-Vinoda (A Sanskrit Treatise on Arbori-Horticulture)*. Indian Research Institute, Calcutta.

11 Dhanikhari Experimental Garden-Cum-Arboretum

Lal Ji Singh, C.P. Vivek, Bishnu Charan Dey, C.S. Purohit, Gautam Anuj Ekka, Mudavath Chennakesavulu Naik, and Fouziya Saleem

CONTENTS

11.1	Introduction	187
11.2	Aquatic Plants	189
11.3	Bambusetum and Section of Rattans	189
11.4	Gymnosperm Conservatory	189
11.5	Fernery	190
11.6	Herbal Garden and Dhanvantri Medicinal Plot	190
11.7	Kurz Plot (Endemic, Endangered, and Threatened Plants)	193
11.8	Navagrah Vatika	193
11.9	Nursery	193
11.10	Orchidarium	193
11.11	Palmetum	196
11.12	Spices Garden	196
11.13	Succulent Plants	196
11.14	Wild Bananas and Zingibers	196
11.15	Conclusion	198
Acknowledgments		199
References		199

11.1 INTRODUCTION

Andaman and Nicobar Islands (ANIs) comprise unique regions of rich plant diversity with a higher rate of endemism (Singh et al., 2014, 2021a, 2021b). It is one of the biodiversity hotspots, situated in the tropical belt between 6° 45'–13° 41' N and 92° 12'–93° 57' E. Total area of the Islands is 8,249 sq km, of which 87% of the land constitute forests. For proper documentation and conservation of the Islands' flora, Botanical Survey of India, Andaman and Nicobar Regional Centre was established in 1972 with Dhanikhari Experimental Garden cum Arboretum (DEGCA) at Nayashahar, South Andaman (Figure 11.1). The laborious task of documentation of Island flora has been carried out by various workers (Parkinson, 1923; Balakrishnan and Rao, 1983; Vasudeva Rao, 1986; Ellis, 1987; Balakrishnan, 1989; Lakshminarasimhan and Rao, 1996; Nayar, 1997; Mathew, 1998; Hajra et al., 1999; Sinha, 1999; Dixit and Sinha, 2001; Rao, 2001; Srivastava and Rao, 2001; Pandey and Diwakar, 2008; Singh, 2012, 2013, 2021; Singh and Murugan, 2013a, 2013b, 2014; Jagadeesh Ram, 2014; Murugan et al., 2014, 2016; Singh and Ranjan, 2015, 2021; Mishra et al., 2016; Naik and Rao, 2016; Singh et al., 2014, 2016a, 2016b, 2016c, 2016d, 2018. 2020a, 2020b, 2020c, 2021a, 2021b; Singh and Misra, 2017, 2020; Naik and Singh, 2020a, 2020b; Naik et al., 2020a, 2020b; Sivaramakrishna et al., 2021, Purti et al., 2022) and the representative plant species have been conserved in the garden.

FIGURE 11.1 Dhanikhari Experimental Garden cum Arboretum (DEGCA), BSI, ANRC.

DEGCA is situated 16 km away from Port Blair with a focus on both *in-situ* and *ex-situ* conservation of plant diversity in the Andaman and Nicobar Islands. Approximately 95% of the area of DEGCA is protected for *in-situ* conservation, and there are 2 ha of cleared land for *ex-situ* conservation of introduced endemic, endangered, and threatened (EET) plant species of the Islands and for raising the nursery toward multiplication and conservation. This garden is recognized in the Islands as a center of excellence in the study of plant diversity of the Islands (Singh et al., 2014). Being the first garden of its kind in the Islands, DEGCA has made great strides toward the conservation of a large number of EET plant species, including ethno-medicinal, economical, and ornamental plants of the Andaman and Nicobar Islands, thereby assisting the government in effective conservation and control of biodiversity loss (Srivastava and Rao, 2001; Singh and Murugan, 2014; Mishra et al., 2016; Singh et al., 2014, 2016a, 2016b, 2016c, 2016d, 2021a, 2021b). The total floristic wealth of the DEGCA represents ca. 600 plant species belonging to angiosperms, gymnosperms, pteridophytes, and bryophytes (Singh et al., 2014, 2021a, 2021b).

The garden has already established large living collections of trees and woody lianas, medicinals, aromatics and spices, orchids, bamboos, ferns and fern allies, palms, rattans, cycads, zingibers, and aquatic plants. The garden also maintains a large number of cultivated plants and economically important plants. Some of the noteworthy plant species successfully propagated in the garden are *Amomum aculeatum* Roxb., *Amomum andamanicum* V.P. Thomas et al., *Angiopteris evecta* (G. Forst.) Hoffm., *Baccaurea ramiflora* Lour., *Bentinckia nicobarica* (Kurz) Becc., *Calamus andamanicus* Kurz, *Carissa andamanensis* L.J. Singh & Murugan, *Dillenia andamanica* C.E. Parkinson, *Etlingera fenzlii* (Kurz) Škornick. & M. Sabu, *Garcinia dhanikhariensis* S.K. Srivast., *Geophila repens* (L.) I.M. Johnst., *Gnetum scandens* Roxb., *Horsfieldia glabra* (Blume) Warb., *Knema andamanica* (Warb.) W.J. de Wilde, *Mangifera andmanica* King, *Mangifera camptosperma* Pierre, *Musa indandamanensis* L.J. Singh, *Nageia wallichiana* (C. Presl) Kuntze, *Podocarpus neriifolius* D. Don, *Psychotria andamanica* Kurz, *Pyrostria laljii* M.C. Naik, Arriola & M. Bheemalingappa, *Semecarpus kurzii* Engl., *Sphaeropteris albosetacea* (Bedd.), R.M. Tryon, *Zeuxine andamanica* King & Pantl., *Vanilla andamanica* Rolfe, *Vanilla sanjappae* R.P. Pandey et al., and *Zingiber pseudosquarrosum* L.J. Singh & P. Singh. For the sake of classifying these plants, specific sections are designed in the Garden, namely Aquatic Plants, Bambusetum, Gymnosperms Conservatory, Fernery, Herbal Garden & Dhanvantri Medicinal Plot, Kurz Plot (Endemic, Endangered, and Threatened Plants), Navagrah Vatika, Orchidarium, Palmetum, Spices Garden, Succulent Plant Sections, and Sections for Wild Bananas and Zingibers.

11.2 AQUATIC PLANTS

This section is designed as an artificial pond near the administrative block for conservation of aquatic plants of the Islands. Floating, submerged, as well as rooted aquatic plants are the main attraction. Some of the species grown in the pond are *Nymphaea nouchali* Burm. f., *Nymphaea omarana* Hort ex Gard., *Nymphaea pubescens* Willd., *Nymphoides indica* (L.) Kuntze, *Limnocharis flava* (L.) Buchenau, *Monochoria vaginalis* (Burm. f.) C. Presl, etc.

11.3 BAMBUSETUM AND SECTION OF RATTANS

Bamboos are the fastest growing woody plants in the world. The bamboo section in the garden has a rich collection of native as well as introduced bamboos. The center of attraction is the endemic species *Dinochloa andamanica* (Kurz) H.B. Naithani and *Schizostachyum andamanicum* M. Kumar & Remesh. The internodal region of *Schizostachyum andamanicum* is used to make reusable drinking straws, which is an invention by the Botanical Survey of India, under consideration for patent by the Government of India. This section also has representatives of other bamboos of the Island flora, such as *Bambusa vulgaris* Schrad., *Dendrocalamus giganteus* Munro, *Dendrocalamus strictus* (Roxb.) Nees, etc.

Thirteen species of canes are represented in the islands' flora (Renuka, 1995). They are used for rafting, house construction, making baskets, as poles carrying goods, etc. in the Islands. Andaman canes are known for their better quality over those of the main lands. DEGCA represents some important species among them, such as *Calamus andamanicus* Kurz, *C. dilaceratus* Becc., *C. longisetus* Griff., *C. palustris* Griff., *C. pseudorivalis* Becc., *C. viminalis* Willd., etc.

11.4 GYMNOSPERM CONSERVATORY

This section showcases the primitive plant group gymnosperms of the Andaman and Nicobar Islands, of which Cycads are a botanical throwback to the Jurassic era. The Island flora has five species of Cycads, namely *Cycas zeylanica* (J. Schust.) A. Lindstr. & K.D. Hill, *C. dharmrajii* L.J. Singh, *C. pschannae* R.C. Srivast. & L.J. Singh, *Dioon edule* Lindl., and *Zamia furfuracea* L.f. ex Aiton. Despite several introductions, *Cycas dharmrajii* and *C. pschannae* are not successfully established in the garden; however, *C. zeylanica* has a lush growth with male and female plants growing together. Due to the straight trunk and unique arrangements of leaves, *C. zeylanica* attracts special attention of visitors. Other gymnospermous plants in the section are *Gnetum latifolium* Blume, *G. montanum* Markgraf, *G. scandens* Roxb., *Nageia wallichiana* (C. Presl) Kuntze, and *Podocarpus neriifolius* D. Don. (Figure 11.2).

FIGURE 11.2 Plants under ex situ conservation at DEGCA: (A) *Etlingera fenzlii* (Kurz) Škornick and M. Sabu; (B) *Eulophia andamanensis* Rchb.f.; (C) *Pandanus leram* Jones ex Voigt; (D) *Musa indandamanensis* L.J. Singh.

11.5 FERNERY

This section includes a glass house for herbaceous ferns and areas surrounding the glass house for giant ferns. This has rare collections like *Angiopteris evecta* (Forst.) K. Hoffm., *Blechnum finlaysonianum* Wall., *Cyathea gigantea* (Wall. ex Hook.) Holttum, *Diplazium proliferum* (Lam.) Thouars, *Sphaeropteris albo-setacea* (Bedd.) Tryon, *S. nicobarica* (Balakr. *et* Dixit) Dixit, *Thelepteris polycarpa* (Blume) K. Iwats., etc.

11.6 HERBAL GARDEN AND DHANVANTRI MEDICINAL PLOT

More than 160 species of medicinal plants of the Islands are planted systematically in this section. Nomenclature and uses of each species are displayed for visitors. Concrete pathways are prepared for easy reach to all the corners of the section. Some of the medicinal plants included here are *Acorus calamus* L., *Aegle marmelos* (L.) Corrêa, *Amomum aculeatum* Roxb., *Andrographis paniculata* (Burm. f.) Wall. ex Nees, *Aristolochia indica* L., *Asparagus racemosus* Willd., *Azadirachta indica* A. Juss., *Bacopa monnieri* (L.) Wettst., *Blechnum finlaysonianum* Wall., *Centella asiatica*

FIGURE 11.3 Plants under *ex situ* conservation at DEGCA: (A) Schizostachyum andamanicum M. Kumar and Remesh; (B) Tacca leontopetaloides (L.) Kuntze; (C) Vanilla andamanica Rolfe; (D) Vanilla sanjappae R.P. Pandey et al.

FIGURE 11.4 Plants under *ex situ* conservation at DEGCA: (A) *Carissa andamanensis* L.J. Singh and Murugan; (B) *Dillenia andamanica* C.E. Parkinson; (C) *Euphorbia epiphylloides* Kurz; (D) *Garcinia dhanikhariensis* S.K. Srivast.

(L.) Urb., *Cerbera odollam* Gaertn., *C. manghas* L., *Cissus quadrangularis* L., *Costus speciosus* (Koen ex. Retz.) Sm., *Curcuma longa* L., *C. manga* Valeton & Zijp, *C. zedoaria* (Christm.) Roscoe, *Cymbopogon citratus* (DC.) Stapf, *Eclipta prostrata* (L.) L., *Eryngium foetidum* L., *Etlingera fenzlii* (Kurz) Škornick. & M. Sabu, *Geophila repens* (L.) I.M. Johnst., *Hibiscus rosa-sinensis* L., *Hyptis suaveolens* (L.) Poit., *Ipomoea aquatica* Forssk., *Justicia adhatoda* L., *Lawsonia inermis* L., *Morinda citrifolia* L., *Moringa oleifera* Lam., *Nyctanthes arbor-tristis* L., *Ocimum tenuiflorum* L., *Opuntia ficus-indica* (L.) Mill., *Phyllanthus amarus* Schumach. & Thonn., *P. emblica* L., *Piper longum* L., *Rauvolfia serpentina* (L.) Benth. ex Kurz, *R. tetraphylla* L., *Ricinus communis* L., *Saraca asoca* (Roxb.) de Wilde, *Sauropus rhamnoides* Blume, *Sida acuta* Burm.f., *Syzygium aromaticum* (L.) Merr. & L.M. Perry, *Tinospora cordifolia* (Willd.) Miers, etc., *Triphasia trifolia* (Burm. f.) P. Wilson, *Vitex trifolia* L., *Zingiber officinale* Roscoe, *Z. pseudosquarrosum* L.J. Singh & P. Singh, etc.

11.7 KURZ PLOT (ENDEMIC, ENDANGERED, AND THREATENED PLANTS)

The Central goal of this section is to create awareness in the researchers and public about the endemic, endangered, and threatened plant species of the Islands. More than 50 species of this category are conserved in the section. Some of them are *Amomum aculeatum* Roxb., *Calamus nicobaricus* Becc. & Hook.f., *Canarium denticulatum* Blume, *Carissa andamanensis* L.J. Singh & Murugan, *Codiocarpus andamanicus* (Kurz) R.A. Howard, *Dillenia andamanica* C.E. Parkinson, *Elaeocarpus floribundus* Blume, *E. rugosus* Roxb. ex G. Don, *Euphorbia epiphylloides* Kurz, *Garcinia andamanica* King, *Knema andamanica* (Warb.) W.J. de Wilde, *Maesa andamanica* Kurz, *Mangifera andamanica* King, *Mangifera griffithii* Hook.f., *Mesua manii* (King) Kosterm., *Mimusops andamanensis* King & Gamble, *Myristica andamanica* Hook.f., *Psychotria andamanica* Kurz, *Pyrostria laljii* M.C. Naik, Arriola & M. Bheemalingappa, *Schizostachyum andamanicum* M. Kumar *et* Ramesh, *Semecarpus kurzii* Engl., *Syzygium andamanicum* (King) N.P. Balakr., *Zeuxine andamanica* King & Pantl., etc. (Figure 11.5).

11.8 NAVAGRAH VATIKA

This section is to depict the representative plant species of nine planets (*Grahas*). As per Hindu mythology, the planets associated with specific plants can control the destiny of a person. The objective of developing *Navagrah Vatika* in the garden is to bring awareness about the cultural system in India by which plants are being conserved in the name of Gods, rituals, and beliefs. The nine plants representing nine planets are *Calotropis gigantea* (L.) Dryand. for Sun, *Butea monosperma* (Lam.) Taub. for Moon, *Achyranthes aspera* L. for Mercury, *Acacia catechu* (L.f.) Willd. for Mars, *Ficus religiosa* L. for Jupiter, *Ficus racemosa* L. for Venus, *Prosopis cineraria* (L.) Druce for Saturn, *Cynodon dactylon* (L.) Pers. for Rahu, and *Desmostachya bipinnata* (L.) Stapf for Ketu.

11.9 NURSERY

Multiplication of plants through, seeds, seedlings, cuttings, and other propagules are done in the Nursery section with emphasis on the multiplication and conservation of EET plant species of the Islands. The multiplied and established seedlings are being reintroduced in their original habitats for population multiplication. Also, the seedlings are distributed to other scientific and non-scientific organizations, schools, colleges, etc. in the Islands.

11.10 ORCHIDARIUM

This section is designed close to the administrative block as a net house to fix the optimum light penetration inside. Richness of orchid flora of the Andaman and Nicobar Islands is well exhibited in the section with more than 100 species in proper classification. Some of them, including rare and endemic species, are *Aerides emericii* Rchb. f., *Agrostophyllum planicaule* (Wall. ex Lindl.) Rchb. f., *Bulbophyllum crassipes* Hook. f., *B. rufinum* Rchb. f., *Coelogyne quadratiloba* Gagnep., *Cymbidium aloifolium* (L.) Sw., *Dendrobium crumenatum* Sw., *D. formosum* Roxb. ex Lindl., *D. shompenii* B.K. Sinha & P.S.N. Rao., *D. tenuicaule* Hook. f., *Eria andamanica* Hook. f., *Eulophia andamanensis* Rchb. f., *E. nicobarica* N.P. Balakr. & N.G. Nair, *Geodorum densiflorum* (Lam.) Schltr., *Luisia balakrishnanii* S. Misra, *Nervilia aragoana* Gaudich., *N. plicata* (Andr.) Schltr., *Pteroceras muriculatum* (Rchb.f.) P.F. Hunt, *Tropidia curculigoides* Lindl., *Vanilla albida* Blume, *V. andamanica* Rolfe, *V. sanjappae* R.P. Pandey et al. and *Zeuxine andamanica* King & Pantl. (Figure 11.6)

FIGURE 11.5 Plants under *ex situ* conservation at DEGCA: (A) *Mangifera andamanica* King; (B) *Mangifera camptosperma* Pierre; (C) *Saraca asoca* (Roxb.) de Wilde; (D) *Pterocarpus dalbergioides* Roxb. ex DC.

Dhanikhari Experimental Garden-Cum-Arboretum 195

FIGURE 11.6 Plants under ex-situ conservation at DEGCA: (A) *Cycas zeylanica* (J. Schuster) Lindsts and K. D. Hill; (B) *Podocarpus neriifolius* D. Don; (C) *Angiopteris evecta* (Forst.) K. Hoffm.; (D) *Sphaeropteris albo-setacea* (Bedd.) Tryon.

11.11 PALMETUM

Bentinckia nicobarica (Kurz) Becc., *Phoenix andamanensis* S. Barrow, and *Pinanga andamanensis* Becc. are endemic palm species that are also endangered or extremely rare in the wild. The section has a lush growth of *Bentinckia nicobarica* and *Pinanga andamanensis*. The hill species *Phoenix andamanensis* S. Barrow is surviving in the section with few individuals. Other species such as *Areca catechu* L., *A. triandra* Roxb. ex Buch.-Ham., *Borassus flabellifer* L., *Caryota mitis* Lour., *Cocos nucifera* L., *Corypha umbraculifera* L., *C. utan* Lam., *Elaeis guineensis* Jacq., *Licuala peltata* Roxb. ex Buch.-Ham., *L. spinosa* Wurmb., *Nypa fruticans* Wurmb., *Phoenix andamanensis* S. Barrow, *P. paludosa* Roxb., *Pinanga andamanensis* Becc., *P. manii* Becc., and *Rhopaloblaste augusta* (Kurz) Moore are also planted in the section in a specific classification (Figure 11.7). A few exotic species of aesthetic values are also planted here.

11.12 SPICES GARDEN

Common spice plants used in the Islands are grown in this section, such as *Cinnamomum verum* J. Presl., *Cinnamomum zeylanicum* Nees, *Myristica fragrans* Houtt., *Piper nigrum* L., *Piper sarmentosum* Roxb., *Syzygium aromaticum* (L.) Merr. & L.M. Perry, etc.

11.13 SUCCULENT PLANTS

As Andaman Nicobar flora is tropical rainforest, a section is designed in the garden to familiarize the plants of xeric adaptations. Succulent plants are poorly represented in the Island flora. However, the occurrence of an endemic cactus plant, *Euphorbia epiphylloides* Kurz, in the Islands' flora is of geographical importance. Therefore, the center of attraction of the succulent plant section is *E. epiphylloides*. Other introduced species like *Agave sisalana* Perrine ex Engelm., *A. vivipara* L., *Euphorbia milii* Des Moul., *E. tithymaloides* L., *E. tirucalli* L., *Hylocereus undatus* (Haw.) Britton & Rose, *Opuntia ficus-indica* (L.) Mill., *Sansevieria trifasciata* Prain, etc. are growing here.

11.14 WILD BANANAS AND ZINGIBERS

This section has good representation of almost all important species of bananas and zingibers of the Islands. Andaman and Nicobar Islands have four wild species of banana, such as *Musa acuminata* Colla, *M. balbisiana* Colla, *M. indandamanensis* L.J. Singh, and *M. paramjitiana* L. J. Singh (Figure 11.6); one cultivated species, *M. paradisiaca* L.; and one ornamental, *M. velutina* H. Wendl. & Dude. The garden has representation of all these. The family Zingiberaceae represents 30 species in the Islands. Of which, the following species are under conservation in the Zingiber section of the garden: *Alpinia luteocarpa* Elmer, *A. manii* Baker, *Amomum aculeatum* Roxb., *A. andamanicum* V.P. Thomas & al., *A. maximum* Roxb., *Boesenbergia albolutea* (Baker) Schltr., *B. siphonantha* (King ex Baker) M. Sabu & al., *Curcuma longa* L., *C. mangga* Valeton & Zijp, *C. petiolata* Roxb., *C. zedoaria* (Christm.) Roscoe, *Etlingera fenzlii* (Kurz) Škornick. & M. Sabu, *Globba pauciflora* King ex Baker, *Hedychium coronarium* J. König, *Zingiber odoriferum* Blume, *Z. officinale* Roscoe, *Z. pseudo-squarrosum* L.J. Singh & P. Singh, *Z. squarrosum* Roxb., and *Z. zerumbet* (L.) Roscoe ex Sm.

Other than the plant species in sections, many plant groups of specific importance are also being conserved in the garden. They are as follows: the furgivorous plants, such as *Aphanamixis polystachya* (Wall.) R. Parker, *Ficus benghalensis* L., *F. benjamina* L., *F. callosa* Willd., *F. exasperata* Vahl, *F. hispida* L.f., *F.religiosa* L., *F.rumphii* Blume, etc.; Wild Rudraksha trees, such as *Elaeocarpus floribundus* Blume, *E. macrocerus* (Turcz.) Merr., *E. petiolatus* (Jack) Wall. ex Steud., and *E. rugosus* Roxb. ex G. Don; various Mistletoe species, such as *Dendrophthoe curvata* (Blume) Miq., *Macrosolen melintangensis* (Kurth) Miq. (Loranthaceae), *Viscum monoicum* Roxb. ex DC., etc.; natural die-yielding plants, such as *Abrus precatorius* L., *Aegle marmelos* (L.) Corrêa, *Aporosa*

Dhanikhari Experimental Garden-Cum-Arboretum

FIGURE 11.7 Plants under *ex-situ* conservation at DEGCA: (A) *Areca triandra* Roxb. ex Buch.-Ham.; (B) *Bentinckia nicobarica* (Kurz) Becc.; (C) *Calamus andamanicus* Kurz; (D) *Licuala peltata* Roxb. ex Buch.-Ham.

octandra (Buch.-Ham. ex D. Don) A.R. Vickery, *Cassia fistula* L., *Curcuma longa* L., *Entada rheedii* Spreng., *Hibiscus rosa-sinensis* L., *Lawsonia inermis* L., *Mallotus philippensis* (Lam.) Müll.Arg., *Mirabilis jalapa* L., *Morinda citrifolia* L., etc.; ornamental plants, such as *Alternanthera brasiliana* (L.) Kuntze, *Araucaria heterophylla* (Salisb.) Franco, *Duranta erecta* L., *Heliconia rostrata* Ruiz & Pav., *Heterotis rotundifolia* (Sm.) Jacq.-Fél., *Pogonatherum paniceum* (Lam.) Hack., *Polyalthia longifolia* (Sonn.) Thwaites, *Ravenala madagascariensis* Sonn., *Zamia furfuracea* L.f. ex Aiton, etc.; wild edible fruit plants, such as *Antidesma bunius* (L.) Spreng., *Aporosa octandra* (Buch.-Ham. ex D. Don) A.R. Vickery, *Artocarpus heterophyllus* Lam., *Dillenia indica* L., *Diospyros ridleyi* Bakh., *Garcinia cowa* Roxb. ex DC., *Grewia calophylla* Kurz ex Mast., *Pandanus leram* Jones ex Voigt, *Syzygium cumini* (L.) Skeels, *Triphasia trifolia* (Burm. f.) P. Wilson, etc.; and economic plants, such as *Dipterocarpus alatus* Roxb. ex G. Don, *D. costatus* C.F. Gaertn., *D. grandiflorus* (Blanco) Blanco *Elaeis guineensis* Jacq., *Korthalsia laciniosa* (Griff.) Mart., *K. rogersii* Becc., *Pterocarpus dalbergioides* Roxb. ex DC., etc.

In addition to the conservation activities and taxonomic research, DEGCA is also playing an important role in public education and developing environmental awareness by organizing programs to promote public understanding of biodiversity.

Though curious plant species of the Islands are being periodically introduced in DEGCA, some are not surviving in the microclimatic condition of the garden. Some plants that were introduced but failed to survive in the garden are *Anoectochilus narasimhanii* R. Sumathi et al. (introduced in 2016), *Arisaema saddlepeakense* Rao *et* Srivastava (introduced in 2015), *Bulbophyllum tenuifolium* (Blume) Lindl. (introduced in 2015), *Calamus basui* Renuka *et* Vijayakumaran (introduced in 2016), *Codiocarpus andamanicus* (Kurz) R.A. Howard (introduced in 2019), *Cyathia gigantia* (introduced in 2012), *Cycas dharmrajii* L.J. Singh (introduced in 2013), *C. pschannae* R.C. Srivast. & L.J. Singh (introduced in 2012), *Cyrtandra burttii* N.P. Balakrishnan (introduced in 2020), *Ficus andamanica* Corner (introduced in 2016), *Desmodium heterocarpon* subsp. *ovalifolium* (Prain) H. Ohashi (introduced in 2018), *Diplazium proliferum* (Lam.) Thouars (introduced in 2012), *Dentella cylindrica* M.C. Naik and L.J. Singh (introduced in 2013), *Gigantochloa nigrociliata* (Buse) Kurz (introduced in 2021), *Musa paramjitiana* L. J. Singh (introduced in 2013), *Portulaca laljii* P. Sivaramakrishna & P. Yugandhar (introduced in 2019), *Rivina andamanensis* L. J. Singh & M.C. Naik (introduced in 2014), *Schizostachyum kalpongianum* M. Kumar & Remesh (introduced in 2016), *Strobilanthes andamanensis* Bor. (introduced in 2021), *Thelypteris polycarpa* (Blume) K. Iwats. (introduced in 2012), *Trichomanes bipunctatum* Poir (introduced in 2012), *T. minutum* Blume (introduced in 2012), and *T. motleyi* (Bosch) Bosch (introduced in 2012).

11.15 CONCLUSION

DEGCA is a natural abode in the Andaman and Nicobar Islands for *ex situ* and *in situ* conservation with a representative collection of the flora of the Islands. This has suitable ambience for lush growth of orchids, ferns, herbs, climbers, shrubs, and trees. This is the lone garden in the Islands participating in conservation activities and taxonomic research of native plants and performing a role in public education and developing environmental awareness. Regular in-house projects on the introduction, multiplication, and conservation of endemic, endangered, and threatened plant species of the Andaman and Nicobar Islands have been carried out in the garden since its establishment. In this way, the garden has enriched about 600 plant species of major plant groups, including, angiosperms, gymnosperms, pteridophytes, and bryophytes. Among them, many are enlisted by IUCN in different categories that might be the choice of genetic base for future studies.

Sufficient seedlings of endemic and rare plant species, such as *Amomum aculeatum, Amomum andamanicum, Calamus andamanicus, Bentinckia nicobarica, Etlingera fenzlii, Garcinia dhanikhariensis, Knema andamanica, Maesa andamanica, Mangifera andamanica, Musa indandamanensis, Pinanga andamanensis, P. manii*, and *Rhopaloblaste augusta* (Kurz) Moore have been raised in the nursery of the garden in order to reintroduce them in their original locality. More than

500 seedlings of *Bentinckia nicobarica* and *Rhopaloblaste augusta* have already been reintroduced in their locality in Nicobar Islands as part of an in-house project of the Botanical Survey of India.

Avenue plants (*Bentinckia nicobarica*, *Cycas zeylanica*, *Rhopaloblaste augusta*, etc.), fruit-yielding curious plants (*Baccaurea ramiflora*, *Garcinia cowa*, *G. dhanikhariensis*, *Mangifera andamanica*, *M. griffithii*, *Pandanus leram*, etc.), medicinal and economical plants (*Elaeocarpus floribundus*, *Horsfieldia glabra*, *Knema andamanica*, *Myristica andamanica*, *Pterocarpus dalbergioides*) are being used for mass plantation programs conducted by the Botanical Survey of India during Van Mahotsav (every year 1–7 July), World Environment Day (every year 5 June), and International Day of Biological Diversity (every year 22 May) and also supplied to different organizations throughout the country on demand.

As Andaman and Nicobar Islands is one of the biodiversity hotspots, DEGCA plays a key role in the Global Strategy for Plant Conservation by making all efforts to control the loss of plant diversity by conserving the EET plant species that are not recorded from any other geographical region.

ACKNOWLEDGMENTS

The authors are grateful to Dr. A.A. Mao, Director, Botanical Survey of India and Ministry of Environment, Forest and Climate Change for constant support and facilities. The authors are thankful to Prof. (Dr.) T. Pullaiah, Editor, and Dr. Galbraith, Co-editor, for critical comments and suggestions that helped to improve the chapter. The authors are also thankful to the scientists and staff of BSI who have always shown readiness to help.

REFERENCES

Balakrishnan, N.P. 1989. Andaman Islands–vegetation and floristics. In: Saldanha, C.J. (ed.), *Andaman, Nicobar and Lakshadweep-an Environmental Impact Assessment*. Oxford & IBH Publishing Co., New Delhi, pp. 55–68.

Balakrishnan, N.P. and M.K.V. Rao. 1983. The dwindling plant species of Andaman and Nicobar Islands. In: Jain, S.K. and R.R. Rao (eds.), *An Assessment of Threatened Plants of India*. Botanical Survey of India, Howrah, pp. 186–201.

Dixit, R.D. and B.K. Sinha. 2001. *Pteridophytes of Andaman and Nicobar Islands*. Bishen Singh Mahendra Pal Singh, Dehra Dun.

Ellis, J.L. 1987. The pteridophytic flora of Andaman and Nicobar Islands. *J. Andaman Sci. Assoc.* 3(2): 59–79.

Hajra, P.K., P.S.N. Rao and V. Mudgal. 1999. *Flora of Andaman-Nicobar Islands, Vol. 1. (Ranunculaceae–Combretaceae)*. Botanical Survey of India, Calcutta, pp. 1–487.

Jagadeesh Ram, T.A.M. 2014. The genus *Herpothallon* (Arthoniaceae) from the Andaman Islands, India. *Lichenologist* 46(1): 39–49.

Lakshminarasimhan, P. and P.S.N. Rao. 1996. A Supplementary list of Angiosperms recorded (1983–1993) from Andaman and Nicobar Islands. *J. Econ. Taxon. Bot.* 20(1): 175–185.

Mathew, S.P. 1998. A supplementary report on the flora and vegetation of the Bay Islands, India. *J. Econ. Taxon. Bot.* 20(2): 249–272.

Mishra, S., G.A. Ekka, R. Vinay and L.J. Singh. 2016. Role of DEGCA, BSI garden in conservation of medicinal plant diversity of Andaman and Nicobar Islands, India. In: Chourasia, H.K. (ed.), *Conservation of Medicinal Plants Conventional and Modern Approaches*. Omega Publications, New Delhi, pp. 95–108.

Murugan, C., S. Prabhu, R. Sathiyaseelan and R.P. Pandey. 2016. A checklist of plants of Andaman and Nicobar Islands. In: Singh, P. and W. Arisdason (eds.). ENVIS Centre on Floral Diversity, Botanical Survey of India, Howrah. Published on the Internet: www.bsienvis.nic.in/Database/Checklist of Andaman-Nicobar Islands_24427.aspx (Accessed 2 June 2018).

Murugan, C., L.J. Singh, S. Prabhu and R. Sathiyaseelan. 2014. The genus *Ruellia* L. (Acanthaceae) in Andaman and Nicobar Islands. *Ind. J. Forestry* 37(4): 425–428.

Naik, M.C. and B.R.P. Rao. 2016. Flora of South Andaman Islands, India. In: Nehra S. et al. (eds.), *Biodiversity in India: Assessment, Scope and Conservation*. Lambert Academic Publishing Heinrick-Booking-Str. Sarbruken, Germany.

Naik, M.C. and L.J. Singh. 2020a. Diversity and Conservation of Rubiaceae of Andaman and Nicobar Islands, India. In: *International Symposium on Plant Taxonomy and Ethnobotany*. BSI, Kolkata. p. 87.

Naik, M.C. and L.J. Singh. 2020b. *Dentella cylindrica* sp. nov. (Rubiaceae: Tribe Spermacoceae) from Andaman and Nicobar Islands, India. *Indian J. For.* 43(4): 308–314.

Naik, M.C., A.H. Arriola and M. Bheemalingappa. 2020b. *Pyrostria laljii*, a new species and first record of *Pyrostria* (Vanguerieae, Rubiaceae) from the Andaman and Nicobar Islands, India. *Ann. Bot. Fennici.* 57: 335–340.

Naik, M.C., L.J. Singh and K.N. Ganeshaish. 2020a. Floristic diversity and analysis of South Andaman Islands (South Andaman District), Andaman and Nicobar Islands, India. *Species* 21(68): 343–409.

Nayar, M.P. 1997. "Hot spots" of plant diversity in India-strategies. In: Pushpangadan, P., K. Ravi and V. Santosh (eds.), *Conservation and Economic Evaluation of Biodiversity*. 1. Science Publishers, Thiruvananthapuram, pp. 59–80.

Pandey, R.P. and P.G. Diwakar. 2008. An integrated check list of Andaman and Nicobar Islands, India. *J. Econ. Taxon. Bot.* 32(2): 403–500.

Parkinson, C.E. 1923. *A Forest Flora of the Andaman Islands*. Bishen Singh Mahendra Pal Singh, Dehra Dun.

Purti, N., L.J. Singh and A.K. Pandey. 2022. New hosts for the cycad blue butterfly, *Luthrodes pandava*, Horsfield (Lepidoptera: Lycaenidae) in an Island ecosystem. *Feddes Repertorium* 133(3): 234–243.

Rao, P.S.N. 2001. Coastal and marine plant diversity of Andaman and Nicobar Islands. *Bull. Bot. Surv. India* 42: 1–40.

Renuka, C. 1995. *A Manual of the Rattans of Andaman and Nicobar Islands*. Kerala Forest Research Institute, Trichur, pp. 1–72.

Singh, L.J. 2012. Mangrove plant diversity in Bay Islands, India and its significance. In: Kumar, P., P. Singh, R.J. Srivastava and R.K. Dubey (eds.), *National Conference on Biodiversity: One Ocean-Many Worlds of Life*. U P State Biodiversity Board, Lucknow, pp. 119–126.

Singh, L.J. 2013. The genus *Macrosolen* (Loranthaceae) in Andaman and Nicobar Islands, with a new record for India. *Rheedea* 23(2): 108–112.

Singh, L.J. 2021. *Septemeranthus* (Loranthaceae), a new monotypic genus from the Andaman and Nicobar Islands, India and its relationship with allied genera. *Feddes Repert.* 132: 193–203.

Singh, L.J., M.D. Dwivedi, S. Kasana, M.C. Naik, G.A. Ekka and R.P. Pandey. 2020b. Molecular systematics of the genus *Musa* L. (Zingiberales: Musaceae) in Andaman and Nicobar Islands. *Biologia* 75: 1825–1843. https://doi.org/10.2478/s11756-020-00552-5

Singh, L.J., G.A. Ekka and S. Mishra. 2016a. Ethnobotanical uses of plants in the Andaman and Nicobar Islands, India. In: Chourasia, H.K. and A.K. Roy (eds.), *Conservation, Cultivation, Disease and Therapeutic Importance of Medicinal and Aromatic Plants*. Today and Tomorrow's, Printers and Publishers, New Delhi, pp. 165–179.

Singh, L.J., G.A. Ekka, S. Mishra, C.P. Vivek, V. Shiva Shankar, M.C. Naik and F. Salee. 2020c. Habitat status of *Musa paramjitiana* L.J. Singh (Musaceae): A critically endangered, endemic species in Andaman and Nicobar Islands, India. *Pleione* 14(1): 121–127.

Singh, L.J., G.A. Ekka, C.P. Vivek and D.R. Misra. 2021a. Gymnosperms of the Andaman and Nicobar Islands: An overview. In: Singh, L.J. and V. Ranjan (eds.), *New Vistas in Indian Flora*. Bishen Singh Mahendra Pal Singh, Dehra Dun, India, pp. 265–278.

Singh, L.J., B.S. Kholia, K.P. Kumar, S. Sharma and A. Ebihara. 2016c. Notes on occurrence and distribution of some filmy ferns in Andaman and Nicobar Islands, India. *Keanean J. Sci.* 5: 61–68.

Singh, L.J., B.S. Kholia, B. Kumar, P. Joshi and S. Sharma. 2016d. On the occurrence of *Thelypteris polycarpa* (Blume) K. Iwats. in Andaman Island with a note its dispersal and distribution in India. *Indian For.* 143(12): 1234–1236.

Singh, L.J., B. Kumar, B.S. Kholia and P. Joshi. 2016b. *Diplazium proliferum*: An addition to the Indian pteridophytic flora from little Andaman. *J. Japanese Bot.* 91: 57–60.

Singh, L.J., S. Mishra, C.P. Vivek and G.A. Ekka. 2018. Systematic account of family Musaceae in the Andaman and Nicobar Islands. In: Chourasia, H.K. and D.P. Mishra (eds.), *Plant Systematics and Biotechnology: Challenges and Opportunities*. Today and Tomorrow's Printers and Publishers, New Delhi, pp. 473–483.

Singh, L.J. and D.R. Misra. 2017. Identity and status of recently described *Cycas pschannae* (Cycadaceae) in the Andaman and Nicobar Islands, India. *Bionature* 37(1): 38–55.

Singh, L.J. and D.R. Misra. 2020. Reappraisal of the genus *Cycas* L. (Cycadaceae) in Andaman and Nicobar Islands, India. *Indian J. Forestry* 43(1): 46–57.

Singh, L.J. and C. Murugan. 2013a. Genus *Dendrophthoe* Mart. (Loranthaceae) from Bay Islands with a new record for India and inventory of host species. *Geophytology* 43(1): 41–49.

Singh, L.J. and C. Murugan. 2013b. Vivpary in *Hibiscus cannabinus* L. (Kenaf.): A potential reproductive strategy in island's ecosystem. *Geophytology* 43(2): 171–175.

Singh, L.J. and C. Murugan. 2014. Seed plant species diversity and conservation in Dhanikhari experimental garden-cum-arboretum in Andaman and Nicobar Islands, India. In: Nehra, S., R.K. Gothwal and P. Ghosh (eds.), *Biodiversity in India: Assessment, Scope and Conservation*. Lambert Academic Publishing, Heinrich-Bocking-Str. Saarbruken, Germany, pp. 253–280.

Singh, L.J., C. Murugan and P. Singh. 2014. Plant genetic diversity of endemic species in the Andaman and Nicobar Islands. National Conference on Islands Biodiversity, U.P. State Biodiversity Board, Lucknow, pp. 49–57.

Singh, L.J. and V. Ranjan. 2015. *Heterotis* Benth. (Melastomataceae): A new addition to Indian flora from Andaman Islands. *Geophytology* 45(1): 101–106.

Singh, L.J. and V. Ranjan. 2021. *New Vistas in Indian Flora*. Vol. 1 & 2: Bishen Singh Mahendra Pal Singh Dehra Dun, Uttarakhand, India, p. 417, p. 819.

Singh, L.J., V. Ranjan, B.K. Sinha, S. Mishra, C.S. Purohit, C.P. Vivek, M.C. Naik and G.A. Ekka. 2021b. An overview of phytodiversity of the Andaman and Nicobar Islands, India. In: Singh, L.J. and V. Ranjan (eds.), *New Vistas in Indian Flora*. vol. 1. Bishen Singh Mahendra Pal Singh, Dehra Dun, Uttarakhand, India, pp. 381–399.

Singh, L.J., C.P. Vivek, G.A. Ekka and B.C. Dey. 2020b. Diversity of family Loranthaceae in Andaman and Nicobar Islands, India. In: *International Symposium on Plant Taxonomy and Ethnobotany*. BSI, Kolkata, p. 219.

Sinha, B.K. 1999. *Flora of Great Nicobar Island* (eds. Hajra, P.K. and P.S.N. Rao). Botanical Survey of India, Calcutta.

Sivaramakrishna, P., P. Yugandhar and G.A. Ekka. 2021. A new species *Dendrophthoe laljii* (Loranthaceae) infesting *Artocarpus heterophyllus* Lam. (Moraceae) in Andaman and Nicobar Islands, India. *J. Asia-Pac. Biodivers*. 14: 452–459.

Srivastava, S.K. and P.S.N. Rao. 2001. Flora of Dhanikhari experimental garden-cum-arboretum, Port Blair, Andaman and Nicobar Islands. *Bull. Bot. Sur. India* 43(14): 1–82.

Vasudeva Rao, M.K. 1986. A preliminary report on the angiosperms of Andaman-Nicobar Islands. *J. Econ. Taxon. Bot*. 8(1): 107–184.

12 M. S. Swaminathan Botanical Garden – A Community Conservation Initiative in the Western Ghats of India

Nadesa Panicker Anil Kumar, V. Shakeela, Merlin Lopus, and Sarada Krishnan

CONTENTS

12.1 Introduction 203
12.2 Status of Global Biodiversity 204
12.3 Biodiversity of the Western Ghats 206
12.4 Botanical Gardens and Biodiversity Conservation 207
12.5 Case Study: M. S. Swaminathan Botanical Garden 208
    12.5.1 The Strategic Approach of MSSBG 210
    12.5.2 Work in Partnership with the Local Self Governments 210
    12.5.3 Work in Partnership with Native Tribal Communities 211
    12.5.4 Looking Ahead 213
12.6 Conclusion 213
References 214

12.1 INTRODUCTION

Botanical gardens have evolved significantly over the past few centuries since their inception in the mid-sixteenth century to today's gardens. Early gardens were associated with European universities for the study of medicinal plants. In the seventeenth to nineteenth centuries, gardens were involved in taxonomy to classify plants, and often plants in gardens were displayed in Linnaean order. Activities involved acquisition, display, study, and distribution of economically important plants as Europeans explored and colonized other parts of the world. During the second half of the twentieth century, realization and acknowledgment about the rapid loss of biodiversity led many gardens to become involved in conservation and sustainable use of biodiversity through *ex situ* and *in situ* conservation. Today, many botanical gardens are tackling global issues of climate change, food security, biodiversity loss, environmental education, and human health, leading to expansion of garden programs beyond the garden walls (Avery, 1957; Borsch and Lohne, 2014; Krishnan and Novy, 2016; Powledge, 2011).

Dodd and Jones (2010) identified seven key areas that are particularly concerned with the outreach strategies of botanic gardens. Among these, the top three are (1) audience development, (2) meeting the needs of communities, and (3) education. Enhancing the relevance of the gardens' activities to meet the needs of the people was reported as the most challenging task for most of the gardens as it needs an inclusive, transformative, multi-disciplinary, and context-specific approach. In the coming decades, human population pressure coupled with food, nutrition, and health needs and developmental aspirations of people in regions of high biodiversity, in particular the tropical regions, are going to result in significant loss of biodiversity, including extinction. Conservation and sustainable use of plant genetic resources have become focal points of many national and international agendas with tremendous success in developing methods for the conservation of genetic resources *ex situ*, while *in situ* conservation is still inadequate (Gole et al., 2002). This inadequacy is mainly due to socio-economic factors, lack of policy and political will, and a lack of scientific understanding of the natural environment and biological characteristics of species (Gole et al., 2002). Biodiversity conservation should be viewed as a social, cultural, and political process undertaken to conserve non-human life by humans for the well-being of themselves as well as for the planet. But, biodiversity is reported to be declining at an unprecedented rate globally despite decades of scholarship and actions for its protection. A major reason for this situation is the poor involvement of the local community in decision-making in conservation.

The Convention on Biological Diversity (CBD) at the tenth meeting of the Conference of the Parties held 18 to 24 October 2010 in Nagoya, Aichi Prefecture, Japan, adopted a revised and updated Strategic Plan for Biodiversity (Convention on Biological Diversity, 2012). The Aichi Biodiversity Targets, effective through 2020, outline five strategic goals to be implemented through 20 major targets, such as addressing the underlying cause of biodiversity loss, reducing direct pressures on biodiversity, safeguarding biodiversity at the ecosystem level, enhancing the benefits provided by biodiversity, and providing for capacity building. At the 15th meeting of the Conference of the Parties (COP 15), held in October 2021, a revised and updated post-2020 Global Biodiversity Framework (Convention on Biological Diversity, 2021) was adopted. This Framework, built around a Theory of Change, advocates for urgent policy actions to bring transformational changes in economic, social, and financial sectors in order to curb the on-going trends in biodiversity loss and to allow the recovery of natural ecosystems. The Theory of Change, with its 4 goals, 8 milestones, and 20 targets, outlines a whole-of-government and society approach to make conservation actions more effective.

The XIV International Botanical Congress held in St Louis, Missouri, USA, in 1999 called for a global initiative for plant conservation. This led to the development of the Global Strategy for Plant Conservation (GSPC), which was adopted by the world's governments as a program under the CBD in 2002. The first GSPC identified 16 targets to be achieved by 2010. A review in 2007 revealed that, even though significant progress had been made, efforts needed to continue beyond 2010. The result was the updated GSPC: 2011–2020, which was adopted at the 10th Conference of Parties (COP) to the CBD in 2010. The mission of the GSPC is to be a catalyst for working together at all levels – local, national, regional, and global – to understand, conserve, and use sustainably the world's immense wealth of plant diversity while promoting awareness and building the necessary capacities for its implementation. Addressing the challenges posed to plant diversity, the GSPC consists of 5 objectives and 16 targets (Convention on Biological Diversity, 2012). A draft post-2020 GSPC has been developed with plant conservation objectives for 2050 and plant conservation targets for 2030. This post-2020 GSPC will be aligned within the post-2020 Global Biodiversity Framework of the CBD (CBD, 2021).

A major lesson from the coronavirus pandemic that began in 2020 is that we must stop exceeding the limits of planetary boundaries and rely on evidence from science to prepare for the future management of the planet's life irrespective of whether it is human or microscopic viral life. The discussion should focus on how to ensure a whole-of-government and society approach and engage the private sector industries and the key non-governmental organizations in achieving the post-COVID development agenda by integrating the post-2020 biodiversity action plans at different levels.

12.2 STATUS OF GLOBAL BIODIVERSITY

The most important challenge to biodiversity in the coming decades will be global climate change causing increasing variability in temperature and precipitation and extreme weather events. This will impact ecosystem services that humans depend upon. In a study to understand the impact of climate change and population growth on biodiversity and ecosystem services and geographic locations of vulnerable human populations, Aukema et al. (2017) report that 40% of rangelands and 30% of rain-fed agriculture have experienced significant changes in precipitation, threatening food security.

The second-greatest threat to biodiversity arises from habitat destruction, alien species invasion, and genetic homogeneity (Swaminathan, 2000). As result of human activities, biodiversity decline in oceans, land, and wetlands ranges from 66%, 75%, and 85%, respectively (Reyers and Selig, 2020; Díaz et al., 2019). All forms of biodiversity, from ecosystems and species to genes across almost all taxa, are threatened and facing an alarming rate of extinction. The Intergovernmental Science-Policy Platform on Biodiversity and Ecosystem Services estimated that 1 million species are threatened with extinction. According to World Wildlife Fund (WWF) reports, 12,505 plant, 1,204, mammal, 1,469 bird, 1,215 reptiles, 2,100 amphibians, 2,386 fish, and 1,414 insect species are considered threatened. According to IUCN, one in every four species of all major taxa is threatened. Nearly 75% of the genetic diversity in crops and animal breeds has been already lost and 90% of the remaining diversity confined to small and family farms. The Intergovernmental Science-Policy Platform on Biodiversity and Ecosystem Services (IPBES, 2019) reported 87% of the world's small farms are in Asia; these farms support a wide range of crop and animal breed genetic diversity, but they are in decline, along with their associated traditional knowledge. Intact natural forest ecosystems and genetically diverse agricultural farms and landscapes are critically vital for providing services like supply of freshwater and protection against storms, floods, and other hazards as well as for resilience to the production landscapes. This massive loss of wild species has now forced environmentalists to think that the planet is entering its sixth great extinction phase.

In the human-dominated Anthropocene epoch where disruptions in the natural environment have become rampant, the emergence of many more pathogens in the future cannot be ruled out. Studies reveal that the source of more than 90% of zoonotic virus transmission is wild animals that are in the range of thousands of species – from rodents, birds, and bats to primates (IPBES, 2019; Jameel, 2020). For the post-coronavirus pandemic development phase, every nation needs to bring out an action plan that has local level actions aimed at protecting and enhancing wild landscapes, ecosystems, and species diversity. Unless there is a paradigm shift in our approach in the management of and investment in our wild biodiversity and the planet's health, we could expect increased health calamities in the future.

The combination of high human population growth with climate variability in critical conservation areas with high biodiversity, such as the Horn of Africa, the Himalayas, the Western Ghats, and Sri Lanka, poses challenges to both humans and sustenance of biodiversity (Aukema et al., 2017). In a study examining the risk of conflict between food security and biodiversity conservation, Molotokos et al. (2017) constructed a combined risk index by overlapping the Global Food Security Index (GFSI) with the National Biodiversity Index (NBI) and ranked various countries. Table 12.1 lists the top 15 countries with highest combined risk, of which India ranks in thirteenth place.

The 2017 State of the World's Plants report provides an optimistic outlook on the effectiveness of conservation actions and policies in securing and protecting some of the most important plant species and communities globally, though there is still more to do (Willis, 2017). Some key findings in the report are as follows:

- Each year, 340 million hectares of vegetated surface burns.
- In total, 452 vascular plant families have been identified across the world.
- In 2016, 1,730 new vascular plant species were documented.
- The genomes of 225 plant species have been completely sequenced.
- At least 28,187 plant species have been documented as having medicinal properties.
- Plants with thicker leaves, denser roots, efficient water use strategies, and higher wood density are adapted to cope better with climate change.
- 6,075 species have been documented as invasive.
- The impact of the spread of invasive pests and pathogens, if not stopped, amount to a cost of approximately US$540 billion per year to agriculture.
- In the CITES Appendices, 31,517 plant species are listed currently.
- There are an estimated 392,000 vascular plant species known to science.

(RBG Kew, 2016; Willis, 2017)

TABLE 12.1
The Top 15 Countries with Highest Combined Conflict Risk between Food Security and Biodiversity Conservation

Rank	Country	GFSI Rank	NBI Rank	Combined Rank
1	Madagascar	98	97	195
2	Burundi	107	86	193
3	Haiti	102	84	186
4	Sierra Leone	106	76	182
5	Democratic Republic of Congo	101	75	176
6	Togo	88	88	176
7	Indonesia	66	107	173
8	Rwanda	82	91	173
9	Tanzania	89	82	171
10	Cameroon	80	87	167
11	Malawi	99	66	165
12	Philippines	69	94	163
13	India	70	92	162
14	Guatemala	68	93	161
15	Bolivia	65	90	155

Source: Molotokos et al. (2017).

Halting the loss of biodiversity that has evolved over millions of years by promoting conservation and sustainable use of wild and agricultural biodiversity for food and agriculture development continues to be the major challenge. In order to achieve results in conservation and enhancement of natural resources, an integrated gene management practice encompassing bio-partnerships, participatory forest management, community gene management, biosphere management, and genetic resources enhancement and sustainable use should be implemented (Swaminathan, 2000). Mainstreaming this approach in critical conservation areas, regions of plant diversity, and biodiversity hotspots through an integrated conservation approach thus becomes a high priority action.

12.3 BIODIVERSITY OF THE WESTERN GHATS

The monsoon forests of the Western Ghats, one of the world's biodiversity hotspots, occur in western India both on the western (coastal) margin of the Ghats and on the drier eastern side, providing habitat for many endemic species of flora and fauna (Athira et al., 2017; IUCN, 1991). For instance, 80% of caecilian diversity (16 out of the 20 known species), 78% of the recorded amphibians, 76% of mollusks, 62% of reptiles, 56% of the 645 evergreen tree species, 43% of liverworts, 41% of fish, and 40% of odonates are endemics to this region (Gunawardene et al., 2007). The floral diversity of the Western Ghats also consists of many wild, unutilized, and underutilized minor fruits that are rich in antioxidants, essential nutrients, and bioactive molecules, offering tremendous potential for future development and commercialization that could result in food and livelihood security of local communities. Among all the 36 global biodiversity hotspots, the Western Ghats are reported with the highest human population density of >300 persons/sq km (Cincotta et al., 2000). Rapid urbanization and deforestation are threatening several plant species with disturbance or change of about 35.3% forest cover between 1920 and 2013 (Mundaragi et al., 2017). Among the biodiversity hotspots, the Western Ghats has high species endemism and is older in terms of geological age. Based on the presence of the fossil plant *Glossopteris* flora throughout the Western Ghat hill ranges, it is considered that peninsular Indian region was a part of the Gondwana supercontinent, which goes back to the time of earth's crust formation (Maithy, 1966; Casshyap and Tewari, 1988; Chandra, 1995).

The Western Ghats have some of the finest non-equatorial tropical evergreen forests in the world with very high levels of speciation and endemism. It is home to more than 5,500 indigenous vascular plant species with about 1,500 endemics (37.5%). Among these, about 700 are tree species, of which nearly 50% (more than 300 species) are endemics, including about 55 species that are globally threatened and critically endangered and listed under IUCN Red Data Book (www.iucnredlist.org/).

The United Nations Educational, Scientific and Cultural Organization's (UNESCO) Man & Biosphere Reserve Program, as well as the ecological and geographical significance of the Western Ghats, led the Ministry of Environment, Forest and Climate Change to create the Nilgiri Biosphere Reserve (NBR) in 1986 with a core area of 1,240 km^2 and a buffer zone of 4,280 km^2. The area lies between 76°–77° 15' E. latitude and 11° 15' – 12° 15' N. longitude, covering the Nilgiri mountain ranges in the geographic territory of Tamil Nadu and the adjoining forests from the states of Kerala and Karnataka. This was the first Biosphere Reserve of the country, and in 2012, the Nilgiri sub-cluster of the Reserve, which includes NBR's nine protected sanctuaries, was declared a world heritage site.

Within the Western Ghats and partly in the Nilgiri Biosphere Reserve, Wayanad District is a hotspot of biodiversity, qualifying as an agrobiodiversity heritage site, notable among which is the genetic diversity of rice with more than 20 landraces cultivated in the district (Anil Kumar et al., 2017). A floristic exploration of the district has documented 2,034 species of angiosperms of which 491 taxa are endemic to the Western Ghats (as cited in Anil Kumar et al., 2017). In Kerala, a network of 23 protected areas (PAs) covering 3,213.24 km^2 and consisting of 17 wildlife sanctuaries, 5 national parks, and 1 community reserve have been established (Athira et al., 2017). Establishment of protected areas, national parks, and sanctuaries are positive trends for curbing deforestation, and the effectiveness of biodiversity conservation will depend on efficient management, which should be done in collaboration with local communities.

12.4 BOTANICAL GARDENS AND BIODIVERSITY CONSERVATION

Through living collections, botanic gardens have been experts in *ex situ* conservation of plants. During the second half of the twentieth century, many botanic gardens became more actively engaged in the *in situ* conservation of rare and threatened plant species (Krishnan and Novy, 2016). Botanic gardens are scientific institutions with the ability to respond to contemporary issues caused by global changes (Donaldson, 2009). To maintain biodiversity and the quality and quantity of ecosystem services and to counter the vast destruction and degradation of the world's terrestrial ecosystems, there is a critical need for ecological restoration, and botanic gardens have the technical know-how, skills, and resources to achieve these goals (Miller et al., 2016).

Fant et al. (2016) identify seven steps by which botanic gardens can implement some of the approaches developed by the zoological community to conserve genetic diversity of threatened plants, especially plants that cannot be conserved as seeds and need to be maintained as living collections, in order to support *in situ* conservation directly. The seven steps are as follows: (1) identifying candidate species for managed breeding, (2) establishing a sponsor institution connected by multiple institutions to maintain and manage *ex situ* collections for a specific species, (3) utilizing available online plant databases to inventory *ex situ* collections to assess genetic diversity of the species of interest, (4) conducting gap analysis between *ex situ* and *in situ* populations, (5) coordinating acquisition of new wild materials to fill the gaps, (6) developing and implementing *ex situ* management plans for the species, and (7) developing collaborations for *in situ* conservation using *ex situ* materials.

To evaluate the effectiveness of botanic garden collections in supporting plant conservation, Maunder et al. (2001) conducted a survey of 119 European botanic gardens in 29 countries. The survey identified 25 gardens in 14 countries taking part in 51 conservation projects focused on threatened plant species listed by the Bern Convention. Of the 573 listed plant species, 308 are being cultivated by 105 gardens. A major finding from this study was that the majority of the species were being cultivated in gardens outside the range of the species and not contributing to their conservation, though these collections represent populations now lost in the wild, with at least nine taxa identified as "extinct in the wild." Recommendations made by this study include increasing applied research in botanic gardens to develop horticultural protocols for threatened taxa; expanding research on storage of germplasms other than as living collections; applying horticultural practices for species recovery, such as propagation, reproductive assistance, reintroduction, weed and pest control, etc.; and linking botanic garden displays for promotion and to raise funds for *in situ* conservation.

Apart from the traditional role of botanic gardens contributing to plant conservation through research and education, the modern-day gardens increasingly recognize the importance of citizen science and involvement (Chen and Sun, 2018; Conrad and Hilchey, 2011). For instance, scientists of many gardens in India partner with local community researchers and volunteers to increase the scope of biodiversity conservation and collection of scientific data through the legally mandated mechanism of People's Biodiversity Registers. A well-prepared Register gives comprehensive information about the biodiversity and the associated people's knowledge of a region, which can be used as a tool in the mainstreaming of biodiversity into development policies, plans, and programs and to facilitate the access and benefit sharing provisions.

With the adoption of the GSPC, remarkable transformation has happened over the past few years in international plant conservation, stimulating new initiatives and programs at all levels and sectors, specifically botanic gardens (Wyse Jackson and Kennedy, 2009). There is still much to do, and botanic gardens have a significant role to play in the conservation of plants in order to stem any new threats of extinction.

12.5 CASE STUDY: M. S. SWAMINATHAN BOTANICAL GARDEN

The biodiversity conservation efforts of M. S. Swaminathan Research Foundation in the Western Ghats resulted in collection and conservation of many endangered plant species and little-known taxa of food, nutrition, and health value found in this region. These collections came from on-farm, wilderness, and forest sites and formed key collections in the establishment of a botanic garden named the M. S Swaminathan Botanical Garden (MSSBG) in the Wayanad district of Kerala. MSSBG is situated in the valley of a natural hillock in an agrarian village traversed with shade-grown coffee groves and paddy/banana fields, near Kalpetta, the district headquarters of Wayanad district. Wayanad is one of the richest centers of biodiversity in India and strategically located in the trans-state corridor of the country. It is the leading district in the state in production of spices and plantation crops, medicinals, and aromatic rice as well as diverse non-wood forest products (NWFP).

The garden occupies about 10 ha and has ten major zones and ten minor components with diverse representative species of tropical flora, consisting of crop wild relatives, wild orchids, medicinal plants, and germplasm of food plant varieties, such as roots and tubers, legumes, citrus, and wild edible greens (Figures 12.1–12.3). All these components are in continuous development. A total of 2,033 live plant species are conserved in the garden, of which 512 are endemic to the Western Ghats and 579 fall under IUCN's threatened plant species category. The rest of the 10 ha of the property is managed as shade-grown coffee, consisting of robusta coffee (*Coffea canephora*) grown under the canopy of wild native tree species. The garden was established as a Resource Centre for supply of seed materials of the "lost crops and species" that were once available to the local community, particularly farmers and the indigenous tribal families, through cultivation and as associated species on-farm. For instance, the garden maintains a germplasm collection with plant genetic resources such as *Dioscorea* – cultivated (22 varieties from 3 species), *Dioscorea* – wild

FIGURE 12.1 Wild edible fruits of the Western Ghats. (a) *Alangium salvifolium* (L. f.) Wang., (b) *Solanum americanum* Mill., (c) *Salacia beddomei* Gamble., (d) *Sarcostigma kleinii* Wight & Arn., (e) *Rourea minor* (Gaertn.) Merr., (f) *Antidesma montanum* Blume., (g) *Baccaurea courtallensis* (Wight) Muell.-Arg. in DC., (h) *Schleichera oleosa* (Lour.) Oken.

FIGURE 12.2 Wild orchids of the Western Ghats. (a) *Bulbophyllum sterile*, (b) *Acanthephippium bicolor*, (c) *Pecteilis gigantea*, (d) *Peristalis porotta*, (e) *Cymbidium aloifolium*, (f) *Dendrobium anilli*, (g) *Oberonia swaminathanii*, (h) *Dendrobium aphyllum*.

(13 varieties from 6 species), *Colocasia* – cultivated (8 varieties of a species), *Colocasia* – wild (4 varieties of a species), Arrowroot (2 varieties of a species), *Amorphophallus* (1 species), *Canna* (2 species), Sweet Potato (3 varieties of a species), *Alocasia* (2 species), Tuberous orchids (2 species), Asparagus (1 species), *Costus* (1 species), *Citrus* (17 varieties from 6 species), Banana (12 varieties), and Legumes (17 varieties from 4 species). It was also established to serve as an educational facility for children, youth, teachers, and parents to learn about the native flora and the value of biodiversity and ecosystem services.

In 2014, the Denver Botanic Gardens in Denver, Colorado, USA, partnered with MSSBG to develop a master plan providing a future vision for growth and development of the M.S. Swaminathan Botanical Garden (Krishnan et al., 2020). Through stakeholder meetings, the following mission was adopted for the garden: "to serve the community as a resource and advocate for botanical and

FIGURE 12.3 RET species conserved at MSSBG. (a) *Actinodaphne malabarica*, (b) *Atuna travancorica*, (c) *Buchanania lanceolata* Wt, (d) *Cynometra beddomei*, (e) *Cynometra travancorica* Beddome, (f) *Hopea ponga* (Dennst.) Mabb., (g) *Humboldtia brunonis*, (h) *Humboldtia decurrens*.

agricultural biodiversity of the Western Ghats and inspire an ethic of conservation." In 2017, an advisory committee was formed to spearhead the implementation of the master plan and develop clear strategies for action. The reinvigorated purpose of the new garden is to promote conservation of endangered plant diversity of the Western Ghats through integrated conservation action and dissemination of scientifically credible knowledge on the critical role of plant diversity in providing ecosystem services, enhancing climate resilience, and supporting human well-being.

12.5.1 The Strategic Approach of MSSBG

MSSBG adopts a "4C" framework where the "C"s represent conservation, cultivation, consumption, and commercialization; the framework operates in a cyclic manner to promote the paradigm of sustainable genetic resource management (Krishnan et al., 2020). By engaging and empowering the "custodian farmers and conservers" to apply this 4C framework, MSSBG proposes to demonstrate sustainable utilization of four key sectors of biodiversity, such as crop diversity, curative diversity (medicinal plants), culinary diversity (traditional and wild food plant diversity), and the rare-endemic-threatened (RET) tree species. The major strategy is to work in partnership with local community families and their representative institutions.

12.5.2 Work in Partnership with the Local Self Governments

MSSBG has formulated an integrated action plan to empower the democratically elected local self governments (LSGs) of Kerala that enjoy a high level of decentralized decision-making power to promote community biodiversity conservation and mainstreaming biodiversity in local development (see Box 12.1). All local bodies of India have the mandate to tailor the biodiversity-related projects toward achieving concurrently the Sustainable Development Goals and the National Biodiversity Targets. Despite the major role played by local communities in conservation, their involvement in research and biodiversity management activities has not received the attention it deserves. There are many farmers in almost every Panchayath (the smallest administrative unit) of the state who are engaged in protection of heritage agriculture and fisheries and biodiverse, climate-smart, nutrition-sensitive farming. MSSBG has built capacity of a few LSGs of Wayanad District to work with such "custodian farmers" and promote the local sustainable agriculture legacy as a sectoral component in the plan. This will lead to strengthening *in situ* and on-farm conservation of the biodiversity with a focus on mainstreaming the genetic diversity of seeds, cultivated plants, and farmed and domesticated animals and their related wild species. These actions are aimed at achieving educated and empowered Biodiversity Management Committees (the legally binding institutional arrangement under the Biodiversity Act 2002 for conservation of biodiversity at local level) working with the respective LSGs to execute and monitor the conservation action.

BOX 12.1 MAINSTREAMING BIODIVERSITY ACTION AREA: CONSERVATION, SUSTAINABLE USE, AND BENEFIT SHARING

1. Organizing studies on the conservation status of different forms of biodiversity found in urban, rural, and tribal settings.
2. Implementing projects for collection and multiplication of the rare, endemic, and threatened plant species.
3. Publication of local-level RED Lists and RED DATA Books.
4. District or Panchayath level GIS and drone imaging of the landscapes and waterscapes with quantifiable information on the land/water use area size and diversity.
5. Organizing preparation of agrobiodiversity registers and conservation corps to enable the farmers to register their seeds/breeds and continue the on-farm conservation.
6. Panchayath or district-level species-specific conservation action projects targeting the critically endangered species across all taxa.
7. Establishing community seed banks for preserving the seeds of useful tree species used by farmers and traditional healers.
8. Ground verification studies of the landscapes and waterscapes outside protected areas in different seasons to understand the present land use and land cover measures.
9. Establishing seed multiplication plots for the varieties and breeds that are unique to the state.
10. Scientific validation of the traditional claims of nutrition and medicinal values of the taxa.
11. Joint publication by botanists and traditional healers/farmers on the food, nutrition, and medicinal values.
12. Databases pertaining to intellectual property rights (IPR) maintained at district and state levels.
13. Establishing community conservation gardens at every district to grow all the plants of direct utility value and to promote the traditional uses for nutrition and health.
14. Promoting eco-technologies and sustainable harvest of the useful plants and plant parts.
15. Building up local level biodiversity funds managed by biodiversity management committees with contributions received from the commercial users of biodiversity.

12.5.3 WORK IN PARTNERSHIP WITH NATIVE TRIBAL COMMUNITIES

MSSBG adopts an approach that assures conservation through the participation of local communities, particularly the native tribal communities. Wayanad district has the largest tribal population of Kerala with the major groups such as *Kurichiya*, *Mullu Kuruma (Kuruma)*, *Paniya*, *Adiya*,

Kattunaikka, Thachinadan Mooppan, Wayanadan Kadar, Karimbalar, Urali Kuruma, and *Pathiyar*. The tribal communities of Wayanad district have vast knowledge on wild plant diversity, and many are included in their food basket. For example, the *Paniya* community uses a large number of plant and small animal diversity, which includes 72 species of leafy vegetables, 25 species of mushrooms, 19 species of tubers, 48 species of fruits and nuts, 36 kinds of native fishes, 8 kinds of crabs, and 5 types of wild honey (Anil Kumar and Ratheesh Narayanan, 2009). Wild leaves are among the most widely consumed wild foods of the district. Most of the leafy wild food plants are locally referred to and classified as weeds, sprouting and flourishing after rains. A study undertaken by MSSBG identified 102 wild edible leaves in Wayanad district. MSSBG studies also identified usage of 40 different types of wild mushrooms in Wayanad (Anil Kumar et al., 2008). The household survey revealed that *Paniya* families consume about 88 species of leafy greens, followed by the *Kattunaikka*, who consume 43 species, the *Kuruma* about 21 types, and the settlers restrict themselves to between 3 and 6 types of leafy greens. Most of these species are herbs (90%), and very few are trees (Ratheesh Narayanan and Anil Kumar, 2007). The *Paniya* community possesses knowledge regarding 136 taxa of wild edible plants, with *Kattunaikka* coming next with knowledge of 97 taxa. The *Kuruma* have knowledge of fewer species, 42 taxa of wild edible plants, which is still much greater than the knowledge of such plants amongst non-tribal communities. Edible roots, tubers, and rhizomes of 24 wild plant species (Table 12.2) are eaten by the tribal communities in Wayanad (Ratheesh Narayanan et al., 2011).

TABLE 12.2
Wild Tubers Consumed by Different Tribes

S/N	Scientific Name	Kattunaikka	Paniya	Kuruma
1	*Adenia hondala*	-	Koombikilangu	-
2	*Aponogeton appendiculatus*	-	Chammikaya	-
3	*Asparagus racemosus*	-	Sathavarikilangu	-
4	*Costus speciosus*	-	Channakoova	-
5	*Dioscorea belophylla*	Hekku	-	-
6	*D. hamiltonii*	Kaluvenni	Bennykilangu	Vennangu
7	*D. hispida*	Kottunoora	-	-
8	*D. intermedia*	Shoddikalasu	-	-
9	*D. kalkapershadii*	Nara	-	-
10	*D. oppositifolia*	Kavalakalasu	Kavalaikilangu	-
11	*D. pentaphylla*	Noorakorana	Noorankilangu	-
12	*D. pentaphylla* var. *communis*	Hendhikorana	-	-
13	*D. pentaphylla* var. *linnaei*	Chenakorana	-	-
14	*D. pentaphylla* var. *rheedii*	Korana	Koranakilangu	-
15	*D. pubera*	Boojikavala	-	-
16	*D. tomentosa*	Salu	-	-
17	*D. wallichii*	Narra	Naraikilangu	Nara
18	*D. wightii*	Narramooyan	Mooyankilangu	-
19	*Dioscorea* sp.	Erekalasu	-	-
20	*Dioscorea* sp.	Moodavenni	Cholabenny	-
21	*Dioscorea* sp.	Hekkuheruman	-	-
22	*Dioscorea* sp.	Heruman	Naravayan	-
23	*Ipomoea mauritiana*	-	Muthukku	-
24	*Pecteilis gigantea*	-	Kundukilangu	-

Wayanad is also known for its medicinal plant wealth, in particular from the ethnic health care system. Unfortunately, due to the changes that have occurred in land use patterns and to some extent the changes in societal values, many of these species, along with the associated traditional knowledge, are in decline. MSSBG has 120 species of wild food plants and more than 300 species of medicinal plants that are used by tribal and rural communities of the district. MSSBG started implementing projects that are aimed at ethnobotanical documentation and village-level herbal knowledge registration. The documentation focuses on local knowledge on aspects such as increase in immunity, seasonal food recipes, disease-specific recipes, health boosting foods, and collection of useful herbals. The garden has initiated steps for building up district-level databases by protecting the IPR of the knowledge providers and ensuring access and benefit-sharing mechanisms are operating in an effective manner. Phytochemical validation of the local registered knowledge of such herbs and recipes will also be undertaken in collaboration with the indigenous knowledge providers and scientists.

12.5.4 Looking Ahead

Differing from the conventional concept of botanical gardens, MSSBG's approach of equal priority to wild, natural, and cultivated traditional and horticultural species and varieties of plants and partnership with local communities helps effective conservation. The garden became a member garden of Botanic Garden Conservation International (BGCI) in 2016, and one of the Lead Botanical Gardens promoted by the Ministry of Environment, Forest and Climate Change, Govt. of India, for conserving RET plants of the Western Ghats. MSSBG was also awarded Level 1 accreditation by the ArbNet Arboretum Accreditation Program for achieving particular standards of professional practices deemed important for arboreta and botanic gardens. Once the master plan is fully implemented, the MSSBG will be the only one of its kind in India, focused on an integrated approach to biodiversity conservation through partnerships, engaging multi-disciplinary technical expertise, and engaging the local community in becoming stewards of their natural heritage.

12.6 CONCLUSION

In the past 30 years, significant progress has been made in plant conservation through stronger infrastructure, institutional capacity, and robust approaches and methodologies, with botanical gardens playing a significant role. Yet, there are still major gaps in achieving the goals set forth in the GSPC and Aichi targets due to knowledge gaps in all aspects of plant life impeding effective conservation (Heywood, 2017). Botanical gardens need to lead the way in curtailing global loss of plant biodiversity. This can be achieved through global partnerships and capacity building and by establishing community-driven gardens with emphasis on location-specific medicinal and nutrition-rich plant species that are managed in partnership with knowledgeable local community members, including traditional healers. Such botanic gardens can promote satellite gardens at the community level. The master garden and its network gardens can ensure the availability of nutrient-rich food species, a refuge for the genetic variability in wild or semi-wild foods, orphan crops, and forest foods, and can engage in educating all the critical stakeholders about the sustainable management of such bio-resources. To our knowledge, there are hardly any organized gardens or efforts in India that are directed toward saving the wild and marginal species of food value in an integrated manner. In the context of the Western Ghats, where tremendous population growth has been inevitable, botanic gardens in general and M.S. Swaminathan Botanical Garden in particular can play a major role in developing best management practices for trade-offs between conservation and development, such as agroforestry, and serving as the resource in imparting this knowledge to the local communities. As recommended by Kremen et al. (1994), in order to achieve sustained conservation of the global plant biodiversity, these natural resources should be conserved in concert with local people through integrated conservation and development projects.

REFERENCES

Anil Kumar, N., P. Prajeesh and K.P. Smiitha. 2017. Grassroots initiatives for sustainable livelihood. In: Laladhas, K.P., P. Nilayangod and V. Oommen (eds.), *Biodiversity for Sustainable Development*. Springer International Publishing, Switzerland.

Anil Kumar, N. and M.K. Ratheesh Narayanan. 2009. *Diversity, Use Pattern and Management of Wild Plants of Western Ghats: A Study from Wayanad District*. MSSRF/RR/04/12.

Anil Kumar, N., M.K. Ratheesh Narayanan and K. Satheesh. 2008. Traditional knowledge of three 'mycophilic' communities on wild edible mushrooms of Wayanad district, Kerala. *Ethnobotany* 20: 41–47.

Athira, K., C. Sudhakar Reddy, K.R. L. Saranya, S. Joseph and R. Jaishankar. 2017. Habitat monitoring and conservation prioritization of protected areas in Western Ghats, Kerala, India. *Environmental Monitoring and Assessment*. 189: 295. DOI: 10.1007/s10661-017-5998-z.

Aukema, J.E., N.G. Pricope, G.J. Husak and D. Lopez-Carr. 2017. Biodiversity areas under threat: overlap of climate change and population pressures on the world's biodiversity priorities. *PLoS One* 12(1): e0170615. DOI: 10.1371/journal.pone.0170615.

Avery, G.S. Jr. 1957. Botanic gardens–what role today? An "operation bootstraps" opportunity for botanists. *Amer. J. Bot*. 44(3): 268–271.

Borsch, T. and C. Lohne. 2014. Botanic gardens for the future: Integrating research, conservation, environmental education and public recreation. *Ethiopian J. Biol.l Sci*. 13: 115–133.

Casshyap, S.M. and R.C. Tewari. 1988. Depositional model and tectonic evolution of Gondwana basins. *Palaeobotanist* 36: 59–66.

Chandra, S. 1995. Bryophytic remains from the Early Permian sediments of India. *Palaeobotanist* 43(2): 16–48.

Chen, G. and W. Sun. 2018. The role of botanical gardens in scientific research, conservation, and citizen science. *Plant Diversity* 40. DOI: 10.1016/j.pld.2018.07.006.

Cincotta, R.P., J. Wisnewski and R. Engelman. 2000. Human population in the biodiversity hotspots. *Nature* 404: 990–992.

Conrad, C.C. and K.G. Hilchey. 2011. A review of citizen science and community-based environmental monitoring: Issues and opportunities. *Environ Monit Assess*. 176: 273–291. https://doi.org/10.1007/s10661-010-1582-5

Convention on Biological Diversity. 2012. *Global Strategy for Plant Conservation: 2011–2020*. Botanic Gardens Conservation International, Richmond, UK.

Convention on Biological Diversity. 2021. The Development of a Post-2020 Global Strategy for Plant Conservation as a Component of the Global Biodiversity Framework. www.cbd.int/doc/c/08a5/5940/83a43eb11e4773bf4f4098bf/sbstta-24-inf-20-en.pdf (Accessed 6 April 2021).

Díaz, S., J. Settele, E.S. Brondízio, H.T. Ngo, J. Agard, A. Arneth, P. Balvanera, K.A. Brauman, S.H.M. Butchart, K.M.A. Chan, L.A. Garibaldi, K. Ichii, J. Liu, S.M. Subramanian, G.F. Midgley, P. Miloslavich, Z. Molnár, D. Obura, A. Pfaff, S. Polasky, A. Purvis, J. Razzaque, B. Reyers, R.R. Chowdhury, Y.J. Shin, I. Visseren-Hamakers, K.J. Willis, and C.N. Zayas. 2019. Pervasive human-driven decline of life on Earth points to the need for transformative change. *Science* 12(13): 366(6471):eaax3100.

Dodd, J. and C. Jones. 2010. *Redefining the Role of Botanic Gardens – Towards a New Social Purpose*. University of Leicester Research Centre for Museums and Galleries, Leicester, UK.

Donaldson, J.S. 2009. Botanic gardens science for conservation and global change. *Trends in Plant Science* 14(11): 608–613. DOI: 10.1016/j.tplants.2009.08.008.

Fant, J.B., K. Havens, A.T. Kramer, S.K. Walsh, T. Callicrate, R.C. Lacy, M. Maunder, A.H. Meyer and P. Smith. 2016. What to do when we can't bank on seeds: What botanic gardens can learn from the zoo community about conserving plants in living collections. *Amer. J. Bot*. 103(9): 1541–1543.

Gole, T.W., M. Denich, D. Teketay, and P.L.G. Vlek 2002. Human impacts on *Coffea arabica* genepool in Ethiopia and the need for its *in situ* conservation. In: Engels, J.M.M., V.R. Rao, A.H.D. Brown and M.T. Jackson (eds.), *Managing Plant Genetic Diversity*. CABI Publishing, New York, pp. 237–247.

Gunawardene, N.R., A.E.D. Daniels, I.A.U.N. Gunatilleke, C.V.S. Gunatilleke, V. Karunakaran, K.G. Nayak, S. Prasad, P. Puyravaud, B.R. Ramesh, K.A. Subramanian and G. Vasanthy. 2007. A brief overview of the Western Ghats – Sri Lanka biodiversity hotspot. *Current Science* 93: 1567–1572.

Heywood, V.H. 2017. Plant conservation in the Anthropocene – challenges and future prospects. *Plant Diversity*. https://doi.org/10.1016/j.pld.2017.10.004.

IPBES. 2019. Summary for policymakers of the global assessment report on biodiversity and ecosystem services of the intergovernmental science-policy platform on biodiversity and ecosystem services. In: S. Díaz, J. Settele, E.S. Brondízio, H.T. Ngo, M. Guèze, J. Agard, A. Arneth, P. Balvanera, K.A. Brauman, S.H.M. Butchart, K.M.A. Chan, L.A. Garibaldi, K. Ichii, J. Liu, S.M. Subramanian, G.F. Midgley, P. Miloslavich, Z. Molnár, D. Obura, A. Pfaff, S. Polasky, A. Purvis, J. Razzaque, B. Reyers, R. Roy Chowdhury, Y.J. Shin, I.J. Visseren-Hamakers, K.J. Willis and C.N. Zayas (eds.), IPBES secretariat, Bonn, Germany. 56 pages. https://doi.org/10.5281/zenodo.3553579

IUCN. 1991. *The Conservation Atlas of Tropical Forests and the Pacific*. Simon & Schuster, New York, USA.

Jameel, S. 2020. *On Ecology and Environment as Drivers of Human Disease and Pandemics*. ORF Issue Brief No. 388, Observer Research Foundation.

Kremen, C., A.M. Merenlender and D.D. Murphy. 1994. Ecological monitoring: A vital need for integrated conservation and development programs in the tropics. *Conservation Biology* 8(2): 388–397.

Krishnan, S., A. Kumar and M. Swaminathan. 2020. Development of a master plan and a fund-raising plan for the M.S. Swaminathan Botanical Garden in Wayanad, India. *Acta Horticulturae*. 1298: 23–27. DOI: 10.17660/ActaHortic.2020.1298.5

Krishnan, S. and A. Novy. 2016. The role of botanic gardens in the twenty-first century. *CAB Reviews* 11(23). DOI: 10.1079/PAVSNNR201611023

Maithy, P.K. 1966. Palaeobotany and stratigraphy of the Karharbari stage with special reference to the Giridih coalfield, India. Symp. Flora. Strat. Gond., Lucknow, pp. 102–109.

Maunder, M., S. Higgens and A. Culham. 2001. The effectiveness of botanic garden collections in supporting plant conservation: A European case study. *Biodiversity and Conservation* 10: 383–401.

Miller, J.S., P.P. Lowry II, J. Aronson, S. Blackmore, K. Havens, and J. Maschinski. 2016. Conserving biodiversity through ecological restoration: The potential contribution of botanical gardens and arboreta. *Candollea* 71(1): 91–98.

Molotokos, A., M. Kuhnert, T.P. Dawson and P. Smith. 2017. Global hotspots of conflict risk between food security and biodiversity conservation. *Land* 6(67). DOI: 10.3390/land6040067.

Mundaragi, A., T. Devarajan, S. Jeyabalan, S. Bhat, and R. Hospet. 2017. Unexploited and underutilized wild edible fruits of Western Ghats in Southern India. *Scientific Papers. Series A. Agronomy* 60: 326–339.

Powledge, F. 2011. The evolving role of botanical gardens. *BioScience* 61(10): 743–749.

Ratheesh Narayanan, M.K. and N. Anil Kumar. 2007. Gendered knowledge and changing trends in utilization of wild edible greens in Western Ghats, India. *Indian J. Trad. Knowl.* 6(1): 204–216.

Ratheesh Narayanan, M.K., N. Anil Kumar, V. Balakrishnan, M. Sivadasan, H. Ahmed Alfarhan and A.A. Alatar. 2011. Wild edible plants used by the Kattunaikka, Paniya and Kuruma tribes of Wayanad District, Kerala, India. *J. Med. Plants Res*. 5(15): 3520–3529.

RBG Kew. 2016. *The State of the World's Plants Report – 2016*. Royal Botanic Gardens, Kew.

Reyers, B. and E.R. Selig. 2020. Global targets that reveal the social–ecological interdependencies of sustainable development. *Nat. Ecol. Evol*. 4: 1011–1019.

Swaminathan, M.S. 2000. Government-industry-civil society partnerships in integrated gene management. *Current Science* 78(5): 555–562.

Willis, K.J. (ed.) 2017. *State of the World's Plants 2017*. Report. Royal Botanic Gardens, Kew.

Wyse Jackson, P.S. and K. Kennedy. 2009. The global strategy for plant conservation: A challenge and opportunity for the international community. *Trends in Plant Science* 14(1): 578–580.

13 Calicut University Botanical Garden (CUBG) and Its Role in Plant Conservation

A.K. Pradeep and Santhosh Nampy

CONTENTS

13.1	Introduction	217
13.2	Aroids	218
13.3	Pteridophytes	219
13.4	Gingers	220
13.5	Cacti and Succulents	220
13.6	Wild Bananas	222
13.7	Aquatic Plants	222
13.8	Medicinal Plants	222
13.9	Bamboosetum	222
13.10	Arboretum	222
13.11	Touch and Feel Garden for Visually-Impaired	223
13.12	Mushrooms and Other Macro Fungi	223
13.13	Trees and Lianas	223
13.14	Trekking Path	223
13.15	Palms	224
13.16	Gymnosperms	224
13.17	Orchidarium	224
13.18	Conservation	224
Acknowledgments		225
References		225

13.1 INTRODUCTION

Calicut University Botanical Garden (CUBG) was established in 1971 in a lush green panoramic, undulating, lateritic land in Malappuram district of Kerala. It is the brainchild of late Prof. B.K. Nayar, the first Head of the Department of Botany and the then Vice Chancellor of the University of Calicut, Prof. M.M. Ghani. It was inaugurated in 1972 by Prof. R.E. Holttum, former Director of the Singapore Botanical Garden. The garden covers an area of 33 acres, including a 13.5-acre arboretum in the Calicut University campus (Figure 13.1). The major thrust area of the garden is the *ex situ* conservation of the rare, endangered, and threatened species of South India, Indian gingers, aroids, leeas, gesneriads, ferns, wild bananas, and medicinal plants. During the last five decades, the garden has developed into an excellent center of biodiversity and *ex situ* conservation of tropical Indian flora and exotic species.

The main area of the garden spreads out in a shallow basin surrounded by sloping terrains except for a narrow gap on the southern side where it slides down to a small transitory reservoir, providing diverse habitats and niches for a variety of plants. The central shallow region is provided with a graceful, placid pool and an octagonal greenhouse and avenues of royal palms and oil palms. The

FIGURE 13.1 A view of Calicut University Botanical Garden.

1 km long ring road from the main entrance on the western side connects the greenhouses, fernery, ginger zone, Victoria Pond, aroid zone, northern gate, bamboosetum, rockery, and medicinal plant conservatory. This enables one to have a quick walk/journey through the exquisite and luxuriant vegetation of myriad hues and fragrance. A map of the garden is provided at the entrance close to the reception office.

The garden also holds many curious plants, such as insectivorous *Nepenthes khasiana*; the upside-down tree, *Adansonia digitata*; the pride of Burma, *Amherstia nobilis*; the giant Victoria lily, *Victoria amazonica*; the cabbage tree, *Andira inermis*; and the cannon ball tree, *Couroupita guianensis*. A number of Gesneriads, Begonias, Commelinas, and Sonerilas collected from different parts of India as part of various research programs are also maintained separately for scientific investigations. Species of *Henckelia*, *Rhynchotechum*, and *Rhynchoglossum* are well displayed in this section. The monotypic and endemic, *Jerdonia indica* is an important addition. Weedy Commelinas along with exotic ornamentals such as *Callisia* and *Tradescantia* enrich the collection. There are also several greenhouses holding different groups of plants, such as aroids, cacti and succulents, pteridophytes, wild gingers, and medicinal plants. Two ponds have been constructed in the garden to harvest rain water and the backwash water from the water treatment plant (Figure 13.2). These water resources are being utilized effectively for developing gardens in the university campus and also partly for drinking purposes.

13.2 AROIDS

The garden holds an excellent collection of wild aroids. Most of these collections are gathered through extensive field exploration in the forests of India. Aroids are planted both in garden beds as well as in nurseries. The wild aroids include aquatic genera such as *Acorus*, *Cryptocoryne*, and *Lagenandra* and root creepers such as *Pothos scandens*, *Epipremnum*, *Monstera*, and *Rhaphidophora* cover the trunks of some trees in the garden. The aquatic aroids are maintained in large shallow ponds filled with mud, mixed with compost and farmyard manure. They are propagated by rhizome cuttings. The

FIGURE 13.2 A pond in the garden during monsoon.

terrestrial species include *Alocasia*, *Amorphophallus*, *Anaphyllum*, *Ariopsis*, *Arisaema*, *Caladium*, and *Philodendron*. Epiphytic *Rimusatia* also thrives in trees and rocks in moist shady areas. Apart from all these, a number of fascinating ornamental aroids are maintained in pots in greenhouses, which include different varieties of *Aglaonema*, *Anthurium*, *Caladium*, *Dieffenbachia*, *Spathiphyllum*, *Xanthosoma*, and *Zamioculcas*. Insects/pests and fungal attack are checked by regular application of bio-pesticides and fungicides.

13.3 PTERIDOPHYTES

The ferns and fern allies are another group of most fascinating plants. They are known for their lush green foliage and magnificent leaf architecture. CUBG has an excellent collection of wild pteridophytes, especially from the southern Western Ghats. The pteridophyte collection includes semi-aquatic species such as *Microsorum pteropus*, *Ceratopteris thalictroides*, *Marsilea pinnata*, *Osmunda hugeliana* (royal fern), and *Acrostichum aureum* (backwater fern). In addition, a variety of ferns across the Western Ghats, such as the edible fern (*Diplazium esculentum*), club mosses (*Lycopodium* spp.), little club mosses (*Selaginella* spp.), maiden hair ferns (*Adiantum* spp.), moon worts (*Botrychium* spp.), birds nest fern (*Asplenium phyllitidis*), stag-horn ferns (*Platycerium* spp.), and tree ferns (*Cyathea* spp.) are grown in the garden. Big clumps of *Angiopteris helferiana*, *Psilotum nudum* (whisk ferns), and *Equisetum ramosissimum* (horse-tail) are some other curious additions (Figure 13.3).

The majority of these ferns are propagated by simple division of the clumps. The separated clumps are either planted in earthen pots or directly in raised beds. Many native species are planted and maintained in environments simulating their natural habitats. Epiphytic species such as *Drynaria quercifolia* (bracket fern) and *Pyrrosia lanceolata* are often found naturally growing on tree trunks in the garden. Another curious creeping epiphytic fern, *Pyrrosia piloselloides*, with its fleshy round necklace-like fronds, blankets tree trunks.

FIGURE 13.3 Fern House.

13.4 GINGERS

A good collection of wild and ornamental gingers (Zingiberaceae) is maintained in different greenhouses, viz. Ginger Villa, Ginger House, and Spices House (Figure 13.4), and in the "Ginger Zone" of the botanical garden. This forms one of the largest collections of live gingers in India. The collection includes gingers from the Western Ghats, Northeast India, and the Andaman and Nicobar Islands and exotics from China, Malaysia, Sri Lanka, and Thailand. About 80% of the live collections of Indian gingers are maintained in the CUBG. It includes more than 2,000 accessions belonging to about 200 taxa, of which more than 60 taxa are endemic and about 30 are endangered. The wild relatives of economically important genera, such as *Alpinia*, *Amomum*, *Curcuma*, *Elettaria*, *Hedychium*, and *Zingiber* are conserved in the field. A good number of exotic ornamentals, as well as potential ornamentals, are also included. The major ornamental ginger genera include *Alpinia* (shell gingers), *Curcuma* (hidden gingers), *Etlingera* (torch gingers), *Globba* (dancing ladies), *Hedychium* (butterfly gingers), *Kaempferia* (peacock gingers), and *Zingiber*. A house for spices of ginger-related species with about 200 accessions from different parts of India is kept here as a separate section.

13.5 CACTI AND SUCCULENTS

Cacti and succulent collections are maintained in one separate greenhouse. These collections are delightful to all those who enjoy rare and exotic xerophytes in idyllic settings. Some large cacti like Euphorbias and Opuntias are maintained as outdoor displays. A wide range of families, such as Apocynaceae, Asparagaceae, Crassulaceae, and Euphorbiaceae, are represented in this garden. The collections cover both native and exotic taxa. The native species are mostly from the drier regions of the Deccan Plateau. *Caralluma*, *Cereus*, *Epiphyllum*, *Sansevieria*, *Sarcostemma*, *Stapelia*, etc. are grown in indoor beds with laterite and pebbles, simulating its natural environs (Figure 13.5).

Calicut University Botanical Garden & Its Role in Conservation

FIGURE 13.4 A view of the Spices House.

FIGURE 13.5 Cacti and Succulents House.

13.6 WILD BANANAS

A unique and fascinating collection of Musaceae comprising 41 taxa of wild bananas from India is beautifully grown in the central low-lying area. The majority of the collections are from Northeast India and gathered as part of the revision of the family Musaceae in India. Some of the wild species have tremendous ornamental potential. Magnificent spadices of Southeast Asian, *Musa haekkinenii*, and *Musa ornata* amidst lush green foliage catch our attention even from a distance. Large bottle-like bases of *Ensete superbum* and *Ensete glaucum* with their large stiff foliage are also an eye-catching sight in the garden. Apart from the wild bananas, 24 Indian cultivars of *Musa* also enrich the banana collections here. These include 28 endemic taxa and 15 endangered species. *Musa* collections when in bloom attract bats to this garden.

13.7 AQUATIC PLANTS

Diverse collections of aquatic plants are maintained in the garden, which include floating *Azolla*, *Eichhornia*, *Lemna*, *Pistia*, *Salvinia*, *Spirodela*, and *Wolffia* and emergents such as *Alisma*, *Nelumbium*, *Ludwigia*, and *Thalia*. Submerged aquatics include *Blyxa*, *Elodia*, *Cabomba*, *Ceratophyllum*, *Cryptocoryne*, *Hydrilla*, and *Vallisneria*. Several endemic species of *Nymphoides* with magnificent white and yellow flowers are another added attraction of this collection. The white flowered *Nymphoides indica*, *N. hydrophila*, *N. macrospermum*, and *N. parvifolia,* and the yellow flowered *N. aurantiacum* are generally called "poor man's water lily" or "flat dwellers water lily." They can be beautifully grown in shallow ponds for their lovely foliage and flowers.

13.8 MEDICINAL PLANTS

One of the major attractions is the rich and varied collection of medicinal plants from the Western Ghats of Kerala. A separate Medicinal Plant House displays more than 200 species of medicinal herbs and shrubs. The collections include *Nalpamara* (comprising four species of *Ficus*), *Dasamoolam* (ten medicinal root plants), and *Dasapushpam* (ten sacred plants of Kerala tradition and culture). *Acorus calamus*, *Rauvolfia serpentina*, *Rubia cordifolia*, *Salacia beddomei*, and *Woodfordia fruiticosa* are some of the important medicinal plants. The rare Maramanjhal (*Coscinium fenestratum*) is grown in the garden. Species such as *Datura metel*, *Ricinus communis*, and *Strychnos nux-vomica*, which are poisonous, are also grown here. Many other plants included in this collection can also be toxic if misused. The "star-plants" are displayed in an open courtyard, which forms another attraction to the visitors.

13.9 BAMBOOSETUM

Bamboosetum is located on the northern boundary. Diverse collections of bamboos, such as the Budda's belly bamboo (*Bambusa ventricosa*), the Giant Bamboo (*Dendrocalamus giganteus*), the thornless yellow bamboo (*Bambusa vulgaris* "Striata"), and hedge bamboo (*B. multiplex*) are some of the major attractions.

13.10 ARBORETUM

The arboretum covers an area of 13.5 acres of land with luxuriant vegetation. This is the core area where visitors are not normally permitted to disturb the vegetation. The tree canopy is almost closed, and a stream passing through the arboretum supports the luxuriant growth of different kinds of shrubs, herbs, and climbers on its sides. Huge trees such as *Antiaris toxicaria*, *Ficus carica*, *Litsea glutinosa*, *Mimusops elengi*, *Sterculia guttata*, and *Terminalia bellirica*; straggling shrubs like *Alangium salvifolium* and *Hugonia mystax*; woody climbers such as *Anamirta cocculus*, *Calycopteris floribunda*, and *Gnetum ula*; and the epiphytic species such as *Pyrrosia lanceolata*, *Drynaria quercifolia*, and *Microsorum punctatum* are some of the main attraction of this tree garden. This area is very fertile, allowing luxuriant growth of understory plants. A number of native epiphytic orchids species, such as *Acampe praemorsa*, *Coelogyne nervosum*, and *Oberonia iridiflora*, are planted in this area. A treetop walk through the arboretum is recently constructed that will provide a bird's eye view of the entire area.

13.11 TOUCH AND FEEL GARDEN FOR VISUALLY-IMPAIRED

A separate section is specially designed for the visually-impaired with provisions to familiarize plants by touch and feel. About 65 species of aromatic and other curious plants and 34 types of seeds and fruits of diverse groups of vegetables, spices, and medicinal plants are displayed here. The audio system and Braille scripts provide basic information, including uses of all the plants exhibited in this garden.

13.12 MUSHROOMS AND OTHER MACRO FUNGI

The garden is remarkable for its high mushroom diversity. It is home to several hundreds of agaric species belonging to genera such as *Agaricus, Amanita, Collybia, Coprinus, Entoloma, Hygrocybe, Lentinus, Marasmius, Mycena, Pleurotus,* and *Termitomyces*. A number of gasteromycetes such as puffballs (*Lycoperdon, Pisolithus*), earth stars (*Geastrum*), bird's nest fungi (*Cyathus*), and stink-horns (*Dictyophora*) are seen here. Several genera of coral fungi (*Clavaria, Ramaria*), jelly fungi (*Auricularia, Tremella*) and bracket fungi (*Ganoderma, Hexagonia, Microporus, Phellinus, Polyporus*) are found in the garden. Ascomycetes genera, such as *Daldinia, Peziza, Rhopalostroma, Trichoglossum,* and *Xylaria* are frequently seen. Specimens of most of these fungi are preserved in the Mycology Laboratory of the Department of Botany.

13.13 TREES AND LIANAS

The majority of the tree species introduced are from the Western Ghats of Kerala. Timber plants such as "*Irul*" (*Xylia xylocarpa*), Mahogany (*Swietenia macrophylla*), "*Maruth*" (*Terminalia aurilata*), Rose wood (*Dalbergia latifolia*), and Teak (*Tectona grandis*) are thriving well in the garden. The African Baobab (*Adansonia digitata*) is a curious tree. Other important trees are: *Adenanthera,* Aeroplane wood tree (*Ochroma*), African tulip tree (*Spathodea*), Camphor tree (*Cinnamomum camphora*), Champaka (*Michelia nilagirica*), Iron wood tree (*Mesua ferrea*), Rudraksha tree (*Elaeocarpus sphaericus*) and Sausage tree (*Kigelia africana*).

The beautiful flowering trees include *Cochlospermum gossypium, Lagerstroemia speciosa,* and *Spathodea campanulata. Adenocalymna alliaceum, Quisqualis indica,* and *Saritaea magnifica* are the attractive climbers with graceful flowers. Magnificent yellow blooms of Cat's claw trumpet, *Doxantha unguis-cati*, is the most striking and perhaps the most spectacular of all bignoniaceous climbers.

13.14 TREKKING PATH

A portion of the garden at its southwestern border is retained to preserve the natural vegetation with a number of trees, palms, lianas, and climbers. The undulating terrain with luxuriant growth of wild climbers such as *Gnetum ula, Hugonia mystax, Calycopteris floribunda,* and the spiny straggling *Hibiscus hispidissimus* and *Calamus thwaitesii* offers visitors a feel of the warmth of the tropical forest while trekking through its narrow trail.

13.15 PALMS

In addition to coconut, avenues of royal palms, and oil palms, several other agri-horticulturally valued palms, such as *Areca*, *Arecastrum*, *Cyrtostachys*, *Licuala*, *Rhapis*, *Phoenix*, *Woodetia*, and *Zalacca* are introduced in the garden. Besides, the common Fish-tail palm, *Caryota urens*, the monocarpic Talipot palm, *Corypha umbraculifera*, runs wild. The Western Ghats endemics such as *Arenga wightii* and *Pinenga dicksonii* also flourish well along the eastern boundary of the garden. A few palms of the Nicobar endemic, *Bentinckia nicobarica*, introduced to the garden a few years back, are also now well established.

13.16 GYMNOSPERMS

In addition to the native *Cycas circinalis*, other cycads, such as *Zamia floridana*, *Z. furfuracea*, and *Dioon mejiae*, and conifers, such as *Agathis*, *Araucaria*, *Cupressus*, *Juniperus*, and *Podocarpus*, are also represented. The *Gnetum ula* climbing over huge trees, bearing male and female cones, is another attraction.

13.17 ORCHIDARIUM

More than 50 species of wild orchids, such as *Acampe*, *Aerides*, *Calanthe*, *Geodorum*, *Habenaria*, *Oberonia*, and *Rhynchostylis*, are grown in the green house. Endemic species like *Acanthephippium bicolor* are conserved in the orchidarium. Ornamental orchids like Dancing girl (*Oncidium*), Dove orchid (*Peristeria*), Soniya (*Dendrobium*), and Spider orchid (*Arachnis*) are the attractions of the orchidarium.

13.18 CONSERVATION

India is very rich in biodiversity and represents 8% of the global biodiversity residing in only 2.4% land area of the world (Raju et al., 2010). This includes 17,053 flowering plants, 1,022 pteridophytes, 2,843 bryophytes, and 64 gymnosperms. According to Hajra and Mudgal (1997), about 32% of flowering plants are endemic. Moreover, the country has a rich medicinal plant flora of about 2,000 species, and most of the drugs prescribed in the native system of medicine, especially ayurveda, have their origin in these plants (Heywood, 1991). A significant role can be played by botanical gardens toward conservation of rare and endangered flora. There is a need to ensure that a wide range of useful germplasm is brought into cultivation for further propagation. Bringing them into cultivation is, however, the first step toward saving them from extinction.

For the great majority of plants, locations inside the botanical garden are identified simulating their natural habitats, or a particular habitat is created artificially for growing them. CUBG takes initiatives in selection, propagation, conservation, and preservation of different groups of plants, such as begonias, leeas, gesneriads, Sonerilas (Figure 13.6), wild gingers and bananas, medicinal plants, and aroids collected from various regions within the country. Emphasis is also given to collect those plants that are known in folk medicine for further study by the research students. The research work in Calicut University Botanical Garden plays an important role in the utilization of its biological resources.

Calicut University Botanical Garden & Its Role in Conservation

FIGURE 13.6 Live collections of Sonerilas in CUBG.

ACKNOWLEDGEMENTS

The authors are thankful to the Head, Department of Botany, and authorities of the University of Calicut for the support. The help rendered by the staff of the botanical garden is also gratefully acknowledged.

REFERENCES

Hajra, P.K. and V. Mudgal. 1997. *Plant Diversity Hotspots in India, an Overview*. Botanical Survey of India, Kolkata.

Heywood, V. 1991. Botanic gardens and conservation of medicinal plants. In: Akerle, O., V. Heywood and H. Synge (eds.), *Conservation of Medicinal Plants*. Cambridge University Press, Cambridge, pp. 213–222.

Raju, V.S., C.S. Reddy and S. Suthari. 2010. Flowering plant diversity and endemism in India, an overview. *Anu J. Nat. Sci.* 1: 27–39.

14 Role of Botanical Garden in Conservation and Citizen Science – A Case Study from Mahatma Gandhi Botanical Garden, University of Agricultural Sciences, Bangalore

A.N. Sringeswara and Sahana Vishwanath

CONTENTS

14.1 Introduction 227
14.2 Mahatma Gandhi Botanical Garden 227
Acknowledgments 231
References 232

14.1 INTRODUCTION

Botanical gardens are gardens holding documented collections of living plants for the purposes of scientific research, conservation, display, and education. Botanical gardens around the world play a pivotal role in conservation of biological diversity. The present concept of a botanical garden has its origin in Europe with the establishment of Italy's Padova Botanic Garden during the year 1545. Botanical gardens around the world are interlinked with an international organization, Botanical Gardens Conservation International (BGCI), established after the launch of the Plant Conservation Programme led by IUCN and WWF (Synge, 1984; Hamann, 1985), which aimed at conservation of plant resources, including public participation, conservation of economically important plants both *ex situ* and *in situ*, and developing awareness among the public in conservation importance. They are the storehouse of genetic diversity required for the crop improvements. The number of botanical gardens has increased from ca. 1,400 to 3,000 from the year 1990 to 2017, respectively (Heywood, 2017), across the world. These botanical gardens represent about one-quarter of the estimated number of vascular plants of the world (Wyse Jackson, 2001; O'Donnell and Sharrock, 2017), one of the important targets under the Global Strategy for Plant Conservation (GSPC) is to have 70% of the world's threatened plants conserved *ex situ* in botanical gardens.

14.2 MAHATMA GANDHI BOTANICAL GARDEN

Mahatma Gandhi Botanical Garden is one such botanical garden established in the year 1973 with the establishment of the new campus of the University of Agricultural Sciences at GKVK,

FIGURE 14.1 (A) View of the entrance of the Mahatma Gandhi Botanical Garden; (B) detailed design of the garden.

Bangalore. The garden spreads over an area of 65 acres (26 ha) and is located at 13° 04' 42.77" North latitude and 77° 34' 39.32" East longitude with an elevation of 925 m above mean sea level and an annual average rainfall of 970 mm. The garden was systematically designed with different blocks aiming to conserve a particular group of plants in each block. The plants conserved in these blocks are in accordance with the Bentham and Hooker system of plant classification, which is one of the best field-oriented classification systems for easy identification of plants in the field with minor modification. Each block of the garden is represented with certain plant families according to the said system. The advantage is that when one walks from the first block to the last, they get a clear idea about the classification system and in some extent about the plant evolution. The entire garden is divided into ten different blocks (Figure 14.1B). The first block is earmarked for conservation of medicinal plants. Thereafter, blocks two to nine are earmarked for the plants from the primitive families under the subclass Polypetalae to advanced families under the subclass Monochlamydae. A minor modification to the system is that the Gymnosperms are conserved at the end, that is, in block 10, along with Monocots for the ease of maintenance.

To date, about 1,000 plants species have been successfully conserved in the garden belonging to 615 genera and 121 families of flowering plants, including 29 threatened species. Collections of live plants of various groups, including conservation dependents, are the main contribution of any botanical garden (Donaldson, 2009). The garden has different conservation sections, such as Threatened and Endemic plants, Balsam and Ginger house, Orchidarium, Medicinal plants (Rajkumar et al., 2011), Gymnosperms, Bambusetum, etc. (Figure 14.2). Among the 121 families represented in the garden, Fabaceae and Orchidaceae dominate with 102 species each, followed by Apocynaceae with 48 species. Dominant families with more than 25 species in the garden are given in Figure 14.3A. *Ficus* is the dominant genus in the garden with 18 species followed by an orchid genus *Dendrobium* with 14 species. Figure 14.3B gives the representation of dominant genera in the garden with more than six species.

Over the past few years, the garden has successfully introduced the giant water lily (*Victoria amazonica* (Poepp.) Klotzsch) from the Acharya Jagadish Chandra Bose Botanical Garden in West Bengal, and that forms one major attraction to the visitors of the garden.

A major focus of the garden is to conserve the native flora of the country. About 68% of the plants conserved in the garden are native to the country, and only 32% of the species are exotic in nature, which comprises mainly ornamental trees, shrubs, and some herbaceous species. Among the species conserved, 123 species are endemic to the country alone. Trees are the dominant life

Role of Botanical Garden in Conservation and Citizen Science   229

FIGURE 14.2 Important conservation sections in the Mahatma Gandhi Botanical Garden. (A) Bambusetum; (B) Threatened and Endemic plants; (C) Orchidarium; (D) Balsam and Ginger house.

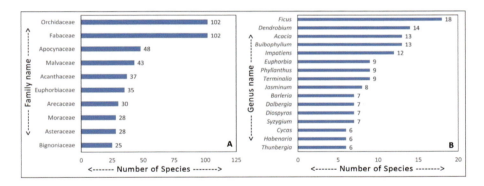

FIGURE 14.3 Dominant family (A) and Genus (B) with more than 25 and 6 Species, respectively, in the botanical garden.

form conserved in the garden (410 species, 40.59%) followed by herbs (323 species, 31.98%), shrubs (163 species, 16.14%), and climbers, including some lianas (114 species, 11.29%).

Along with the flora, the garden harbors a rich faunal diversity represented with 150 birds, 50 butterflies, many insects, and some reptiles. With this rich faunal diversity, the garden serves as one of the important centers for studying plant–animal interactions.

The garden has a well-established herbarium housing more than 12,000 specimens comprising about 3,500 species of the flora of the state in particular and India in general. The collection in the herbarium dates back to the 1890s, and it forms one of the oldest herbaria in the state. The herbarium has the acronym *UASB* from the Index Herbariorum of the New York Botanical Garden (Thiers, 2016).

FIGURE 14.4 Important threatened plants conserved in the Botanical Garden: (A) *Vateria indica* L.; (B) *Cynometra travancorica* Bedd.; (C) *Dipterocarpus indicus* Bedd.; (D) *Dysoxylum malabaricum* Bedd. ex C.DC.; (E) *Hildegardia populifolia* (DC.) Schott & Endl.; (F) *Myristica malabarica* Lam.; (G) *Kingiodendron pinnatum* (Roxb. ex DC.) Harms.

Conservation activities in the garden include *ex situ* collection of threatened and endemic plants of the Western Ghats and developing the propagation protocols for them. Among the 29 species of threatened plants conserved, there is 1 critically endangered species, 16 endangered species, and 12 vulnerable species as per the IUCN categories. Out of the 29 threatened species, 17 of them are endemic to the Western Ghats, and the rest of the 12 species are distributed elsewhere also (Figure 14.4). Out of the 123 endemic species conserved in the botanical garden, 93 of them are endemic to the Western Ghats only.

Role of Botanical Garden in Conservation and Citizen Science 231

FIGURE 14.5 Awareness creation on the conservation aspects: (A) drawing competition to school children and environmental education; (B) educating school students; (C) imparting training to the college students on plant identification; (D) training to officials of the Forest Survey of India on tree identification.

The garden has a well-maintained nursery with more than 100 species, including threatened and endemic plants, economically important species, medicinal plants, and agroforestry species. Conservation can be best achieved with the exchange of conservation-dependent plants among the similar organizations and also among the institutions who are willing to propagate in their campuses. Major objectives of the botanical gardens are to support the collections of conservation-dependent taxa for *ex situ* conservation through maintenance of stocks and utilizing them sustainably to achieve the conservation goal (Cibrian-Jaramillo et al., 2013). The major aim of the Mahatma Gandhi Botanical Garden is to distribute important plant species to the needy people, including the farmers to propagate in their farm land under the agroforestry concepts, and to exchange species among similar organizations.

With this collection and conservation activities taking place at the garden, the Ministry of Environment, Forest and Climate Change (MoEF & CC), Govt. of India, recognized this garden as one of the Lead Botanical Gardens in the country and funded for the overall development.

The garden is actively involved in environmental education among school children, college students, and the general public. Education programs have been conducted regularly to impart technical knowledge on conservation education and sustainable use of bioresources (Figure 14.5).

The botanical garden, with a focus on science and conservation, should have a comprehensive collection policy to enrich their collections for achieving the effective conservation goal and for the sustenance of the botanical garden.

ACKNOWLEDGMENTS

Authors acknowledge the authorities of the University of Agricultural Sciences, Bangalore, for the facility and the Ministry of Environment, Forests and Climate Change (MoEF&CC) for financial support for development of the garden under the Assistance Botanical Garden scheme.

REFERENCES

Cibrian-Jaramillo, A., A. Hird, N. Oleas, H. Ma, A.W. Meerow, J. Francisco-Ortega and P. Griffith. 2013. What is the conservation value of a plant in a botanic garden? Using indicators to improve management of ex situ collections. *Bot. Rev.* 79: 559–577.

Donaldson, J.S. 2009. Botanic gardens science for conservation and global change. *Trends Plant Sci.* 14: 608–613.

Hamann, O. 1985. The IUCN/WWF plants conservation programme 1984–85. *Vegetatio* 60: 147–149.

Heywood, V.H. 2017. Plant conservation in the Anthropocene e Challenges and future prospects. *Plant Diversity* 39: 314–330.

O'Donnell, K. and S. Sharrock. 2017. The contribution of botanic gardens to *ex situ* conservation through seed banking. *Plant Divers* 39: 373–378.

Rajkumar, M.H., A.N. Sringeswara and M.D. Rajanna. 2011. Ex-situ conservation of medicinal plants at University of Agricultural Sciences, Bangalore, Karnataka. *Recent Research in Science and Technology* 3(4): 21–27.

Synge, H. 1984. *The IUNC/WWF Plants Conservation Programme 1984–85*. Gland, Switzerland.

Thiers, B. 2016. *Index Herbariorum: A Global Directory of Public Herbaria and Associated Staff*. New York Botanical Garden's Virtual Herbarium.

Wyse Jackson, P.S. 2001. An international review of the *ex situ* plant collections of the botanic gardens of the world. *Bot. Gard. Conserv. News* 3: 22–33.

15 The Role of the Kuzbass Botanical Garden in Solving Environmental Problems and in Plant Conservation *in situ* and *ex situ*

Andrey Kupriyanov

CONTENTS

15.1 Introduction	233
15.2 The Location of the Kuzbass Botanical Garden	234
15.3 Botanical Research	236
15.4 Herbarium	236
15.5 Results of Botanical Research	237
15.6 Creation of Collections and Displays	238
15.7 Biodiversity Conservation	240
15.7.1 *Ex situ* Conservation of Plants	240
15.7.2 *In situ* Conservation of Plants	241
15.7.3 Red Data Books	241
15.7.4 Environmental Education and Upbringing	242
15.7.5 Revegetation on Rock Dumps	243
15.8 Conclusion	244
References	245

15.1 INTRODUCTION

The scientific concept of *introduction* has been used since the sixteenth century. It originated from the Latin word of the same form: *introduction*. Introduction is a purposeful human activity aimed at bringing plants (genera, species, subspecies, varieties, forms) under cultivation in a particular natural-historical area where they have not previously grown. Additionally, the term means bringing into cultivation plants from the indigenous flora. Introduction is part of the system of botanical sciences. Institutions dealing with this science are botanical gardens. Collections of living plants that are grown in open field conditions and in greenhouses and used for research and for arranging displays form the basis of botanical gardens. The fundamental nature of introduction lies in studying the adaptive processes that occur in plants at all levels of life systems, namely from biochemical processes to changes in growth and development rhythms. This process is referred to as plant acclimatization; it ensues in nature due to global climate changes or when plants spread to new territories. Acclimatization takes an extremely long time. In introduction experiments, botanists-introducers model this process, reducing to a great extent the natural timescale that spans geological periods. Studying the processes occurring in plants during their acclimatization is the fundamental aim of introduction. Practical implementation of introduction leads to the enrichment of mankind with new useful plants. The global success of introduction provides an opportunity to overcome the world food crisis. The sharp increase in the productivity of agricultural crops that were bred by virtue of the mobilization of the world plant resources made it possible to accomplish the "Green Revolution" in the middle of the twentieth century and thus feed humanity. The success of introduction disproved the grim predictions of an inevitable food crisis made by Malthus. Hunger on Earth is a social, not a biological, phenomenon that is possible thanks to the progress of introduction. The inexhaustibility of introduction arises from the unlimited resources of intraspecific and form diversity of plants in wildlife. It only remains for humanity to exploit the natural potential of the plant world to prevent food and environmental crises and to provide the population with safe medicinal drugs. The innovativeness of introduction developments should be revealed in integration projects and consist in both bringing a new plant into cultivation and in unlocking its useful properties. In Siberia and Russia, with their extreme climatic features, the search for new cultivated plants for many sectors of the national economy is not over yet. In the twenty-first century, introduction is as innovative as physics and chemistry.

In 1989, Botanic Gardens Conservation International (BGCI) compiled a list of characteristic features distinguishing a botanical garden from other institutions. The list includes the presence of research collections; the registration of collections, which provides data about the natural habitat of species; labeling of plants; accessibility to visitors; and an opportunity to perform research on the plant collections. Thus, the basic object of activity in any botanical garden is collections; the main methods employed in its work are collecting and introducing new plants.

Botanical gardens have unique, unparalleled opportunities to unite the traditional taxonomy and the needs of agriculture, forestry, and medicine as well as the study and conservation of biological diversity.

Currently, there are more than 2,200 botanical gardens in 153 countries of the world. The number of collections in them vary from hundreds to dozens of thousands of taxa. It is known that there are more than 400 botanical gardens and dendrological parks (arboretums) in Europe, about 200 in the USA, 150 within the territory of the USSR, and 107 in modern Russia. (Kuzevanov, 2013). Siberia boasts eight large botanical gardens (Figure 15.1).

The Kuzbass Botanical Garden is one of the youngest academic botanical gardens in Russia. It was laid out on 28 December 1991 in the south of Western Siberia (Kemerovo) covering an area of 186.3 hectares.

15.2 THE LOCATION OF THE KUZBASS BOTANICAL GARDEN

The site of the Kuzbass Botanical Garden is on the eastern outskirts of Kemerovo; its area is 186.3 hectares. The territory of the garden is confined to the flood plain and the first terrace above the flood plain of the Tom. The absolute elevation of the river surface varies from 117 to 132 m above sea level. The soil cover is represented by meadow chernozem gley soils, meadow marsh soils, and drained podzolic soils.

The climate is continental: winters are cold and protracted; summers are short and warm. The average annual air temperature is 0.9°C. The warmest month is July with an average temperature of +24.5°C. The coldest month is January with an average temperature of –24°C. The record cold maximum is –57°C. The first autumn ground frosts are observed from 26 August to 14 September; spring ground frosts are observed from 28 May to 11 June. The average frost-free period is 110 days. Autumn is short, with a rather warm September and cold October. By mid-October, cool temperatures above zero gradually change to cold temperatures below zero. November is fully included in the winter months with an average air temperature of –9.7°C. Such climatic conditions are a significant obstacle to the introduction of plants in the Kuzbass Botanical Garden.

Role of the Kuzbass Botanical Garden in Plant Conservation

FIGURE 15.1 The most significant botanical gardens in Siberia: (1) Tomsk (the Botanical Garden of Tomsk State University; (2) Novosibirsk (Central Siberian Botanical Garden of the Siberian Branch of the RAS); (3) Kemerovo (the Kuzbass Botanical Garden of the Federal Research Centre of Coal and Coal Chemistry of the Siberian Branch of the RAS); (4–5) Barnaul (the Botanical Garden of Altai State University; the Botanical Garden of the Institute of Horticulture of Siberia); (5) Irkutsk (the Botanical Garden of Irkutsk State University); Siberian Branch of the RAS); (7–8) Yakutsk (Botanical Garden of the North-Eastern Federal University; Yakutsk Botanical Garden of the Institute of Biological Problems of the Cryolithozone" SB RAS).

FIGURE 15.2 General view of the Kuzbass Botanical Garden.

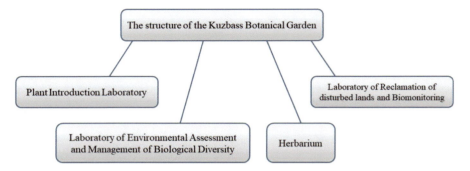

FIGURE 15.3 The structure of the Kuzbass Botanical Garden.

15.3 BOTANICAL RESEARCH

Botanical studies of the Kuzbass area have had a rich history since the eighteenth century; such outstanding Russian botanists as I. Gmelin (eighteenth century), P. N. Krylov and V. V. Reverdatto (nineteenth century), and N. N. Laszczynski and V. P. Sedelnikov (twentieth century) worked here. The materials of botanical expeditions are stored in the herbaria of Moscow (MW), Saint Petersburg (LE), Tomsk (TK), and Novosibirsk (NS).

Present-day research conducted by the botanical garden staff is aimed at studying the flora of southern Siberia, including Kemerovo Oblast and Kazakhstan. At present, floristic research is focused on a thorough study of the floristic composition of the specially protected natural areas (SPNA). A complete inventory of the flora of the Kemerovo Oblast was carried out.

The Kuzbass Botanical Garden has strong ties with Kazakhstan where joint research centered on the topic "A study of endemic plants of particular floristic regions of Kazakhstan and the creation of a database on the herbarium fund" has been repeatedly performed in liaison with the joint-stock company "IRPH 'Phitochemistry'" (Director General, Academician of the National Academy of Sciences of the Republic of Kazakhstan S. M. Adekenov). Scientists of the Kuzbass Botanical Garden participated in the implementation of the research grant of the Ministry of Education and Science of the Republic of Kazakhstan for 2018–2020 (No. AP05132458, "Molecular genetic analysis of the gene pools of the rare plant species populations of Northern Kazakhstan") as well as the grant of the Ministry of Education and Science of the Republic of Kazakhstan for 2018–2020 (state registration No. 0118RK00404; No. AP0513246, "Research and protection of ornamental and rare plants of the Syrdarya-Turkestan Natural Park").

The Kuzbass Botanical Garden has organized 26 expeditions to Kazakhstan since 2006: 11 expeditions to the area of the Kazakh Uplands; 9 expeditions to the Syrdarya Karatau and the West Tian Shan; 3 to the East Kazakhstan region (Kazakhstan Altai, the Saur Mountains, and the Zaysan Basin); 3 to the North Kazakhstan region; 2 expeditions to the West Kazakhstan region; and 1 to the Dzungarian Alatau. I. A. Khrustalyova, O. A. Kupriyanov, T. O. Strelnikova, and Yu. A. Manakov took an active part in them.

15.4 HERBARIUM

The emergence of the Kuzbass Botanical Garden Herbarium is associated with the first collections assembled by T. Ye. Buko in the Kuznetsk Alatau in the 1990s, totaling 1,500 sheets.

At present, the Herbarium contains about 60,000 sheets. It consists of four sections: Kemerovo region, Kazakhstan and Central Asia, Siberia, and Bryophytes. The Herbarium was entered into the international database Index Herbariorum under the acronym KUZ.

The garden has collections of the genera *Stipa*, *Festuca*, and *Artemisia*. Some of the materials were deposited for storage by Siberian and Kazakh scientists: A. L. Ebel, Yu.A. Kotukhov, N. N. Lashchinsky, V. P. Amelchenko, O. M. Maslova, D. V. Zolotov, and others.

The section "Kemerovo region" contains 19,654 herbarium sheets of vascular plants that belong to 1,584 species, 548 genera, and 123 families. Almost all staff members of the botanical garden became the main collectors: S. A. Sheremetova, T. Ye. Buko (about 7,000 herbarium sheets), A. N. Kupriyanov (more than 4,000 sheets), A. L. Ebel, Yu. A. Manakov (more than 2,000 sheets), Ye. A. Kuzmina, T. O. Strelnikova (about 1,500 sheets), L. A. Gorshkova, I. A. Khrustalyova, G. I. Yakovleva, O.A. Kupriyanov (about 1,000 sheets), Yu. V. Morsakova, D. V. Chusovlyanov (about 500 sheets), V.N. Bersenyova, O. V. Barysheva, S. A. Skoblikov (more than 300 sheets), N.V. Demidenko, R. T. Sheremetov, D. A. Sidorov (more than 200 sheets), and others.

The section "Kazakhstan and Central Asia" (KAZ) of the Herbarium contains more than 20,000 sheets of inserted herbarium that are included in the main collection fund. Most herbarium collections were carried out from 1996 to 2019. At present, there are herbarium materials from nine areas: Akmola Region (Akmolinskaya Oblast), Almaty Region (Almatinskaya Oblast), Aktobe Region (Aktyubinskaya Oblast), East Kazakhstan Region (Vostochno-Kazakhstanskaya Oblast), Jambyl Region, Karaganda Region, Kostanay Region, Pavlodar Region, and Turkistan Region. The herbarium collection of Kazakh species was most intensively enlarged from 2010 to 2019.

The section "Siberia" of the Herbarium has about 20,000 sheets and covers the areas of the south of Western and Eastern Siberia, Khakassia, and Tyva.

The section "Bryophytes" of the Herbarium was put together in 2005 by A.Ye. Nozhinkov, PhD (Biol.). The collection of bryophytes comprises 1,760 specimens of mosses and liverworts represented by 157 genera and 345 species that were collected on the territory of Siberia and Kazakhstan (Sheremetova et al., 2021a).

A typical herbarium contains 64 taxa: Asteraceae: *Artemisia* (8), *Galatella* (1), *Hieracium* (1), *Pilosella* (1), *Achillea* (2), *Rhaponticoides* (1), and *Cousinia* (1); Boraginaceae: *Myosotis* (1); Caryophyllaceae: *Gypsophila* (1); Euphorbiaceae: *Euphorbia* (1); Poaceae: *Agrostis* (1), *Agrotrigia* (2), *Elymotrigia* (7), *Elymus* (5), and *Stipa* (19); Ranunculaceae: *Pulsatilla* (1), *Ranunculus* (3), and *Thalictrum* (2); and Rosaceae: *Potentilla* (1).

The Herbarium created the integrated information system titled the Kuzbass Digital Herbarium, which made it possible to acquire the most in-depth scientific information about the regional flora diversity and conduct its documentation in digital form in accordance with the standards of global information databases. The research permitted the generalization of all available data; the collections provided insight into the actual diversity of the flora of Kemerovo Oblast. The state of the collections in the Kuzbass municipalities was determined, and those areas where additional research was necessary (blind spots) were identified. The data obtained from the texts of original labels and the geographical coordinates of the areas where the plants were collected (reflected electronically on maps) will allow researchers to compile complete lists related to individual districts of Kemerovo Oblast automatically.

15.5 RESULTS OF BOTANICAL RESEARCH

In the twenty-first century, botanical research performed by KuzBG staff members is progressing to fundamentally new perspectives of understanding and generalizing previously obtained and accumulated materials. The number of publications devoted to the issues of studying the Kemerovo Oblast vegetation cover has significantly increased. The number of herbarium collections from different Kuzbass districts has increased dramatically since the beginning of the twenty-first century thanks to the expeditionary research pursued by KuzBG staff during the expeditions organized in collaboration with scientists from the Central Siberian Botanical Garden of the Siberian Branch of the Russian Academy of Sciences and Tomsk State University (A. L. Ebel, N. N. Lashchinsky, V. A. Cheryomyshkina, O. Yu. Pisarenko, N.V. Makunina, N. V. Shchegoleva, A. A. Zverev, A. V. Klimov,

G. I. Yakovleva, and others). In the last 20 years, evidence of more than 150 new species not found in Kuzbass earlier has been provided, and data about new locations of rare plants have been obtained. Additionally, the specific features of rare species populations have been studied; the bibliography of research papers on flora and vegetation has increased by more than 350 publications (Sheremetova et al., 2021b).

The *Black Data Book of Siberian Flora* (2016) was published on the basis of joint botanical research performed in collaboration with other Siberian botanists. This monograph provides the first consolidated data on invasive plants in the territory of Siberia. It includes 58 higher vascular plants, which are adventive for the local flora and pose an economic and environmental threat. The morphological description provided for each plant is supplemented with the history of the species naturalization in Siberia. In addition, a distribution map with the assessment of its aggressiveness is created. Data on the biology and ecology of such plants are supplied. The monograph explores the consequences of their introduction and possible practical applications. The data obtained for Siberia for the first time make it possible to understand the consequences of human economic activity as well as the distribution, abundance, and number of invasive species, their reproduction, and the rate of distribution. The book proposes recommendations for curbing the spread of these plants and outlines preventative measures that can be adopted to combat invasive species.

The monograph *The River Tom Basin (Floristic and Physiographic Specific Features)* (2020) by S. A. Sheremetova and R. T. Sheremetov is the most significant floristic generalization of numerous botanical studies performed in the territory of Kemerovo Oblast. The book presents the results of 15 years of floristic research. The primary purpose of the monograph was to assess and compare the current state and trends in the development of the flora of the River Tom basin based on the basin approach. The authors substantiate the use and appropriateness of the basin approach for comparative floristic research.

The results of botanical research in the territory of Kazakhstan are presented in numerous publications and monographic reviews: *A guide to plants of Karkaraly National Park* (Kupriyanov et al., 2008); *A guide to vascular plants of Bayanaul National Park* (Kupriyanov et al., 2013); *The flora of Burabay National Park* (Sultangazina et al., 2014); *The floristic diversity of Boralday* (Kupriyanov et al., 2017a); *Pyrogenous successions in the pine forests of the Kokshetau Uplands after fires* (Sultangazina et al., 2017); *The flora of the Big Ulutau* (Kupriyanov et al, 2017b); *A compendium of the flora of the Kazakh Uplands* (Kupriyanov, 2020); and "Rare plant species of Northern Kazakhstan" (Sultangazina et al., 2020). Herbarium collections served as the basis for the description of new species from the territory of Kazakhstan: *Achillea* × *kasakhstanica* Kupr. et Alibekov, *A. kamelinii* Kupr., *A. tianschanica* Kupr. et Kulemin, *Artemisia saurense* Kupr. *Rhaponticoides zaissanica* Kupr., A.L. Ebel et Khrustaleva, *Cousinia* × *pavlovii* Kupr., Lashchinskiy et A.L. Ebel.

15.6 CREATION OF COLLECTIONS AND DISPLAYS

At present, scientific collections and displays comprising 1,630 species are assembled in the 40-hectare area of the botanical garden. The species include medicinal plants – 157 species. Collections of the genera *Lilium* L. – 190 species and varieties; *Hemerocallis* L. – 80 species and varieties; *Hosta* Tatt – 65 species and varieties; *Allium* L. – 66 species, *Bergenia* Moench – 37 taxa; *Astilbe* Hamilt. – 21 species and varieties; *Paeonia* L. – 50 species and varieties; *Iris* L. – 40 species and varieties; *Phlox* – 40 species and varieties; *Salix* L. – 122 species, hybrids, and varieties; *Syringa* L. – 33 species and varieties. A collection of the natural flora of – 504 species, rare and endangered plants – 75 species, the arboretum – 150 species.

The first display, "Systematikum," was created in 2005. It aims to demonstrate natural flora plants, mainly from the south of Siberia, and to assess the primary introduction of wild plants in cultivation. Plants are arranged in a systematic order in accordance with A. Takhtajan's system. In total, more than 450 species of the natural flora of the south of Western Siberia were planted in the display. The Kuzbass Botanical Garden creates such conditions for plants that are as close as possible to their natural habitats. The collection of natural flora plants delights visitors and staff with continuous flowering throughout the growing season. *Erythronium sibiricum* (Fisch. et Mey.) Kryl., *Tulipa patens* C. Agardh ex Schult. & Schult. f., *Adonis vernalis* L., *Brunnera sibirica* Steven, and others start blooming in mid-April. *Allium microdictyon* Prokh., *Bupleurum longifolium* ssp. *aureum* (Fisch. ex Hoffm.) Soó), *Linum perenne* L., and others bloom in June and July. *Sanguisorba officinalis* L., *Eryngium planum* L., and others bloom in August.

The apothecary garden of the Kuzbass Botanical Garden was laid out in 2006 and covers an area of 0.9 hectares. Its purpose is to reflect the diversity of wild-growing and cultivated official medicinal plants in Kemerovo Oblast as fully as possible and to apply the grown material in the educational process at schools and universities as well as in research. At present, the collection of medicinal plants comprises more than 150 species (curated by I. N. Yegorova). Plants in the display are arranged in groups depending on the content of biologically active substances, including plants containing vitamins, terpenoids, bitterness, polysaccharides, saponins and phytoecdysones, cardiac glycosides, simple phenols and phenol glycosides, phenylpropanoids and lignans, anthracene derivatives, tannins, flavonoids, coumarins and chromones, alkaloids, and biologically active substances of various groups. The apothecary garden has become a platform for educational and outreach activities as vocational and university students have their internships here.

The garden of topiary forms was created in 2007 and covers an area of 1 hectare. It aims to demonstrate the potential of clipping woody plants and plant plasticity under systematic clipping conditions (curated by O. O. Vronskaya). This part of the botanical garden always attracts and surprises visitors with its variety of options for shaping Siberian plants.

O. O. Vronskaya created a collection of lilies and day lilies in 2008. It comprises 190 species and varieties of lilies (*Lilium* L.) and 80 species and varieties of day lilies (*Hemerocallis* L.). The phenological spectra and the flowering time of species and varieties of the genus *Lilium*, the biomorphological parameters of vegetative organs (shoot height, number of flowers in an inflorescence, diameter of flowers), the flowering duration of one flower, and inflorescence were studied. An assessment of both the decorative indicators of species and varieties as well as the prospects of the introduction of some varieties was undertaken. The indicator role of the basic pigments of the photosynthetic apparatus (chlorophyll a and b, carotenoids) in the leaves of lily species and varieties in different vegetation phases was revealed to determine their resistance.

A collection of willows (*Salix* L.). was put together in 2010 (curated by O. O. Vronskaya, PhD, Biol.). The role of willows in soft landscaping and planting greenery in Siberian cities and towns is significantly belittled. The Kuzbass Botanical Garden holds a unique collection totaling about 120 forms, varieties, and species of the genus *Salix* L.

Alliums grow on almost all continents. The genus is characterized by an amazing variation in the flower shapes, colors, and sizes. Flowers are arranged in globular, cone-shaped, umbellate, and panicle inflorescences. They differ in flowering time and duration and are low-maintenance plants. Thus, alliums are an attractive option for novice gardeners. Nevertheless, this species is not in demand with flower growers and landscapers. The allium collection (curated by A. Ye. Nozhinkov) comprises 56 species. The collection aims to demonstrate the variety of *Allium* species and forms as well as their uses in landscape design. The collection includes species with high decorative qualities: *Allium caeruleum* Pall., *Allium schoenoprasum* L., *Allium sphaerocephalon* L., *Allium cristophii* Trautv., and others.

The collection of the *Paeonia* L. varieties and species dates back to 2008 and 45 varieties resistant to Siberian weather conditions. The planting material was provided by the Altai Botanical Garden. In subsequent years, it was enriched with the varieties brought from the Research Institute of Horticulture of Siberia named after M. A. Lisavenko (Barnaul) and amateur flower growers' private collections.

The KuzBG collection of the genus *Paeonia* L. currently holds 45 varieties that vary in the flower structure and shape. Japanese peonies have very unusual flowers. They can open quite easily and

FIGURE 15.4 Expositions and collections of the botanical garden: (1) exposition of medicinal plants; (2) collection of *Lilium* L. species; (3) exposition of the *Salix* L. species; (4) sistematicum; (5) exposition of topiary forms.

quickly. Additionally, most beautiful flower arrangements can be made with them. Japanese peonies are highly frost resistant in the conditions of the south of Siberia.

The phlox collection was restored in 2016 on the initiative of T. A. Sukovatitsina. The pride of the collection is the old varieties created at the beginning of the twentieth century: Europe (1910), Viking (1914), Schneepyramide (1918), and many others. The variety Europe belongs to the widespread Continental Series created by Pfitzer. It is resistant to diseases and pests. The variety is decorative – no wonder it has been in demand for 100 years.

Moss gardens are elite displays in botanical gardens. The moss garden of the Saihō-ji Temple, which was laid out by the famous Japanese gardener Musō Soseki in the middle of the fourteenth century, is the gold standard of sophistication and elegance. The KuzBG Moss Garden was laid out in 2015 (curated by Dr. A. Ye. Nozhinkov).

15.7 BIODIVERSITY CONSERVATION

Biodiversity conservation is the foundation for humanity's habitat. All mankind strives for sustainable development, but it cannot be supported without preserving the entire complex of floral diversity as it plays the role of a buffer in the noosphere–biosphere interaction. One of the main tasks the Kuzbass Botanical Garden faces is biodiversity conservation (Global Strategy for Plant Conservation, 2011–2020; Strategy and Action Plan, 2014).

15.7.1 *Ex situ* Conservation of Plants

The Convention on Biological Diversity (1992) identifies two main ways to conserve the elements of biodiversity. *Ex situ* management means off-site conservation of plants, that is, outside their natural habitats. Article 8 of the Convention states that when a significant adverse impact on biological diversity is established, government agencies shall regulate and control the relevant processes and categories of activities. The role of botanical gardens in *ex situ* management becomes strategic as it is only possible to save populations of rare and endangered plants, study their biology, and ensure the preservation of populations in such gardens.

When designing a new coal deposit, it is mandatory to perform a pre-design study of the territories to identify rare and endangered species. In case such plants are discovered, the Kuzbass Botanical Garden commits to preserving the reserve part of the populations in the Garden territory (Kupriyanov et al., 2017c). This methodology was reflected in Strategies for biodiversity conservation in policies and in programs for the development of the energy sector in Russia and also in the United Nations Development Programmes (Kupriyanov et al., 2017c). At present, the following populations of rare and endangered plants of southern Siberia are conserved in the botanical garden: *Adonis villosa* Ledeb., *Allium vodopjanovae* Friesen, *Astragalus follicularis* Pall., *Epipactis helleborine* (L.) Grantz, *Erythronium sibiricum* (Fisch. et C.A. Mey.) Kryl., *Gypsophila patrinii* Ser., *Linum perenne* L., *Stipa pennata* L., *Thymus marschallianus* Willd., and *Ziziphora clinopodioides* Lam. (Klimova et al., 2021).

15.7.2 IN SITU CONSERVATION OF PLANTS

The modern theory of biodiversity conservation is based on the ecoregional approach. A concerted effort of a great number of specialists united by the World Wide Fund for Nature resulted in identifying 200 ecoregions with high biological diversity. They are key to global biodiversity conservation. This program was called Global-200. The Kuzbass Botanical Garden took an active part in the implementation of the WWF project "Ensuring long-term biodiversity conservation in the Altai-Sayan ecoregion." A.N. Kupriyanov was the scientific editor of the monographs *A System of Specially Protected Areas in the Altai-Sayan Ecoregion* (Kemerovo, 2001) and *Biological Biodiversity of the Altai-Sayan Ecoregion* (Kemerovo, 2003).

The necessity to identify the most important botanical sites in Europe (key botanical territories) was proclaimed at the first conference of the public European organization *Planta Europa* (France) in 1995 (Anderson, 2003). The only project to identify key botanical territories in Russia was conducted in Kemerovo Oblast. This process was accelerated significantly after *Plantlife International* became involved in the work. This wild plant conservation charity financed expeditionary research in 2007–2008 within the framework of the project "A Strategy for the Conservation of Plants in the Altai-Sayan Ecoregion." Twenty-one key botanical territories (KBT) that are of paramount importance for plant conservation in Kuzbass were identified in Kemerovo Oblast. The KBT locations, the main types of habitats, unique plant species, and vegetation types requiring conservation and monitoring were specified in accordance with the criteria complying with the Global Strategy for Plant Conservation (Buko et al., 2009).

The botanical garden has proposed and created seven regional zakazniks (nature reserves) in Kemerovo Oblast since 2012 (the Archekassky Ridge, the Bachatsky Hills, Karakansky, the Kokuiskoye Swamp, Relict, the Luchshevo Uvaly, Thaidon) and two botanical nature monuments (Artyshta and the Kostenkovsky Rocks). A list of prospective natural areas was prepared at the suggestion of the Kuzbass Botanical Garden at the beginning of 2019. The list included 16 sites for creating specially protected natural areas (SPNA). The inclusion of new areas in the regional SPNA system will increase the area by 165,000 hectares, thus making the proportion of protected areas equal to 15%.

15.7.3 RED DATA BOOKS

The Russian Red Data Book (2008) is a list of rare and endangered organisms. This annotated list of species and subspecies indicates their current and past distributions, number and reasons for

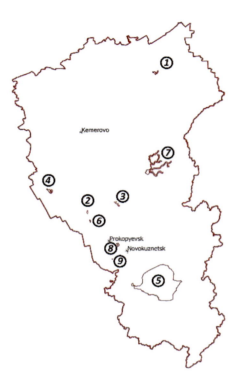

FIGURE 15.5 Nature reserves: (1) Archekassky Ridge; (2) Bachatsky Hills; (3) Karakansky; (4) Kokuiskoye Swamp; (5) Relict; (6) Luchshevo Uvaly; (7) Thaidon. Botanical nature monuments: (8) Artyshta; (9) Kostenkovsky Rocks.

its reduction, reproduction features, the measures already taken, and the measures necessary to conserve the species. The book advances the thesis that it is a document containing a data set concerning the protected flora and fauna and identifies measures to conserve them. Russian legislation makes provisions for publishing national and regional Red Data Books.

The Red Book of the Kemerovo Region – Kuzbass. (2021) is an integral part of the regional component of environmental education and upbringing. It is essential to enhance the population's local history knowledge and to shape public opinion concerning the necessity to conserve floral diversity, especially the rarest and most vulnerable plants living in the area. In this case, the Red Data Book is considered a source of local history information that public education specialists lack. Botanical research performed by the botanical garden staff in many ways contributes to the formation of a list of the species to be conserved. A.N. Kupriyanov is the executive editor of the second (2012) and third (2021) editions of the Kuzbass Red Data Book.

It is difficult to overestimate the importance of the Red Books. Since the plants included in them have state and regional protection, mining companies are obliged to carry out their activities taking into account the conservation of these species and their places of growth. Red Books are the most important factor in environmental education and education of the population and help to organize plant protection in certain areas. The presence of species listed in the Red Book provides justification for the creation of new SPNA.

15.7.4 Environmental Education and Upbringing

The threat of a global environmental crisis on the cusp of the twentieth and twenty-first centuries determines the necessity to develop strategies for building optimum relationships between man and nature. The United Nations Conference on Environment and Development (Rio de Janeiro, 1992) made several important decisions related to the environment. Many countries, including Russia, signed the Convention on Biological Diversity. These events marked the key boundary in the history of human civilization. Almost 30 years have passed since that momentous forum held in Rio de Janeiro, but there has been no significant paradigm change from the extensive to intensive economic management yet. The attempt to combat global warming through a global concerted effort of mankind focused on the implementation of treaties was not a success as the Kyoto Protocol (1997) did not lead to the desired results, and the 2009 Copenhagen accord also failed.

The reason for such slow progress toward greening, energy saving, and the modernization of both technologies and lifestyles is in the first place the human factor. Economic, political, and national interests are of secondary importance. The human egocentric world view prevents the "resetting" of people's morality. There are too few proponents of environmental ethics and environmental morality; they do not constitute a critical mass capable of changing the existing paradigm. The necessity to engage in environmental and botanical education and upbringing is determined by the position of botanical gardens between the research into fundamental processes in plants and the activity that is open to the public and whose results are shown to society.

Shaping an environmental worldview is a complex, multifaceted process that, in principle, must begin at an early age. It is based on the following three pillars: environmental pragmatism and practicability, environmental ethics, and environmental local history education. The principle of environmental local history education rests on several axioms: you cannot love what you do not know; you cannot save what you do not love; you cannot save nature globally. Thus, the understanding of the global nature of the processes taking place on the planet should be balanced with the responsibility for the environmental well-being of a particular person's place of residence. The botanical garden staff developed an environmental and local history methodology of environmental education and upbringing. It is based on the axioms mentioned earlier.

The environmental and local history methodology involves the maximum proximity of biological objects under study to the real place of students' residence. The clarification of general biological and environmental laws should be based on specific plants and animals living in students' residence.

The implementation of a large-scale project aimed at the introduction of this method was undertaken in Tashtagolsky and Belovsky districts of Kuzbass through the preparation and publication of a textbook for the regional component. For relatively small areas, *The Red Book (rare and endangered plants and animals of the Belovsky district of the Kemerovo region,* 2001) and *The Red Book (rare and endangered plants and animals of the Tashtagolsky district of the Kemerovo region,* 2014), textbooks on Botany (*Travelling with Plants Around Mountainous Shoria, The Green World of the Steppe Kuzbass*), textbooks on Ecology (*My House Mountainous Shoria, The Steppe Kuzbass Environment*), and textbooks on tourism, as well as posters and handouts for the aforementioned textbooks, were published and distributed to schools.

The botanical garden traditionally hosts ecological and biological excursions and runs sessions with school students on its premises.

Over the past 15 years, about 480,000 residents of six municipal districts participated in the experiment on the introduction of the environmental and local history methodology in the Kemerovo region, which is 18% of the total number of residents. The work was carried out within the framework of the United Nations Development Program and the Global Environment Facility (UNDP/GEF) "Conservation of biodiversity in the Russian part of the Altai-Sayan Ecoregion." Testing of students' knowledge at the end of the experiment showed the high effectiveness of this method. The students' level of knowledge about biological diversity increased by 1.7 points (from 2.1 to 3.8 points on a five-point scale).

15.7.5 REVEGETATION ON ROCK DUMPS

The Kuzbass Botanical Garden is located within the area of the Kuznetsk Basin, which is one of the most extensive coal deposits in the world with estimated overall coal reserves of about 580 billion tons. Despite the progress of the "green economy," coal production is steadily growing and amounts to 230–250 million tons per year. About 9 billion tons of coal have been mined from the bowels of the Kuznetsk Basin over the whole period of mining in the area.

The downside of the coal industry development in the Kuznetsk Basin and of the increase in the share of open-pit coal mining is a rise in the area of disturbed lands. At present, they comprise 150,000 hectares. Such territories are characterized by a completely changed general profile of the earth surface and entirely or partly destroyed biological diversity.

The paradigm of sustained development of human civilization changed at the end of the twentieth century. The state of biodiversity and its conservation were chosen as the basic criterion for the necessary development. The pragmatic technologies of the fertility restoration of disturbed lands are being replaced by nature-like technologies that aim at the following: the restoration of the ecological functions of rock dumps; the restoration of the high level of biological diversity on rock dumps; the introduction of the environmental approach to natural resources management; and the creation of new nature-like technologies for the restoration of the vegetation cover of disturbed lands. To restore the biodiversity of waste rock dumps, nature-like technologies for the reconstruction and restoration of the vegetation cover on rock dumps were introduced. Such technologies accelerate the natural processes of the overgrowth of rock dumps; they ensure the creation of multi-species plant communities that approach natural communities in a short period (3–6 years).

The botanical garden staff members are developing nature-like technologies to restore the natural vegetation cover on coal mine rock dumps (Kopytov and Kupriyanov, 2019). The technology used to restore vegetation cover consists in applying a grass-seed mixture to a dump. Such mixtures are collected in areas of native vegetation similar to zonal steppe vegetation and are applied to the surface of a dump. When this method is employed, 30–40 zonal meadow-steppe species settle on the experimental site, including typical steppe species: *Artemisia austriaca*, *Dianthus versicolor*, *Elisanthe noctiflora*, *Galium verum*, *Goniolimon speciosum*, *Gypsophila patrinii*, *Hedysarum gmelinii*, *Medicago falcata*, *Seseli ledebourii*, and *Stipa capillata*. A nature-like plant community is formed in six years. By contrast, a similar community is only formed in several decades in case of natural vegetation overgrowth.

The technology employed to reconstruct vegetation cover consists in applying a soil layer containing plant seeds and rhizomes to a dump. Such soil layers are removed during quarrying for coal and are not kept in storage clamps for long. Implementing this technology makes it possible to create a nature-like plant community with rich species diversity (30–50 species) on a dump, which cannot be achieved within several decades in the case of self-organized vegetation. The development of such technologies is in accord with the global agenda to conserve biodiversity (Kupriyanov et al., 2021).

Two patents (Ufimtsev et al., 2019, Ufimtsev, 2020) have been obtained for the development of new nature-like technologies for restoring floral diversity in landfills, which are actively used by coal companies. The results of the restoration of the dumps of the coal company are presented in Figure 15.6

15.8 CONCLUSION

The role of botanical gardens in the twenty-first century will become even more critical. The global challenges include the search for new plants to cure new diseases, the development of new technologies to accelerate carbon dioxide sequestration and combat global warming, and the conservation of floristic diversity *in situ* and *ex situ*. In the context of global urbanization, the role of botanical gardens is also becoming more significant in environmental education and upbringing.

The Kuzbass Botanical Garden is one of the youngest academic botanical gardens in Russia. Thirty years of evolution is a short time for a botanical garden. Nevertheless, the challenges the Kuzbass Botanical Garden is embracing are extremely significant. They concern the study of the Siberian flora, the introduction of plants of the natural flora of southern Siberia, the digitization of

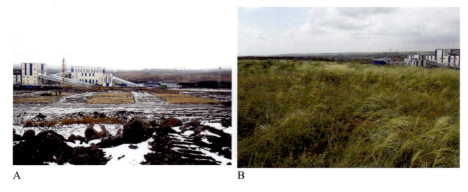

FIGURE 15.6 Reconstruction of meadow-steppe plant communities on dumps: (A) 2014; (B) 2020.

the herbarium collections, the development of nature-like technologies to restore the vegetation cover of waste rock dumps, the *in situ* and *ex situ* conservation of plants, and the environmental education and upbringing of the population.

REFERENCES

Anderson, S. 2003. Identifying Important Plant Areas: A Site Selection Manual for Europe, and a Basis for Developing Guidelines for Other Regions of the World. M. *Izd-vo Predstavitel'stva Vsemirnogo soyuza okhrany prirody (IUSN) dlya Rossiii stran SNG* [In Russian].
Biological Biodiversity of the Altai-Sayan Ecoregion. 2003. Russian Representative Office of the WWF. (Sci. ed. A. Kupriyanov). Kemerovo. [In Russian].
The Black Book of the Flora of Siberia. 2016. (edited by Yu. Vinogradova & A. Kupriyanov). Novosibirsk. 440 p. [In Russian].
Buko T., S. Sheremetova, N. Lashchinsky et al. 2009. *Key Botanical Territories of the Kemerovo Region.* Kemerovo. [In Russian].
Convention on Biological Diversity. Rio de Janeiro, 5 June 1992. Convention on Biological Diversity.
Global Strategy for Plant Conservation. 2011–2020. Global Strategy for Plant Conservation: 2011–2020. Botanic Gardens Conservation International, Richmond, UK.
Klimova, O.A., V.A. Latokhin and A.N. Kupriyanov. 2021. Preservation of rare and endangered plants in conditions of intensive coal mining. *Problems of Industrial Botany of Industrially Developed Regions.* Kemerovo, pp. 53–56 [In Russian].
Kopytov, A.I. and A.N. Kupriyanov. 2019. A new strategy for the development of the coal industry in Kuzbass and solving environmental problems. *Coal.* 11: 89–93. [In Russian].
Kupriyanov, A.N. 2020. Synopsis of the flora of the Kazakh Upland. Novosibirsk. 423 p. [In Russian].
Kupriyanov, A.N., A.L. Ebel, N.N. Laschinsky and B.M. Moshkalov. 2017a. Floristic diversity of Boroldai. Shymkent. 237 p. [In Russian].
Kupriyanov, A.N., I.A. Khrustaleva, S.M. Adekenov and E.M. Gabidullin. 2017b. *Flora of Big Ulutau.* Novosibirsk. 184 p. [In Russian].
Kupriyanov, A.N., I.A. Khrustaleva, Yu. A. Manakov and S.M. Adekenov. 2008. *Plant Determinant of the Karkaraly National Park.* Karaganda. 264 p. [In Russian].
Kupriyanov, A.N., I.A. Khrustaleva, Yu. A. Manakov and S.M. Adekenov. 2013. *Determinant of Vascular Plants of Bayanaul National Park.* Novosibirsk. 215 p. [In Russian].
Kupriyanov, A.N., O.A. Kupriyanov, Yu. A. Manakov, V.I. Ufimtsev. 2021. Effects of the growth substrate on the restoration of *Stipa capillata* L. populations on refuse dumps. *Contemporary Problems of Ecology* 2021 (2): 193–200. DOI: 10.1134/S1995425521020062]. [In Russian].
Kupriyanov, A.N., O.A. Kupriyanov and A. Yu. Ovchinnikov. 2017c. Methodological recommendations for the conservation of rare plant species in the implementation of coal mining projects on the example of the conservation of *Glycyrrhiza uralensis* Fisch. and *Epipactis helleborine* (L.) Crantz. for subsequent

introduction/reintroduction to places suitable for growth./Collection of methodological recommendations for the conservation and restoration of biological diversity in the implementation of economic activities of energy sector enterprises. United Nations Development Programme, Global Environment Facility: 1–12.

Kuzevanov, V. Ya. 2013. Botanic gardens and arboreta as ecological resources for sustainable development. *Role of the Arboretum and Botanical Garden against Climate Change of the East Asia (International Symposium)*, Seoul, KNA. 2013. p. 24. http://bogard.isu.ru/articles/2013_koreakuzevanov_korea_2013.pdf.

Kyoto Protocol to the United Nations Framework Convention on Climate Change. Adopted on December 11, 1997

The Red Book (rare and endangered plants and animals of the Belovsky district of the Kemerovo region) 2011. (Sci. ed. A. Kupriyanov). Kemerovo. 110 p. [In Russian].

The Red Book (rare and endangered plants and animals of the Tashtagolsky district of the Kemerovo region) 2014. (Sci. ed. A. Kupriyanov). Kemerovo. 120 p. [In Russian].

The Red Book of the Kemerovo Region – Kuzbass. Vol. 1. Rare and endangered species of plants and fungi. 2021. Kemerovo. 245 p. [In Russian].

The Red Book of the Kemerovo Region. Vol. 1. Rare and endangered species of plants and fungi. 2012. Kemerovo. 206 p. [In Russian].

The Red Book of the Russian Federation. (Plants and fungi). 2008. M. 855 p. [In Russian].

Sheremetova, S.A., I.A. Khrustaleva, A.N. Kupriyanov, T.O. Strelnikova, G.I. Yakovleva and E.B. Rotkina. 2021a. Additions to the flora of Kemerovo Region (2010–2020). *Bot. Zhurn.* 106(7): 696–702. [In Russian].

Sheremetova, S.A., I.A. Khrustaleva and P. Stieglbauer. 2021b. Biodiversity of vascular plants of Kuzbass. *Problems of Industrial Botany of Industrially Developed Regions*. Kemerovo. 2021. DOI: 10.1051/bioconf/20213100023

Sheremetova, S.A. and R.T. Sheremetov. 2020. The Tom River basin: Floristic and physico-geographical features. Novosibirsk 323 p. [In Russian].

Strategy and Action Plan for the conservation of biological diversity of the Russian Federation. 2014. M. 257 p. [In Russian].

Sultangazina, G.Zh., I.A. Khrustaleva, A.N. Kupriyanov and S.M. Adekenov. 2014. Flora of the National Natural Park "Burobai". Novosibirsk. 242 p. [In Russian].

Sultangazina, G.Zh. and A.N. Kupriyanov. 2017. Pyrogenic successions in the pine forests of the Kokshetau upland after fires. Novosibirsk. 174 p. [In Russian].

Sultangazina, G.Zh., A.N. Kupriyanov and S.V. Boronnikova (eds.). 2020. Rare plant species of Northern Kazakhstan. Kostanay. 2020. 260 p. [In Russian].

The system of specially protected natural territories of the Altai-Sayan ecoregion. 2001. Russian Representative Office of the WWF. (Sci. ed. A. Kupriyanov). Kemerovo. 173 p. [In Russian].

Ufimtsev, V.I. 2020. A method for creating sustainable forest ecosystems on dumps formed during the development of open-pit coal deposits. *Patent of the Russian Federation*. № 12831345.

Ufimtsev, V.I., A.N. Kupriyanov, Yu. A. Manakov and O.A. Kupriyanov. 2019. A method for restoring ecosystems that were disrupted during the development of mineral deposits by an open method. *Patent of the Russian Federation*. №. 2862040.

16 Contribution of Botanic Garden to Plant Conservation
233 Years of Conservation History and Actions of CSIR-NBRI Botanic Garden

J.S. Khuraijam, S.K. Tewari, and S.K. Barik

CONTENTS

16.1	Introduction	247
16.2	Germplasm Collections	249
16.3	Conservatory	249
16.4	Cactus and Succulent House	250
16.5	Palm House	250
16.6	Fern House	251
16.7	Moss House	251
16.8	Orchidarium	252
16.9	Cycads	253
16.10	Touch and Smell Garden	254
16.11	Waterlily and Lotus Collection	254
16.12	Arboretum	254
16.13	Conservation of Threatened Plants	255
16.14	CSIR-NBRI Web Portal on Threatened Plants	257
16.15	R&D for New Varieties/Improvement of Floral and Ornamental Crops	257
16.16	Avian Diversity	257
16.17	Societal Services	259
16.18	Ecosystem Services	259
16.19	Conclusion	259
References		259

16.1 INTRODUCTION

Considering the high rate of species extinction due to various anthropogenic causes, it is essential to conserve threatened plants on a war footing basis before they become extinct. Therefore, conservation biology has become an extremely important discipline in today's context, and it is truly a "mission-oriented crisis discipline" (Soulé, 1986). The discipline is an interdisciplinary subject including diverse areas of biology, such as dispersal, migration, demographics, effective population size, inbreeding depression, and minimum population viability of threatened species. The principles underlying each of these disciplines have immediate implications for the management of species and ecosystems, captive breeding and reintroduction, genetic analyses, and habitat restoration (Hunter and Gibbs, 2011; Primack, 2008; Van Dyke, 2008). Botanic gardens have the most important role in conservation action as they often serve as the last refuge of many threatened plants and provide the much-needed propagation materials for multiplication, the first step to recover a threatened species. Besides, the botanic gardens have the requisite infrastructure and expertise to initiate conservation action for the plant species, which are under imminent threat of extinction.

The Botanic Garden of CSIR-National Botanical Research Institute, Lucknow is one of the oldest and historical botanic gardens of India. Established as *Sikander Bagh* in AD 1789 as a royal garden by the then Nawab of Oudh, Saadat Ali Khan, the garden has transformed itself into CSIR-National Botanical Research Institute (CSIR-NBRI), an active full-fledged premier plant science research institute of the country. The present name CSIR-NBRI was acquired in 1978. Prior to that it was christened as *Sikandar Bagh*, Government Horticultural Garden, and National Botanic Garden. The National Botanic Garden of the then Government of United Province (since 1948) came under the umbrella of the Council of Scientific and Industrial Research (CSIR), Government of India, in 1953.

The CSIR-NBRI garden is well known for its rich germplasm and plant varieties, conservation activities, and sustainable utilization of plant resources having economic, taxonomic, ornamental, horticultural, biological, ecological, educational, and recreational values (Figure 16.1). It is located in the heart of Lucknow, the capital city of the Indian state of Uttar Pradesh. Spread over an area of 65 acres with dense tree canopy and rich plant diversity, the garden is a national facility with

FIGURE 16.1 Views of CSIR-NBRI Botanic Garden: (A) Banyan tree, a heritage tree of Uttar Pradesh in the Botanic garden; (B) Water lily and lotus pond with diverse aquatic plant species.

four main functions, viz. conservation, education, scientific research, and recreation. It is designed to conserve the indigenous and exotic flora and makes those available to various stakeholders for study, research, and industrial use.

The Botanic Garden is a living repository of more than 6,500 taxa/cultivars of various groups of native and exotic plants. The garden has an excellent collection of ornamental crops, trees, houseplants, medicinal plants, cycads, palms, ferns, bryophytes, bonsai, water lilies, cacti, and succulents. There are several thematic gardens and plant houses where plants are displayed for educational and aesthetic purposes. Considering the rich plant diversity of the Botanic Garden, the Institute has been designated as a National Repository by the National Biodiversity Authority, Chennai. Besides, this Botanic Garden has also been recognized as a Lead Botanical Garden by the Ministry of Environment, Forest and Climate Change, Government of India, for enhancing the *ex situ* conservation activities of threatened species. The CSIR-NBRI Botanic Garden is a part of the Botanic Gardens Conservation International (BGCI) network. The garden is recognized as a DUS center for three floral crops viz., *Chrysanthemum*, *Canna*, and *Gladiolus*, by the Protection of Plant Varieties and Farmers' Rights Authority, Government of India.

16.2 GERMPLASM COLLECTIONS

The Botanic Garden is known for varietal development of ornamental plants and conservation of threatened plants. It has an excellent germplasm collection of 200 cultivars of *Bougainvillea*, 100 cultivars of *Gladiolus*, 200 cultivars of *Chrysanthemum*, 150 cultivars of *Canna*, 200 cultivars of rose, a Fern House with 65 species of ferns and fern allies, an arch-shaped conservatory for tropical and subtropical plants with 300 species/cultivars, a Cactus and Succulent House with 200 species/varieties, a Palm House with 52 species, a Cycad House with 71 species, an Orchidarium with 120 species, a Moss House with 20 species, and an Arboretum with 400 species of trees and shrubs.

16.3 CONSERVATORY

A horseshoe-shaped plant house, with an area of 1,370 sq m, serves as a conservatory for house plants and species from tropical and sub-tropical climates (Figure 16.2). There are more than 300 species/cultivars of ornamental, wild, and cultivated plants in the conservatory. Some interesting plants are *Bambusa ventricosa*, *Costus pictus*, *Encephalartos villosus*, *Microcycas calocoma*, *Hoya carnosa*, *Fatsia japonica*, and *Ginkgo biloba*. There is also a large collection of *Aglaonema*, *Alocasia*, *Anthurium*, *Asparagus*, *Calathea*, *Chlorophytum*, *Codiaeum*, *Dieffenbachia*, *Dracaena*, *Peperomia*, *Philodendron*, *Pandanus*, and *Syngonium*.

FIGURE 16.2 Conservatory.

16.4 CACTUS AND SUCCULENT HOUSE

A pagoda-shaped glass house, with an area of 284 sq. m, holds the germplasm of about 200 species or varieties of cacti and succulent plants from arid regions (Figure 16.3). Some of the important genera under *ex situ* conservation in this glass house are *Adenium, Agave, Aloe, Astrophytum, Cereus, Cissus, Cleistocactus, Coryphantha, Dasylirion, Disocactus, Echinocactus, Echinocereus, Echinopsis, Epiphyllum, Euphorbia, Ferocactus, Gasteria, Graptopetalum, Haworthia, Huernia, Hylocereus, Kalanchoe, Lithops, Mammillaria, Opuntia, Pereskia, Sansevieria, Sedum, Selenicereus, Stapelia, Stenocereus,* and *Yucca*. The plants are aesthetically laid out in informal beds and landscaped with pebbles and rocks in a fascinating way. The plant house also conserves threatened species such as *Aloe harlana, Coryphantha maiz-tablasensis, Dracaena draco, Echinocactus grusonii,* and *Euphorbia cylindrifolia*. At least 25 species in the plant house are in the IUCN Red List. The major attraction of the Cacti and Succulents House is the majestic living gymnosperm *Welwitschia mirabilis*. CSIR-NBRI Botanic Garden is the only custodian of this bizarre plant in the entire South Asia (Figure 16.4). Several Indian succulents with high ornamental value have also been conserved in the plant house. Medicinally important succulents such as *Euphorbia neriifolia, Plectranthus amboinicus,* and *Aloe vera* have been displayed and conserved.

16.5 PALM HOUSE

The Palm House is of 765 sq. m area, and contains the germplasms of Arecaceae. More than 52 species of palms are displayed in pots of various sizes and on the ground (Figure 16.5). Some interesting species are *Arenga pinnata, Areca catechu, Caryota urens, C. mitis, Chamaedorea elegans, C. stolonifera, Chrysalidocarpus lutescens, Daemonorops kunstleri, Elaeis guineensis, Licuala grandis, L. spinosa, Livistona chinensis, L. cochin-chinensis, Mascarena verschaffeltii, Phoenix reclinata, Ptychosperma macarthuri, Thrinax barbadensis,* and *T. excelsa*.

FIGURE 16.3 Cactus and Succulent House.

FIGURE 16.4 *Welwitschia mirabilis* in the Cactus and Succulent House.

FIGURE 16.5 Palm House.

16.6 FERN HOUSE

This pyramidal house, with an area of 400 sq m, houses the germplasm of 65 fern species from India and abroad. Some of the species are *Adiantum capillus-veneris* (Figure 16.6), *A. hispidulum*, *Blechnum occidentale*, *Bolbitis heteroclita*, *Diplazium esculentum*, *Drynaria quercifolia*, *Equisetum debile*, *Lygodium flexuosum*, *Microsorium alternifolium*, *Nephrolepis cordifolia*, *N. duffii*, *N. tuberosa*, *Ophioglossum reticulatum*, *Psilotum nudum*, *Pteris cretica* "albolineata," and *P. vittata*.

16.7 MOSS HOUSE

The Moss House is the only plant house of its kind in the country. It houses 20 species of mosses that are beautifully displayed on mounds (Figure 16.7).

FIGURE 16.6 Fern House.

FIGURE 16.7 Moss House.

16.8　ORCHIDARIUM

About 120 species of orchids are conserved in this plant house (Figure 16.8). Some notable species are *Paphiopedilum spiceranum, Paphiopedilum insigne, Vanda coerulea, Renanthera imschootiana, Dendrobium aphyllum, Vanda tessellata,* and *Eulophia nicobarica*. The orchids were collected from natural habitats of eight states, viz., Assam, Bihar, Jharkhand, Meghalaya, Manipur, Nagaland, Odisha, and Uttarakhand. Some of the species conserved in the plant house are *Acampe praemorsa, Arundina graminifolia, Bulbophyllum crassipes, Coelogyne cristata, Cymbidium*

Contribution of Botanic Garden to Plant Conservation

FIGURE 16.8 Orchidarium.

bicolor, Dendrobium aphyllum, Dendrobium herbaceum, Dendrobium moschatum, Dendrobium polyanthum, Gastrochilus inconspicuous, Pelatantheria insectifera, and *Rhynchostylis retusa*.

16.9 CYCADS

The Institute has the largest collection of cycads (71 species) in South Asia (Figure 16.9), and in terms of the number of Cycads conserved, it ranks ninth in the world. The collection includes ten species of Indian *Cycas*, viz. *Cycas annaikalensis, C. circinalis, C. beddomei, C. divyadarshanii, C. nayagarhensis, C. orixensis, C. pectinata, C. sphaerica, C. swamyi*, and *C. zeylanica*. The Institute has successfully developed seed germination technique for six species of Indian *Cycas*, and efforts are being made to develop the technique for all the species.

Besides these Indian species, 19 other species of cycads have been successfully propagated in the Cycad House through seed germination, viz. *Cycas revoluta, C. micronesica, C. nayagarhensis, C. nongnoochiae, C. orixensis, C. pachypoda, C. pectinata, C. riuminiana, C.*

FIGURE 16.9 Cycad House.

schumanniana, C. seemannii, C. silvestris, C. tansachana, C. wadei, C. zeylanica, Macrozamia moorei, Zamia erosa, Z. furfuracea, Z. loddigesii, and *Z. pumila.* Seedlings of *Zamia loddigesii, Z. pumila,* and *Z. furfuracea* were raised through germination of viable seeds developed by artificial pollination. Artificial pollination of *Microcycas calocoma, Dioon spinulosum,* and *Cycas revoluta* is being carried out. Besides these, seven species were propagated through vegetative propagation.

16.10 TOUCH AND SMELL GARDEN

This garden is unique in India and one of the six such gardens in the world. The other gardens are in Berlin, Kyoto, New York, Nanjing, and Warsaw. The garden facilitates the differently abled persons to enjoy the plant kingdom and its diversity. The garden is equipped with a Braille system that describes the plants in the garden (Figure 16.10).

16.11 WATER LILY AND LOTUS COLLECTION

Twenty-five varieties of water lilies, including giant Amazonian water lilies, viz. *Victoria amazonica, V. cruzianiana,* and *Eurayle ferox,* are conserved in water lily ponds, tubs, and pots. Twenty varieties and 14 land races of lotus are also conserved (Figure 16.11). The rare 108 petal lotus from Manipur is the major attraction. Varietal development of water lilies and lotus are in progress.

16.12 ARBORETUM

More than 400 species of trees are conserved in the arboretum. These are planted in a systematic order and properly labeled. Some of the indigenous and exotic tree species are, *Adansonia digitata, A. macrophylla, Annona muricata, Boswellia serrata, Bischofia javanica, Butea monosperma, Chorisia insignis, Cinnamomoum camphora, Coccoloba uvifera, Dillenia indica, Diospyros malabarica, Dalbergia lanceolata, Ficus benghalensis, F. krishnae, Litsea glutinosa, Oroxylum*

FIGURE 16.10 Touch and Smell Garden.

FIGURE 16.11 Water lily and lotus collection in pots.

indicum, *Pterocarpus marsupium*, *Saraca declinata*, *Shorea robusta*, *Santalum album*, *Strychnos nux-vomica*, *S. potatorum*, *Syzygium jambos*, *Tabebuia palmeri*, *Tecomella undulata*, and *Tectona grandis*.

16.13 CONSERVATION OF THREATENED PLANTS

The conservation of threatened plants has been a priority area of the Botanic Garden. As a Lead Botanical Garden under the Assistance to Botanic Garden Scheme of the Ministry of Environment, Forest and Climate Change, Government of India, conservation of threatened plants is a mandatory activity of CSIR-NBRI Botanic Garden (Figure 16.12). The conservation actions include collection of threatened plants from the wild, often supported by ecological niche modeling, standardization of their propagation protocols through seeds, macro/micro-vegetative propagation methods, and multiplication in large numbers for distribution to different botanic gardens of universities, institutions, and colleges for *ex situ* conservation. The garden also provides planting materials to state forest departments for *in situ* conservation of these threatened plants. The forest departments plant these threatened species in different agroclimatic regions/locations in protected areas and reserved forests identified through ecological niche modeling for achieving *in situ* conservation success. CSIR-NBRI has established a field germplasm bank of threatened plants at its Banthra field station. The plant collections of the garden also include many threatened and taxonomically important plants from different countries. Currently, CSIR-NBRI leads a pan-India CSIR program on conservation of 500 threatened plants of the country following a tested 10-point conservation protocol (Adhikari et al., 2018; Barik et al., 2018a, 2018b, 2018c; Chrungoo et al., 2018; Haridasan et al., 2018; Lyngdoh et al., 2018; Panda et al., 2018), which is as follows:

1. Population inventory, characterization, and mapping using ecological niche modeling (ENM);
2. Meta population modeling of selected species populations to determine the conservation status, minimum viable population size, and to assess extinction risk;

FIGURE 16.12 Threatened plants are being multiplied in the CSIR-NBRI garden for conservation: (A) *Paphiopedilum spiceranum*, (B) *Nepenthes khasiana*, (C) *Hoya pandurata*, (D) *Indopiptadenia oudhensis*, (E) *Cycas beddomei*, and (F) *Trachycarpus takil*.

3. Identification of factors responsible for depleting species populations and developing a species-specific recovery strategy;
4. Molecular characterization of the selected species populations to identify those with greater diversity for genetic enrichment based on source–sink concept;
5. Characterization of active principles in selected species in different habitats/populations;
6. Standardizing the macro- and micro-propagation techniques for mass multiplication;
7. Reproductive biology of the selected species to address the regeneration failure;
8. Production of planting materials for reintroduction of the species in the areas identified through ENM;
9. Herbarium and establishment of field gene banks at appropriate ecological zones; and
10. MoU with Forest Department and communities, and reintroduction with post-introduction monitoring protocol.

16.14 CSIR-NBRI WEB PORTAL ON THREATENED PLANTS

CSIR-NBRI also maintains a web portal entitled, *Threatened Plants of India* (http://www.threatenedplantsindia.in) with detailed species information, distribution pattern, and threat status of 2,346 threatened plants of India (following IUCN categorization; IUCN, 2012). The portal is regularly updated and is a dynamic database of threatened plants (Figure 16.13).

16.15 R&D FOR NEW VARIETIES/IMPROVEMENT OF FLORAL AND ORNAMENTAL CROPS

Since the 1960s, CSIR-NBRI Botanic Garden has been regularly developing and releasing trait-specific new varieties of *Gladiolus*, *Bougainvillea*, *Chrysanthemum* and *Tuberose* following hybridization, and mutagenesis through chemical and gamma irradiation approaches. It has developed more than 30 varieties of *Bougainvillea*, 70 varieties of *Chrysanthemum*, two varieties of *Tuberose*, and 15 varieties of *Gladiolus*. All these varieties are extremely popular among the gardeners and nurseries.

16.16 AVIAN DIVERSITY

The Botanic Garden is also home to 61 species of birds (Figure 16.14). Most of the species are resident, and a few are migratory species. There are more than 100 Indian peafowls in the garden. These birds provide important ecosystem services like pest control, pollination, and seed dispersal.

FIGURE 16.13 CSIR-NBRI website, *Threatened Plants of India*.

FIGURE 16.14 Some of the avian species of the garden: (A) *Pavo cristatus*, (B) *Milvus migrans*, (C) *Athene brama*, (D) *Amaurornis phoenicurus*, (E) *Turdoides striata*, (F) *Halcyon smyrnensis*, (G) *Pycnonotus jocosus*, (H) *Oriolus xanthornus*, (I) *Vanellus indicus*, (J) *Dendrocitta vagabunda*, (K) *Cinnyris asiaticus*, (L) *Merops orientalis*, (M) *Eudynamys scolopacea*, (N) *Acridotheres tristis*, (O) *Psittacula krameri*.

16.17 SOCIETAL SERVICES

The Botanic Garden organizes three annual flower shows, viz. Chrysanthemum and Coleus Show, Rose and Gladiolus Show, and Bougainvillea Show, for creating awareness among the masses regarding the importance and usefulness of plants, promotion of floriculture, and outreach of the garden's research activities. The Botanic Garden also organizes several training programs on horticulture to enhance entrepreneurship capabilities, especially among youths and women.

16.18 ECOSYSTEM SERVICES

Besides research, education and recreation services, the garden also hosts hundreds of morning walkers everyday who get the benefits of health care ecosystem service from the garden.

16.19 CONCLUSION

The CSIR-NBRI Botanic Garden has come a long way in providing diverse services such as conservation, education, recreation, and health care ecosystem services. Continuous enrichment of its germplasm collections and mentoring small botanic gardens are its mandated activities, which are religiously performed regularly. More such efforts from different stakeholders associated with infrastructure improvement are expected to take the age-old garden to a new level.

REFERENCES

Adhikari, D., Z. Reshi, B.K. Datta, S.S, Samant, A. Chettri, K. Upadhaya, M.A. Shah, P.P. Singh, R. Tiwary, K. Majumdar, A. Pradhan, M.L. Thakur, N. Salam, Z. Zahoor, S.H. Mir, Z.A. Kaloo and S.K. Barik. 2018. Inventory and characterization of new populations through ecological niche modelling improve threat assessment. *Current Science* 114: 519–531.

Barik, S.K., N.K. Chrungoo and D. Adhikari. 2018b. Conservation of threatened plants of India. *Current Science* 104: 468–469.

Barik, S.K., B.R.P. Rao, K. Haridasan, D. Adhikari, P.P. Singh and R. Tiwary. 2018c. Classifying threatened species of India using IUCN criteria. *Current Science* 114: 588–595.

Barik, S.K., O.N. Tiwari, D. Adhikari, P.P. Singh, R. Tiwary and S. Barua. 2018a. Geographic distribution pattern of threatened plants of India and steps taken for their conservation. *Current Science* 114: 470–503.

Chrungoo, N.K., G.R. Rout, S.P. Balasubramani, P.E. Rajasekharan, K. Haridasan, B.R.P. Rao, R. Manjunath, G. Nagduwar, P. Venkatasubramanian, A. Nongbet, M. Hynniewta, D. Swain, S. Salamma, K. Souravi, S.N. Jena and S.K. Barik. 2018. Establishing taxonomic identity and selecting genetically diverse populations for conservation of threatened plants using molecular markers. *Current Science* 114: 539–553.

Haridasan, K., A.A. Mao, M.K. Janarthanam, A.K. Pandey, S.K. Barik, S.K. Srivastava, P.C. Panda, S. Geetha, S.K. Borthakur, B.K. Datta and B.R.P. Rao. 2018. Contributions of plant taxonomy, herbarium and field germplasm bank to conservation of threatened plants: Case studies from the Himalayas and Eastern and Western Ghats. *Current Science* 114: 512–518.

Hunter, M.L. and J.P. Gibbs. 2011. *Fundamentals of Conservation Biology*, 3rd ed. Blackwell Science, Malden, MA.

IUCN. 2012. *IUCN Red List Categories and Criteria, Version 3.1*, 2nd ed. IUCN, Gland, Switzerland.

Lyngdoh, M.K., A. Chettri, D. Adhikari and S.K. Barik. 2018. Metapopulation modelling of threatened plants to assess conservation status and determine minimum viable population size. *Current Science* 114: 532–538.

Panda, P.C., S. Kumar, J.P. Singh, P. Gajurel, P.K. Kamila, S. Kashung, R.N. Kulloli, P.P. Singh, D. Adhikari and S.K. Barik. 2018. Improving macropropagation and seed germination techniques for conservation of threatened species. *Current Science* 114: 562–566.

Primack, R.B. 2008. *A Primer of Conservation Biology*, 4th ed. Sinauer, Sunderland, MA.

Soulé, M.E. 1986. *Conservation Biology: The Science of Scarcity and Diversity*. Sinauer Associates, Inc., Sunderland, MA.

Van Dyke, Fred. 2008. *Conservation Biology: Foundations, Concepts, Applications*, 2nd ed. New York: Springer.

Index

A

Abies spectabilis, 13
Abrus precatorius, 180, 184
Abutilon neelgherrense, 179, 184
Abutilon ranadei Woodrow & Stapf, 146
Acacia catechu (Cutch tree), 193
Acampe praemorsa, 160, 223, 252
Acanthephippium bicolor, 209
Acer oblongum, 13
Achillea × *kasakhstanica* Kupr. et Alibekov, 238
A. kamelinii Kupr., 238
A. tianschanica Kupr. et Kulemin, 238
Achyranthes aspera L., 193
Aconitum kusnezoffii Rchb., 118
Acorus calamus, 14, 118, 190, 222
Acrostichum aureum, 219
Acrotrema arnottianum Wight, 145
Actiniopteris radiata, 157
Actinodaphne madraspatana, 179
Actinodaphne malabarica, 210
Adansonia digitata, 218, 223, 254
A. macrophylla, 254
Adansonia grandidieri, 54
Adenanthera pavonina, 184
Adenia hondala, 212
Adenocalymma alliaceum, 223
Adiantum capillus-veneris, 157, 251
A. caudatum, 157
A. hispidulum, 251
Adonis mongolica Simonov, 115, 118
Adonis vernalis L., 239
Adonis villosa Ledeb., 241
Aechmea bracteata (Sw.) Griseb., 147
Aechmea gamosepala Wittm, 147
Aegle marmelos (L.) Corrêa, 190
Aerides crispa, 160
Aerides emericii Rchb. f., 193
A. maculosa, 160
A. ringens, 160
Aesculus indica, 15
Afzelia xylocarpa, 56, 57
Agapanthus africanus, 14
Agave angustifolia, 184
Agave cantula, 14
A. sisalana, 184
Aglaia malabarica Sasidh., 146
Agrostophyllum planicaule (Wall. ex Lindl.) Rchb. f., 193
Ailanthus triphysa, 165
Alangium salvifolium (L.f.) Wang., 164, 209, 223
Allium caeruleum Pall., 239
Allium cristophii Trautv, 239
Allium macrostemon Bunge, 118
Allium microdictyon Prokh., 239
Allium obliquum L., 118
Allium schoenoprasum L., 239
Allium sphaerocephalon L., 239
Allium vodopjanovae Friesen, 241
Alnus nepalensis, 13

Aloe harlana, 250
Aloe vera, 158
Alphonsea madraspatana, 179
Alsophila nilgirensis (Holttum) R.M.Tryon, 145
Alstonia scholaris (L.) R.Br., 30, 145
Alstonia venenata R.Br., 145
Amherstia nobilis Wall., 138, 218
Amomum aculeatum Roxb., 188, 190, 193
Amomum andamanicum V.P. Thomas et al., 188
Amorphophallus interruptus, 57
Amorphophallus konkanensis, 179, 184
Amorphophallus paeoniifolius, 157
Amorphophallus titanum, 28
Anacolosa densiflora Bedd., 146
Anamirta cocculus, 223
Ancillary Botanic Gardens, 83, 85, 90
Andaman and Nicobar Islands, 187–189, 193, 196, 198–199
Andira inermis, 218
Andrographis beddomei, 179, 181, 184
Andrographis glandulosa, 178
Andrographis nallamalayana, 178
Andrographis paniculata (Burm.f.) Wall. ex Nees, 190
Andrographis serpyllifolia, 179
Anemarrhena asphodeloides Bunge, 118
Angiopteris evecta (G. Forst.) Hoffm., 188, 190
Angiopteris helferiana C. Presl., 145, 219
Anisoptera costata, 56
Annona muricata, 254
Annona reticulata, 184
Anogeissus latifolia, 184
Antiaris toxicaria, 222
Antidesma montanum Blume., 209
Aphyllorchis montana, 180
Aponogeton appendiculatus, 212
Aponogeton crispus, 157
A. natans, 157
A. undulatus, 157
Aporosa cardiosperma (Gaertn.) Merr., 145
Aquatic plants, 222
Aquilaria crassna, 57
Arboreta, 95–96, 98, 100–101
Arboretum, 67–70, 72, 76, 95, 157, 160–161, 164, 254
Areca catechu, 250
Areca triandra, 164
Arenga pinnata, 67, 250
Arenga wightii Griff., 143, 146, 159, 164
Aristolochia indica, 180, 190
Artabotrys hexapetalus, 180
Artanema longifolium (L.) Vatke, 145
Artemisia austriaca, 244
Artemisia saurense Kupr., 238
Artocarpus heterophyllus Lam., 145
Artocarpus hirsutus, 164–165
Arundina graminifolia, 252
Asparagus filicinus, 14
Asparagus gonoclados, 184
Asparagus racemosus Willd., 157, 190, 212
Asplenium phyllitidis, 219

261

Aster altaicus, 96
Astragalus boreali-mongolicus Y.Z. Zhao, 119
Astragalus follicularis Pall., 241
Atalantia wightii, 164
Atuna travancorica, 210
Azadirachta indica, 180, 190
Azanza lampas (Cav.) Alef., 145

B

Baccaurea courtallensis (Wight) Müll. Arg., 145, 209
Baccaurea ramiflora Lour., 188
Bacopa monnieri, 180, 190
Bambusa bambos, 184
Bambusa multiplex (Lour.) Raeusch. ex Schult., 143, 222
Bambusa ventricosa, 222, 249
Bambusa vulgaris Schrad., 22, 143, 189
Bambusetum, 189, 229
Barleria acanthoides, 162
B. acuminata, 162
B. buxifolia, 162
Barleria courtallica Nees, 145, 162
B. cristata L., 162–163
Barleria cuspidata, 162, 180
B. durairajii, 162
B. gibsonii, 162–164
B. grandiflora, 162–164
B. hochstetteri, 162
B. involucrata, 162–163
B. lawii, 162
B. longiflora, 162
B. lupulina, 162
B. montana, 162
B. morrisiana, 163
B. mysorensis, 162
B. nitida, 162–163
B. noctiflora, 162
B. pilosa, 162
B. prattiana, 162–163
Barleria prionitis L., 162–163
B. repens, 162–163
B. sepalosa, 162–163
Barleria stocksii, 162, 179, 181
Barleria strigosa Willd 145, 162–163
B. terminalis, 162
B. tomentosa, 162–163
Bauhinia variegata, 15
Begonia dipetala Graham, 145
Bentinckia condapanna Berry ex Roxb., 143, 146, 161, 164
Bentinckia nicobarica (Kurz) Becc., 161, 164, 188, 224
Berchemia berchemiifolia, 97
Berchemia racemosa Siebold & Zucc., 96
Betula utilis, 20
Bidoup–Nui Ba Botanic Garden, 57
Billbergia pyramidalis (Sims) Lindl., 147
Billbergia zebrina (Herb.) Lindl., 147
Biodiversity Conservation, 240
Biodiversity Hotspot, 135, 147
Bischofia javanica, 254
Black Data Book, 238
Blechnum brasiliense Desv., 145
Blechnum finlaysonianum Wall., 190
Blechnum occidentale, 251

Blepharistemma membranifolium, 164
Bogor Botanic Gardens, 26
Bolbitis heteroclita, 251
Bolbitis presiliana (Fée) Ching., 145
Bombax ceiba, 13, 160
Borassus flabellifer, 164, 184
Boswellia ovalifoliolata, 178, 182, 184
Boswellia serrata, 160, 184, 254
Botanical garden of MAS, 105–109, 111–116, 118, 122, 125–127
Botanic/Botanical gardens, 1–16, 19–21, 29–30, 39, 47, 54–56, 58, 60, 66, 70–71, 77, 79, 81–92, 95–96, 100–101, 124
Botanic Garden of CSIR-National Botanical Research Institute, 248
Botanic Gardens Conservation International (BGCI), 2, 11, 13, 21, 54, 60–62, 66, 122, 213, 227, 234, 249
Boucerosia diffusa, 179–181
Boucerosia frerei (G.D.Rowley) Meve & Liede, 146
Boucerosia indica, 179
Bouea macrophylla, 67
Brachystelma annamacharyae, 178, 181
Brachystelma ciliatum, 179, 180
Brachystelma kadapense, 178
Brachystelma kolarense, 179, 181
Brachystelma maculatum, 178, 179, 181
Brachystelma nigidianum, 178
Brachystelma pullaiahii, 178
Brachystelma seshachalamense, 178
Brachystelma vemanae, 178, 181
Brassica oleracea, 26
Breynia vitis-idaea, 184
Brocchinia reducta, 138
Brunnera sibirica Steven, 239
Buchanania barberi Gamble, 146
Buchanania cochinchinensis, 164
Buchanania lanceolata Wt., 210
Bu Gia Map Botanic Garden, 55, 57
Bulbophyllum crassipes Hook.f., 193, 252
B. rufinum Rchb. f., 193
Bulbophyllum sterile, 209
Bupleurum longifolium, 239
Butea buteiformis, 15
Butea monosperma, 20, 145, 180, 184, 193, 254
Byttneria herbacea, 184

C

Cabomba furcata Schult. & Schult.f., 146
Cacti and succulens, 220–221
Caesalpinia bonduc, 184
Caesalpinia sappan, 180
Calamus andamanicus Kurz, 143, 188, 189
Calamus brandisii Becc., 146
C. dilaceratus Becc., 189
C. longisetus Griff., 189
Calamus manan Miq., 30
Calamus nicobaricus Becc. & Hook.f., 193
C. palustris Griff., 189
C. pseudorivalis Becc., 189
Calamus thwaitesii, 223
Calamus vattayila Renuka, 143

Index

C. viminalis Willd., 189
Calicut University Botanical Garden (CUBG), 217–220, 224
Calliandra cynometroides Bedd., 146
Calophyllum apetalum, 164–165
Calotropis gigantea (L.) Dryand., 193
Calycopteris floribunda, 223
Camellia bugiamapensis, 57
Camellia cucphuongensis, 57
Camellia dalatensis, 61
Canarium commune, 67
Canarium denticulatum Blume, 193
Canavalia gladiata, 180
Canna indica, 180
Capparis grandis, 180
C. zeylanica, 180
Caragana tibetica Kom, 118
Carallia brachiata, 164–165
Caralluma stalagmifera, 180
Carex duriuscula C.A.May, 111
Carissa andamanensis L.J. Singh & Murugan, 188, 193, 195
Carya sinensis, 57
Caryota mitis, 164, 250
Caryota urens, 164, 224, 250
Cassia fistula, 15
Cassia roxburghii, 184
Cassine glauca, 180
Castanopsis cerebrina, 58
Castanopsis lecomtei, 57
Castanopsis tribuloides, 15
Cedrus deodara, 15
Centella asiatica (L.) Urb., 180, 190
Cephalotaxus hainanensis, 56
Ceratophyllum demersum L., 146
Ceratopteris thalictroides, 219
Cerbera manghas L., 192
Cerbera odollam Gaertn., 192
Cereus pterogonus, 184
C. repandus, 184
Ceropegia anantii, 166
C. anjanerica, 166
C. attenuata, 166
C. bhatii, 166
Ceropegia candelabrum 179
C. concanensis, 166
C. evansii, 166
C. fantastica, 166
Ceropegia hirsuta, 180
C. huberi, 166
C. jainii, 166
Ceropegia juncea, 180
C. lawii, 166
C. maccannii, 166
C. mahabalei, 166
C. media, 166
C. mohanramii, 166
C. noorjahaniae, 166
C. oculata, 166
C. odorata, 166
C. panchganiensis, 166
Ceropegia pullaiahii, 178, 181
C. rollae, 166

C. sahyadrica, 166
C. santapaui, 166
Ceropegia spiralis, 179, 181
C. vincifolia, 166
Chamaedorea elegans, 250
C. stolonifera, 250
Chionanthus mala-elengi, 164
Chlorophytum arundinaceum, 157
C. belgaumense, 157
C. bharuchae, 157
C. borivilianum, 157
C. breviscapum, 157
C. glaucoides, 157
C. gothanense, 157
C. heynei, 157
C. indicum, 157
C. kolhapurense, 157
C. laxum, 157
C. malabaricum, 157
C. nepalense, 157
C. nilgheriensis, 157
C. tuberosum, 157, 180
Choerospondias axillaris, 13, 15
Chorisia insignis, 254
Chrysalidocarpus lutescens, 250
Chukrasia tabularis, 54
Cibinong Botanic Gardens, 28
Cibodas Botanic Gardens, 28
Cinchona calisaya, 26, 28, 32
Cinnamomoum camphora, 223, 254
Cinnamomum balansae, 58
Cinnamomum chemungianum M.Mohanan & A.N.Henry, 146
Cinnamomum culilawan, 67
Cinnamomum goaense, 164
Cinnamomum riparium Gamble, 146
Cinnamomum tamala, 15
Cissus discolor Blume, 145
Cissus quadrangularis L., 180, 192
Cissus rotundifolia, 184
Cleidiocarpon cavaleriei, 56
Cleistanthus collinus, 184
Coccoloba uvifera, 254
Cochlospermum gossypium, 223
Cochlospermum religiosum (L.) Alston, 145
Codiocarpus andamanicus (Kurz) R.A. Howard, 193
Coelogyne cristata, 252
Coelogyne nervosum, 223
Coelogyne quadratiloba Gagnep., 193
Coffea canephora, 26
Combretum malabaricum (Bedd.) Sujana, Ratheesh & Anil Kumar, 145
Commiphora caudata, 180
Conservatory, 249
Convallaria keiskei Miq., 118
Convention on Biological Diversity (CBD), 95, 204, 240, 242, 243
Cordia dichotoma, 184
Corylus ferox, 20
Coryphantha maiz-tablasensis, 250
Corypha umbraculifera L., 138, 143, 159, 164
C. utan, 159, 164
Coscinium fenestratum, 222

Costus comosus, 143
Costus pictus, 249
Costus speciosus (Koen ex. Retz.) Sm., 180, 192, 212
Costus stenophyllus Standl. & L.O.Williams, 143
Couroupita guianensis, 218
Cousinia × *pavlovii* Kupr., Lashchinskiy et A.L. Ebel., 238
Crinum asiaticum, 157
C. brachynema, 157
C. latifolium, 157
C. malabaricum, 157
C. pratense, 157
C. solapurense, 157
C. viviparum, 157
C. woodrowii, 157
Crotalaria sandoorensis, 179, 184
Crotalaria shevaroyensis, 164
Croton scabiosus, 178, 183–184
CSIR-NBRI botanic garden, 248–249
Cu Chi Botanic Garden, 55, 57
Cuc Phuong Botanic Garden, 55, 57
Cupressus torulosa, 14
Curcuma aromatica, 14, 157
Curcuma longa L., 192
C. mangga Valeton & Zijp, 192
Curcuma neilgherrensis, 157, 184
C. zedoaria (Christm.) Roscoe, 192
Cyathea gigantea (Wall. ex Hook.) Holttum, 190
Cyathea spinulosa, 14
Cycads, 253
Cycas annaikalensis, 253
Cycas beddomei, 158, 178, 183, 184, 253, 256
Cycas circinalis, 165, 184, 224, 253
C. dharmrajii L.J. Singh, 189
C. divyadarshanii, 253
C. micronesica, 253
C. nayagarhensis, 253
C. nongnoochiae, 253
C. orixensis, 253
C. pachypoda, 253
Cycas pectinata, 20, 253
C. pschannae R.C. Srivast. & L.J. Singh, 158, 189
C. revoluta, 158, 184, 253–254
C. riuminiana, 253
C. rumphii, 158, 184
C. schumanniana, 254
C. seemannii, 254
Cycas seshachalamensis, 178, 182, 184
C. silvestris, 254
Cycas sphaerica, 179, 183–184, 253
C. swamyi, 253
C. tansachana, 254
C. wadei, 254
Cycas zeylanica (J. Schust.) A. Lindstr. & K.D. Hill, 182, 184, 189–190, 253–254
Cymbidium aloifolium (L.) Sw., 193, 209
Cymbidium bicolor, 252–253
Cymbidium lancifolium Hook., 96
Cymbopogon citratus (DC.) Stapf, 180, 192
Cynanchum acidum, 180
Cynanchum thesioides (Freyn) K. Schum, 118
Cynodon dactylon (L.) Pers., 193
Cynometra beddomei Prain, 146, 210
Cynometra bourdillonii Gamble, 146
Cynometra cauliflora, 67
Cynometra travancorica Beddome 210, 230
Cypripedium calceolus L., 118
Cypripedium japonicum, 97
Cypripedium macranthos Sw., 118

D

Dacrycarpus imbricatus (Blume) de Laubenf., 30
Dactylorhiza hatagirea, 20
Daemonorops kunstleri, 250
Dalbergia cochinchinensis, 57
Dalbergia lanceolata, 254
Dalbergia latifolia, 20, 180, 182, 184, 223
Dalbergia oliveri, 56, 57
Dalbergia paniculata, 180
Dalbergia sissoo, 15
Dalbergia tonkinensis, 54, 56
Datura metel, 222
Daucus carota, 26
Decalepis hamiltonii, 179, 184
Decaschistia cuddapahensis, 178, 181, 184
Delonix regia, 184
Delphinium malabaricum, 164, 166
Dendrobium anilli, 209
Dendrobium aphyllum, 209, 252–253
Dendrobium crumenatum Sw., 30, 160, 193
D. formosum Roxb. ex Lindl., 193
Dendrobium herbaceum, 160, 253
Dendrobium moschatum, 253
Dendrobium polyanthum, 253
D. shompenii B.K. Sinha & P.S.N. Rao., 193
D. tenuicaule Hook. f., 193
Dendrocalamus giganteus Munro, 143, 189, 222
Dendrocalamus strictus (Roxb.) Nees, 189
Desmostachya bipinnata (L.) Stapf, 193
Dhanikhari Experimental Garden, 187–188
Dianthus versicolor, 244
Dichrostachys cinerea (L.) Wight & Arn., 145
Dicliptera cuneata, 184
Dillenia andamanica C.E. Parkinson, 188, 193, 195
Dillenia indica, 254
Dimocarpus longan Lour., 146
Dinochloa andamanica (Kurz) H.B.Naithani, 143, 189
Dionaea muscipula, 138
Dioon edule Lindl., 189
Dioon mejiae, 224
Dioon spinulosum, 254
Dioscorea belophylla, 212
D. hamiltonii, 212
D. hispida, 212
D. intermedia, 212
D. kalkapershadii, 212
D. oppositifolia, 212
Dioscorea pentaphylla, 180, 212
D. pubera, 212
D. tomentosa, 212
D. wallichii, 212
D. wightii, 212
Diospyros buxifolia (Blume) Hiern, 145
Diospyros ebenum J.Koenig ex Retz., 145
Diospyros macrophylla Blume, 30

Index

Diospyros malabarica, 254
Diospyros melanoxylon, 184
Diospyros mollis, 54
D. mun, 54
Dipcadi concanense, 167
Dipcadi krishnadevarayae, 179
Diplazium esculentum, 219, 251
Diplazium proliferum (Lam.) Thouars, 190
Dipterocarpus alatus, 54
Dipterocarpus indicus Bedd., 230
Dipterocarpus intricatus, 57, 61
Dipterocarpus retusus, 56
Doxantha unguis-cati, 223
Dracaena draco, 250
Drimia nagarjunae, 179
Drynaria quercifolia, 219, 223, 251
Dyckia brevifolia Baker, 147
Dysoxylum beddomei Hiern, 146
Dysoxylum malabaricum Bedd. ex C. DC., 164, 230

E

Eastern Ghats, 173
Echinocactus grusonii., 250
Echinosophora koreensis, 100
Eclipta prostrata (L.) L., 192
Ehretia aspera, 184
Eka Karya Bali Botanic Gardens, 28
Elaeagnus conferta Roxb., 145
Elaeis guineensis, 26, 32, 159, 250
Elaeocarpus angustifolius, 20
Elaeocarpus floribundus Blume, 193
Elaeocarpus grandiflorus J.E.Sm., 30
E. rugosus Roxb. ex G. Don, 193
Elaeocarpus serratus L., 145, 164
Elaeocarpus sphaericus, 15, 223
Elisanthe noctiflora, 244
Encephalartos villosus, 249
Ensete glaucum, 222
Ensete superbum, 222
Environmental education, 242
Ephedra gerardiana, 14, 20
Epipactis helleborine (L.) Grantz, 241
Equisetum debile, 251
Equisetum ramosissimum, 219
Eranthemum capense L., 145
Eria andamanica Hook. f., 193
Erinocarpus nimmonii, 164, 166
Eriolaena lushingtonii, 179, 182, 184
Eryngium foetidum L., 192
Eryngium planum L., 239
Erythronium sibiricum (Fisch. et Mey.) Kryl., 239, 241
Etlingera elatior (Jack) R.M.Sm., 143
Etlingera fenzlii (Kurz) Škornick. & M. Sabu, 188, 191–192
Eulophia andamanensis Rchb.f., 160, 191, 193
Eulophia nicobarica N.P. Balakr. & N.G. Nair, 193, 252
Euphorbia antiquorum, 184
E. caducifolia, 184
Euphorbia cylindrifolia, 250
Euphorbia epiphylloides Kurz, 193, 195
Euphorbia neriifolia, 250
Euphorbia prostrata, 57
Euphorbia royleana, 14
Euphorbia seshachalamensis, 179
E. tirucalli, 184
E. tortilis, 184
Euphorbia vajravelui Binojk. & N.P. Balakr., 146
ex situ conservation, 2–3, 10–11, 20–21, 26, 29, 32, 34, 60, 62, 78, 100, 136–138, 147, 173–174, 176, 184, 203–204, 207, 217, 240–241, 245, 249–250, 255

F

Fernarium, 71
Fernary, 190
Fern House, 220, 251, 252
Ferula ferulioides (Steud.) Korovin, 118–9
Ficus benghalensis, 184, 254
Ficus carica, 222
Ficus dalhousiae, 179, 182
Ficus heterophylla, 181
Ficus hispida, 180
Ficus krishnae, 254
Ficus mollis, 184
Ficus racemosa L., 180, 193
Ficus religiosa L., 193
Filicium decipiens (Wight &Arn.) Thwaites, 145
Flacourtia indica (Burm.f.) Merr., 145
Flacourtia inermis, 67
Flacourtia montana J.Graham, 145, 164
Flemingia chappar, 162
F. lineata, 162
F. nana, 162
F. paniculata, 162
F. praecox, 162
F. procumbens, 162
F. semialata, 162
F. sootepensis, 162
F. stricta, 162
F. strobilifera, 162
F. wallichii, 162
F. wightiana, 162
Forsythia ovata, 100
Fragaria x ananassa, 26

G

Galium verum, 244
Garcinia andamanica King, 193
Garcinia cambogioides, 164
Garcinia dhanikhariensis S.K. Srivast., 188, 195
Garcinia gummi-gutta (L.) Roxb., 145, 164
Garcinia indica, 164
Garcinia talbotii, 164
Garcinia travancorica Bedd., 146, 164
Garcinia wightii T. Anderson, 146
Gastrochilus flabelliformis, 160, 166
Genetic resources, 99
Geodorum densiflorum (Lam.) Schltr., 160, 193
Geophila repens (L.) I.M. Johnst., 188, 192
germplasm collections, 249
gingers, 220
Ginkgo biloba, 15, 184
Givotia moluccana, 180
Global Biodiversity, 204, 206

Global Strategy for Plant Conservation (GSPC), 2, 95–97, 122, 204, 208, 213, 240, 241
Glochidion talakonense, 179
Gloriosa superba, 14
Gluta travancorica Bedd., 146, 164
Glycyrrhiza uralensis Fisch., 119
Gnetum edule (Willd.) Blume, 145
Gnetum gnemon L., 138, 145
G. montanum Markgraf, 20, 189
Gnetum scandens Roxb., 188–189
Gnetum ula Brongn., 138, 145, 158, 223–224
Goniolimon speciosum, 244
Goniothalamus rhynchantherus Dunn, 146
Goniothalamus wynaadensis (Bedd.) Bedd., 146
Grammatophyllum speciosum Blume, 138
Guaiacum officinale, 54
Gymnacranthera canarica (Bedd. ex King) Warb., 146
Gymnema sylvestre, 180
Gymnosphaera gigantean (Wall. ex Hook.) S.Y. Dong, 145
Gymnostachyum febrifugum Benth., 145
Gynostemma pentaphyllum, 57
Gypsophila patrinii Ser., 241, 244

H

Habenaria longicornu, 179
Habenaria panigrahiana, 179
Habenaria rariflora, 179
Habenaria roxburghii, 160
Hanabusaya asiatica, 100
Hanoi Botanic Garden, 55, 56
Hardwickia binata, 164, 180
Harpullia arborea, 164–165
Hedysarum gmelinii, 244
Herbal Garden, 190
Herbarium, 236
Heritiera percoriacea Kosterm., 30
Hevea brasiliensis, 26
Hibiscus hispidissimus, 223
Hibiscus rosa-sinensis L., 192
Hildegardia populifolia (DC.) Schott & Endl., 164, 179, 182–184, 230
Holigarna arnottiana, 164
H. grahamii, 164
Holoptelea integrifolia, 184
Hopea chinensis, 61
Hopea cordata, 57
Hopea odorata, 54
Hopea parviflora, 164
Hopea ponga (Dennst.) Mabb., 146, 210
Horsfieldia glabra (Blume) Warb., 188
Horsfieldia kingii, 20
Hoya carnosa, 249
Hoya pandurata, 256
Hubbardia diandra, 166
Hubbardia heptaneuron, 166–167
Hugonia mystax, 223
Humboldtia bourdillonii Prain, 146
Humboldtia brunonis Wall., 146, 164–165, 210
Humboldtia decurrens Bedd.ex Oliv., 145, 146, 164, 210
Humboldtia sanjappae, 164
Humboldtia vahliana Wight, 145
Hydnocarpus macrocarpus (Bedd.) Warb., 146
Hydnocarpus pentandrus, 164
Hydrilla verticillata (L.f.) Royle, 146
Hygrophila balsamica (L.f.) Raf., 146
Hygrophila difformis (L.f.) Blume, 146
Hymenodictyon orixense, 184
Hyphaene dichotoma, 159, 164
Hyptis suaveolens (L.) Poit., 192

I

Index Seminum, 34
Indigofera mysorensis, 184
Indonesia, 25–36, 38–44, 46–48
Indopiptadenia oudhensis, 256
in-situ conservation, 3, 10–12, 20–21, 29, 31–32, 100, 137, 147, 173–174, 178, 203–204, 207–208, 241, 255
International Union for the Conservation of Nature (IUCN), 2, 20–21, 66, 96–97, 115, 118, 136–138, 227, 230
Intsia bijuga (Colebr.) O. Kuntze, 30, 54
Iphigenia ratnagirica, 167
Ipomoea aquatica Forssk., 192
Ipomoea mauritiana, 212
Iris dichotoma Pall., 96
Iris odaesanensis, 100
Iris pallida, 14
Iris tenuifolia Pall., 116
Ixora brachiata Roxb., 145
Ixora polyantha Wight, 145

J

Jasminum flexile Vahl, 145
Jasminum sambac, 184
Jawaharlal Nehru Tropical Botanic Garden and Research Institute (JNTBGRI), 135–138, 145, 147–148
Jerdonia indica, 218
Johor Botanic Gardens, 70
Juniperus horizontalis, 14
Justicia adhatoda L., 192
Justicia gendarussa, 67

K

Kalanchoe daigremontiana, 180
Kepong Botanic Gardens, 68–69, 78
Keteleeria davidiana, 56
Keteleeria evelyniana, 57
Khaya senegalensis, 54, 57
Kigelia africana, 223
Kingiodendron pinnatum (Roxb. ex DC.) Harms., 230
Knema andamanica (Warb.) W.J.de Wilde, 188–189, 193
Knema attenuata (Wall. ex Hook.f. & Thomson) Warb., 146, 164
Korea, 95–101
Korthalsia rogersii Becc., 143
Kuzbass Botanical Garden, 234–236, 239–241, 243–244
Kyoto protocol, 243

L

Lagerstroemia macrocarpa, 164
Lagerstroemia speciosa, 223

Index

Lake Baikal, 104, 108, 127
Lancea tibetica Hook. f. & Thomson
Lang Hanh Arboretum, 55, 56
Lansium domesticum, 67
Lawsonia inermis L., 192
Lead Botanical Garden (LBG), 149, 151–154, 156, 160, 162, 164, 168, 173
Lepidagathis rajasekharae, 179, 181
Lepisanthes tetraphylla, 180
Licuala grandis, 250
L. spinosa, 250
Lilium martagon L., 118
Lilium oxypetalum, 14
Lilium pensylvanicum Ker Gawl., 118
Limnocharis flava (L.) Buchenau, 189
Limnophila heterophylla (Roxb.) Benth., 146
Limnophila indica (L.) Druce, 146
Limonia acidissima, 180
Linum perenne L., 239, 241
Liparis nervosa, 160
Litchi chinensis, 32
Lithocarpus amygdalifolius, 57
Lithocarpus elegans, 58
Litsea glutinosa, 181, 222, 254
L. cochin-chinensis, 250
Livistona chinensis, 250
Livistona jenkinsiana, 164
local botanic gardens, 39, 40, 43–45
Ludwigia adscendens (L.) H. Hara, 146
Luisia balakrishnanii S. Misra, 193
Lutheria splendens (Brongn.) Barfuss & W. Till, 147
Lycium ruthenicum Murray, 116
Lygodium flexuosum, 251

M

Madhuca bourdillonii (Gamble) H.J.Lam, 146
Madhuca indica, 184
Maesa andamanica Kurz, 193
Magaleranthis saniculijolia, 100
Magnolia bidoupensis, 58
Magnolia champaca, 15
Magnolia hodgsonii, 20
Mahatma Gandhi Botanical Garden, 227–229, 231
Mahonia napaulensis, 14
Malaysia, 65, 68, 70–73, 77–79
Malus domestica, 26
Mangifera andamanica King, 189, 193–194
Mangifera camptosperma Pierre, 189, 194
Mangifera griffithii Hook.f., 193
Mangifera indica L., 145, 189
Mangolia, 103–108, 111–115, 117–128
Manihot utilissima, 26
Marsilea pinnata, 219
Mascarena verschaffeltii, 250
Medicago falcata, 244
Melaka Botanical Garden, 72
Melaleuca cajuputi, 67
Melastoma malabathricum L., 145
Memecylon lushingtonii, 179
Mentha arvensis, 180
Mesua ferrea L., 146, 182, 223
Mesua manii (King) Kosterm., 193

Metroxylon sagu, 67
Meyenia hawtayneana, 179
Michelia champaca L., 30
Michelia nilagirica, 223
Microcycas calocoma, 249, 254
Microsorium alternifolium, 219, 251
Microsorum punctatum, 157, 223
Millenium Seed Bank, 34
Millingtonia hortensis, 56
Mimusops andamanensis King & Gamble, 193
Mimusops elengi, 222
Monochoria vaginalis (Burm. f.) C. Presl, 189
Monoon fragrans, 164
Morinda citrifolia L., 192
Moringa oleifera Lam., 192
Moss house, 251–252
Moullava spicata, 164
Mount Kinabalu Botanical Garden, 73
M. S. Swaminathan Botanical Garden, 203, 208–209, 213
Murraya paniculata (L.) Jack, 145
Musa haekkinenii, 222
Musa indandamanensis L.J. Singh, 189, 191
Musa ornata, 222
Myristica andamanica Hook.f., 146, 193
Myristica malabarica Lam., 146, 164, 230

N

Nageia wallichiana (Presl) Kuntze, 143, 189
Nanophyton erinaceum (Pall.) Bunge, 116
Narcissus tazetta, 14
National Botanical Garden, 4–6, 13, 72
Nelumbo nucifera, 180
Neomarckia gracilis, 14
Neoregelia spectabilis (Antoine) L.B.Sm., 147
Neottianthe cucullata (L.) Schltr., 118
Nepal, 3, 10–11, 13–14, 16, 20–21
Nepenthes khasiana, 218, 256
Nephrolepis biserrata (Sw.) Schott, 145
Nephrolepis cordifolia, 251
N. duffii, 251
N. tuberosa, 251
Nervilia aragoana Gaudich., 193
N. plicata (Andr.) Schltr., 160, 193
Nitraria sibirica Pall., 116
Nothopegia aureofulva Bedd. ex Hook.f., 146
Nothopegia castaneifolia, 164
Nyctanthes arbor-tristis L., 192
Nymphaea nouchali, 180, 189
Nymphaea omarana Hort ex Gard., 189
N. pubescens, 180, 189
N. rubra, 180
Nymphoides indica (L.) Kuntze, 158, 189, 222
N. hydrophila, 222
N. macrospermum, 222
N. parvifolia, 222
Nypa fruticans Wurmb., 143, 159, 164

O

Oberonia iridiflora, 223
Oberonia swaminathanii, 209
Ochna obtusata, 184

Ochreinauclea missionis (Wall. ex G.Don) Ridsdale, 146
Ocimum tenuiflorum L., 192
Ophioglossum reticulatum L., 145, 252
Ophiorrhiza chandrasekharanii, 179
Opuntia dillenii, 184
Opuntia ficus-indica (L.) Mill., 14, 192
Orchidarium, 193, 224, 229, 252–253
Orchis militaris L., 118
Oroxylum indicum, 20, 180, 183, 254–255
Osbeckia aspera (Meerb. ex Walp.) Blume, 145
Osbeckia virgata D.Don ex Wight & Arn., 145
Osmunda hugeliana, 219
Otonephelium stipulaceum, 164
Oxystelma esculentum, 180

P

Paeonia lactiflora Pall., 115, 118
Palmetum, 196
Palm house, 250–251
Pandanus furcatus, 20
Pandanus leram Jones ex Voigt, 191
Pandanus nepalensis, 13
Paphiopedilum appletonianum, 59
Paphiopedilum concolor, 57
Paphiopedilum druryi (Bedd.) Steinthe, 138
Paphiopedilum insigne, 252
Paphiopedilum malipoense, 58, 61
Paphiopedilum spiceranum, 252, 256
Parashorea chinensis, 56
Parashorea stellata, 56
Paris polyphylla, 14
Paris verticillata M. Bieb., 118
Parkia timoriana (DC.) Merr., 30
Pecteilis gigantea, 209, 212
Pedana Botanical Gardens, 71
Peganum harmala L., 118
Penang Botanical Gardens, 72
Pentalinon luteum, 184
Perak Botanical Gardens, 71
Peristalis porotta, 209
Phoenix reclinata, 250
Phu An Bamboo Village, 58
Phyllanthus amarus Schumach. & Thonn., 192
Phyllanthus emblica L., 145, 180, 192
Phyllanthus gageanus (Gamble) M.Mohanan, 145
Phyllanthus indofischeri, 179, 182
Phyllanthus narayanswamii, 179, 181, 184
Physic Garden, 13, 14
Piliostigma foveolatum, 164
Pimpinella tirupatiensis, 179, 181, 184
Pinanga dicksonii (Roxb.) Blume, 143, 145, 164
Pinanga manii, 159, 164
Pinanga javana, 30
Pineatum, 158
Pinenga dicksonii, 224
Pinus kesiya, 58
Pinus krempfii, 58, 61
Piper longum L., 192
Piper nigrum, 180
Pittosporum dasycaulon, 164
Plant conservation, 1–3, 12, 21, 26–31, 35, 44–48, 62, 66, 77–79, 95–97, 100, 204, 207–208, 213

Platycerium bifurcatum (Cav.) C.Chr., 145
Podocarpus neriifolius D.Don, 13, 15, 20, 143, 189–190
Podophyllum hexandrum, 20
Poeciloneuron indicum, 164
Poeciloneuron pauciflorum Bedd., 146
Polyspora huongiana, 58, 61
Pongamia pinnata, 184
Pontederia crassipes Mart., 146
Pothos scandens, 218
Prioria pinnata (Roxb. ex DC.) Breteler, 146, 164
Prosopis cineraria (L.) Druce, 193
Protected areas, 98
Psilotum nudum (L.) P.Beauv., 145, 219, 251
Psychotria andamanica Kurz, 189, 193
Pteridophytes, 219
Pteris argyraea T.Moore, 145
Pteris cretica, 251
Pteris ensiformis Burm. f., 145
P. vittata, 251
Pterocarpus dalbergioides Roxb. ex DC., 194
Pterocarpus indicus, 57
Pterocarpus macrocarpus, 56, 57
Pterocarpus marsupium, 20, 146, 255
Pterocarpus santalinus, 164, 179, 182–184
Pteroceras muriculatum (Rchb.f.) P.F. Hunt, 193
Pterospermum reticulatum, 164–165
Ptychosperma macarthuri, 250
Purwodadi Botanic Gardens, 28
Putrajaya Botanic Garden, 72
Putranjiva roxburghii, 20
Pyrostria laljii M.C. Naik, Arriola & M. Bheemalingappa, 189, 193
Pyrrosia piloselloides, 219
Pyrrosia lanceolata, 219, 223
Pyrus communis, 26

Q

Quercus platycalyx, 58
Quercus semecarpifolia, 15
Quercus setulosa, 57
Quisqualis indica, 223

R

Radermachera xylocarpa, 183
Rainforest Discovery Centre, 73, 75
Ranunculus kadzusensis Makino, 96
Rare and Endangered species, 116
Rauvolfia hookeri S.R.Sriniv. & Chithra, 146
Rauvolfia serpentina (L.) Benth. ex Kurz, 14, 180, 192, 222
R. tetraphylla L., 180, 192
Red book, 114, 118–119, 121
Red data book, 53, 211, 241
Red List, 2, 20–21, 53–54, 57–58, 60, 96–97, 115, 118, 121, 137–138, 143, 146–147, 211
Renanthera imschootiana, 252
Revegetation, 243
Rhaponticoides zaissanica Kupr., A.L. Ebel et Khrustaleva, 238
Rheum nobile, 20
Rhizophora mucronata, 162

Index

Rhododendron cowanianum, 20
Rhododendron garden, 13
Rhododendron lapponicum (L.) Wahlenb., 118
Rhododendron lowndesii, 20
Rhododendron variegata, 13
Rhynchosia beddomei, 179, 182, 184
Rhynchosia ravii, 179
Rhynchospora corymbosa (L.) Britton, 145
Ricinus communis L., 192, 222
Rimba Ilmu Botanic Gardens, 71
Rosa laxa Retz., 118
Rourea minor (Gaertn.) Merr., 209
Royal Palace Botanical Garden, 71
Rubia cordifolia, 222

S

Sabah Agriculture Park, 73, 75
Sagaraea laurina, 164
Sageraea thwaitesii Hook.f.& Thomson, 146
Saigon Zoo and Botanical Garden, 54, 55
Salacia beddomei Gamble, 145, 209, 222
Salacia chinensis, 180
Salacia fruticosa Wall. ex M.A. Lawson, 145
Salsola laricifolia Turcz. ex Litv 116
Sambucus manshurica Kitag., 118
Sambucus williamsii Hance, 118
Sanguisorba officinalis L., 239
Santalum album, 179, 255
Sapindus mukorossii, 15
Saposhnikovia divaricata (Turcz.) Schischk., 116, 119
Saraca asoca (Roxb.) W.J.de Wilde, 146, 180, 183, 192, 194
Saraca declinata (Jack) Miq., 138, 255
Sarawak Biodiversity Centre, 76
Sarawak Botanical Garden, 76
Sarcostigma kleinii Wight & Arn., 209
Saritaea magnifica **223**
Sauropus rhamnoides Blume, 192
Saussurea involucrata Matsum. & Koidz., 118
Saxifraga hirculus L., 118
Schizostachyum andamanicum M. Kumar & Remesh, 189, 192–193
Schleichera oleosa (Lour.) Oken, 209
Scutellaria baicalensis Georgi, 119
Sedum roseum (L.) Scop., 118
Seed bank, 151–152
Selaginella erythropus (Mart.) Spring, 145
Selaginella willdenowii (Desv.) Baker, 145
Semecarpus kurzii Engl., 189, 193
Semecarpus travancoricus, 164
Senegalia catechu, 15, 20
Senna siamea, 184
Seseli ledebourii, 244
Shorea falcata, 57
Shorea robusta, 164, 255
Shorea roxburghii, 57, 179, 183
Shorea tumbuggaia, 179, 183–184
Siberia, 234–241, 244
Sida acuta Burm.f., 192
Sindora tonkinensis, 56
Solanum americanum Mill., 209
Solanum melongena, 26

Solanum tuberosum, 26
Solenocarpus indicus, 164
Solidago virgaurea, 118
Sophora flavescens Aiton, 118
Sophora interrupta, 179
Sophora tonkinensis, 54
Spathodea campanulata, 184, 223
species recovery, 164
Sphaeropteris albosetacea (Bedd.), R.M. Tryon, 189–190
S. nicobarica (Balakr. *et* Dixit) Dixit, 190
Spices house, 221
Spondias pinnata, 182
Stelechocarpus burahol (Blume) Hook.f. & Thomson, 30
Stemonia tuberosa, 180
Stenosiphonium wightii Bremek., 145
Sterculia guttata, 222
Sterculia urens, 184
Stereospermum tetragonum, 184
Stipa capillata, 244
Stipa pennata L., 241
Strobilanthes barbatus Nees, 145
Strobilanthes ciliates Nees, 145
Strobilanthes cordifolia, 184
Strobilanthes lawsonii Gamble, 145
Strobilanthes lupulina Nees, 145
Strobilanthes wightii (Bremek.) J.R.I.Wood, 146
Strychnos nux-vomica, 222, 255
S. potatorum, 255
Suregada multiflora, 182
Suriana Botanic Conservation Gardens, 72–73
Swietenia macrophylla, 57
Swietenia mahagoni, 184
Syzygium alternifolium, 179, 183–184
Syzygium andamanicum (King) N.P. Balakr., 193
Syzygium aromaticum (L.) Merr. & L.M.Perry, 192
Syzygium caryophyllatum (L.) Alston, 145
Syzygium cumini (L.) Skeels, 145
Syzygium jambos, 13, 255
Syzygium laetum (Buch.-Ham.) Gandhi, 146, 164
Syzygium palghatense Gamble, 146
Syzygium stocksii (Duthie) Gamble, 146, 164
Syzygium zeylanicum (L.) DC., 145

T

Tabebuia palmeri, 255
Tabernaemontana alternifolia, 164
Tabernaemontana gamblei Subr. & A.N.Henry, 146
Tacca leontopetaloides (L.) Kuntze, 192
Tamarindus indica, 184
Taxus mairei, 3, 20
Tecomella undulata 255
Tectona grandis, 184, 223, 255
Tephrosia calophylla, 179
Terminalia arjuna, 180
Terminalia bellirica, 184, 222
Terminalia bialata, 164
Terminalia catappa, 184
Terminalia chebula, 56, 180
Terminalia pallida, 179
Terminalia paniculata, 223
Thelypteris polycarpa (Blume) K. Iwats., 190

Theobroma cacao, 26
Threatened Plants, 255
Thrinax barbadensis 250
T. excelsa, 250
Thrixspermum japonicum (Miq.) Rchb.f., 96
Thunbergia fragrans Roxb., 145
Thunbergia mysorensis (Wight) T. Anderson, 145
Thymus marschallianus Willd., 241
Thyrsostachys siamensis Gamble, 143
Thysanolaena latifolia, 14
Tillandsia ionantha Planch., 147
Tillandsia usneoides (L.) L., 147
Tinospora cordifolia (Willd.) Miers, 192
Touch and smell garden, 254
Trachycarpus fortunei, 164
Trachycarpus takil, 159, 164, 256
Traixspermum japonicum (Miquel) Rchb. f.
Trang Bom Arboretum, 55, 56
Trapa natans L., 146
Trewia nudiflora, 182
Triphasia trifolia (Burm. f.) P. Wilson, 192
Tripogon tirumalae, 179
Trollius sajanensis Sipliv., 118
Tropidia curculigoides Lindl., 193
Tulipa patens C. Agardh ex Schult. & Schult. f., 239
Tulipa uniflora (L.) Besser ex Baker, 18
Tylophora indica, 180

U

Ulmus wallichiana, 20
Utricularia aurea Lour., 146
Utricularia gibba L., 146

V

Vachellia eburnea, 184
V. nilotica, 184
Valeriana jatamansi, 14
Vanda coerulea, 252
Vanda tessellata, 252
Vanilla albida Blume, 193
Vanilla andamanica Rolfe, 189, 192–193
Vanilla sanjappae R.P. Pandey et al., 189, 192–193
Vanilla wightii Lindl. ex Wight, 145
Vateria indica L., 146, 164–165, 230

Vatica bantamensis (Hassk.) Benth. & Hook.ex Miq., 30
Vatica chinensis, 164
Victoria amazonica (Poepp.) Klotzsch, 146, 158, 162, 180, 185, 218, 228
Vietnam, 51–58, 60, 62
Vitex trifolia L., 192

W

Wallichia disticha, 164
Wallisia cyanea Barfuss & W.Till., 147
Welwitschia mirabilis, 250–251
Western Ghats, 135, 137–138, 147, 154, 157, 160–162, 164, 166–168, 203, 205–207, 209–210, 213
Wolffia globosa (Roxb.) Hartog & Plas, 146
Woodfordia fruiticosa, 222
Wrightia tinctoria, 180

X

Xylia xylocarpa, 223

Y

Yogi Vemana University, 173
Yucca aloifolia, 184
Yucca gloriosa, 14

Z

Zamia erosa, 254
Zamia floridana, 224
Zamia furfuracea L.f. ex Aiton, 180, 189, 224, 254
Z. loddigesii, 254
Z. pumila, 254
Zanthoxylum armatum, 14
Zea mays, 26
Zeuxine andamanica King & Pantl., 189, 193
Zingiber officinale Roscoe, 192
Zingiber pseudosquarrosum L.J. Singh & P. Singh, 189, 192
Zingiber zerumbet, 158
Ziziphora clinopodioides Lam., 241
Ziziphus mauritiana, 161, 184
Ziziphus oenoplia, 184
Zygophyllum potaninii Maxim., 119